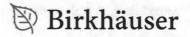

Birkhäuser Advanced Texts Basler Lehrbücher

Series Editors

Steven G. Krantz, Washington University, St. Louis, USA

Shrawan Kumar, University of North Carolina at Chapel Hill, Chapel Hill, USA

Jan Nekovář, Sorbonne Université, Paris, France

More information about this series at http://www.springer.com/series/4842

Christophe Cheverry · Nicolas Raymond

A Guide to Spectral Theory

Applications and Exercises

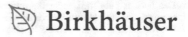 Birkhäuser

Christophe Cheverry
University of Rennes 1
Rennes, France

Nicolas Raymond
University of Angers
Angers, France

ISSN 1019-6242   ISSN 2296-4894   (electronic)
Birkhäuser Advanced Texts Basler Lehrbücher
ISBN 978-3-030-67464-9   ISBN 978-3-030-67462-5   (eBook)
https://doi.org/10.1007/978-3-030-67462-5

Mathematics Subject Classification: 35Pxx

This book is published under the imprint Birkhäuser, www.birkhauser-science.com by the registered
company Springer Nature Switzerland AG
The registered company address is: Gewerbestrasse 11, 6330 Cham, Switzerland

Bientôt, tu auras tout oublié ;
bientôt, tous t'auront oublié.

FOREWORD

I am happy to warmly recommend this very nice volume by Christophe Cheverry and Nicolas Raymond on the spectral theory of linear operators on Hilbert spaces and its applications to mathematical quantum mechanics. This book will provide the reader with a firm foundation in the spectral theory of linear operators on both Banach and Hilbert spaces. The text emphasizes key components of the theory and builds on them in its exploration of several examples and applications. Throughout, the authors emphasize the connections between functional analysis and partial differential equations.

Several unique features make this text particularly appealing. For example, the notion of a Fredholm operator is introduced early in Chapter 3 on the spectrum of linear operators. There, in addition to the fundamental properties of Fredholm operators, the Fredholm nature of certain operators is used as a tool in the decomposition of the spectrum into essential and discrete parts and developed further in Chapter 6. A clear explanation of the Grushin method and its relation to Fredholm operators is presented in Chapter 5. The Grushin method is used in the text to develop the analytic perturbation theory of isolated eigenvalues. It is also applied to the analysis of Toeplitz operators.

Other modern tools developed in the text include the functional calculus based on the Fourier transform and the development of spectral measures in Chapter 8.

In Chapter 9, the notion of a local trace of an operator, that is, the trace of an operator restricted to a finite region by cutoff functions, is discussed. Not only is this an elegant application of the ideas developed in preceding chapters, but is also an introduction to the modern topic of eigenvalue asymptotics, nicely described in the notes to the chapter. The authors also use this topic to provide the reader with an introduction to semi-classical analysis.

This readable text concludes with applications to variational principles for density matrices, an application to local trace estimates, a proof of Stone's formula for spectral projections using the Fourier functional calculus, and a presentation of the Mourre estimate with an application to the Limiting Absorption Principle and to the study of the absolutely continuous spectrum of Schrödinger operators.

A Guide to Spectral Theory: Applications and Exercises invites you to take a step into the world of spectral theory, and, to quote many ancient philosophers; the reader should remember that a walk of thousand miles begins with one step.

Peter D. Hislop
Professor of Mathematics
University Research Professor
University of Kentucky
Lexington, KY, USA
July, 3 2020

PROLEGOMENA

The origin of this book and what it is.—This textbook was born from lectures, which we gave to master and Ph.D. students at the universities of Nantes and Rennes (France) during the fall semesters 2019 and 2020. It is first and foremost meant to be a graduate level textbook introducing readers to the spectral analysis of unbounded linear operators and its applications. We tried to keep it self-contained, relatively short, and accessible to non-specialist readers. As such, it can be used by students for self-study, and by teachers for lecturing on spectral theory. In the last chapters, perspectives for young (and maybe seasoned) researchers are proposed. More advanced applications are presented, and some background ideas for in-depth courses and seminars are given. From this point of view, this book might go beyond what is usually covered in short textbooks. As its title might suggest, this book is an invitation to a guided walk. All the possible directions are not explored, but we only go through selected paths, and sometimes, indicate elegant shortcuts. Our text should be considered a manual, or (as the ancients said) an *enchiridion*, whose ambition is to help the reader emancipate herself/himself, and travel through other mathematical worlds where spectral theory (and its applications) plays a role. In order to ease self-training, it contains many exercises (often with solutions and almost always with hints). The specificity and interest of this little book is to bring closer and conciliate, in the prism of spectral theory, various subjects as, *e.g.*,

- partial differential equations,
- variational methods,
- compact and Fredholm operators (via « Grushin » reductions),

- spectral theorem (with the help of basic measure theory),

- Mourre theory (thanks to elementary non-self-adjoint coercivity estimates).

Prerequisites.—Of course, when opening this book, the reader should not have just fallen off the turnip truck. She/He is expected to have a basic understanding of linear functional analysis and partial differential equations (PDE). Still, many classical results are recalled in Appendix A, and, as often as necessary, we recall basic PDE technics when discussing some spectral issues. As we will see all along this book, many spectral problems have strong connections with PDEs: most of our examples and applications are clearly oriented in this direction. Indeed, we believe that it is pedagogically important to exemplify abstract statements about operator and spectral theory to understand what they can mean in practice. This book is also focused on unbounded operators. We will see that the description of the domains of such operators can lead to elliptic PDE questions, which are often not discussed in books on spectral theory. A basic knowledge of bounded linear operators could also help the reader even though we tried to make it unnecessary.

Selected references.—*Spectral theory* was born in the early twentieth century from David Hilbert in his original and implicit « Hilbert space theory » which was developed in the context of integral equations; see [21, p. 160]. At that time, mathematics experienced many fertile developments, which left a lasting mark on our way of teaching mathematics today. A brief look at the mathematical works of this « world of yesterday » never fails to surprise us: From the ideas and theorems of that time, there emanates a subtle scent of modernity. We may even feel close to these mathematicians through the motivations that we still share with them; see [4, Preface, p. v]. Among these motivations, we should emphasize the will to build bridges between different areas of science. In particular, in [4], R. Courant felt the need to stimulate the dialog between Mathematics and Physics. Even though we will illustrate as often as possible the abstract theorems by means of explicit examples or detailed exercises inspired by Quantum Mechanics, our purpose is somehow more modest.

We can talk about spectral theory in many ways, which are often scattered in various books or articles (see below). There are many contributions that investigate the spectral theory of the Schrödinger operators, as well as their applications. Probably, one of the most prominent ones is the textbook series by Reed and Simon [35], which covers all the material required to study quantum mechanical systems, and extends the original ambition of Courant and Hilbert [4]. It is also worthy to mention the two very nice books [6, 7] by B. Davies, the first for a concise reference and the second for a more specific and developed version. Regarding some aspects, the reader can find useful information in the work of D. E. Edmunds and W. D. Evans [10], as well as in the contribution of Einsiedler and Ward [11]. To discover the general underlying PDE context, the reader can consider the book [44] by M. E. Taylor. Since our book is focused on unbounded operators, a preliminary knowledge about bounded operators might be useful; see for instance [28].

Our text is freely inspired by many other books [3, 20, 22, 25, 27, 31, 34, 37, 45, 49]. It also owes very much to various lecture notes by mathematicians of « the world of today ». We are especially grateful to our colleagues Z. Ammari, C. Gérard, F. Nier, S. Vũ Ngọc, and D. Yafaev. May also P. D. Hislop be thanked for his many valuable comments.

Structure of the book.—The book is organized as follows.

Chapter 1 begins with a short introduction to the relation between Spectral Theory and Quantum Physics. It also contains a discussion about a quite basic but very instructive spectral problem. As such, it may be viewed as a warm up for the students. It introduces some important ideas and landmarks with a concrete example. The presentation is voluntarily purified from the general formalism that will be introduced later in the book.

Chapter 2 is devoted to unbounded operators, which are generalizations of bounded operators. Along this path, the different notions of closed, adjoint, and self-adjoint operators are defined, and they are illustrated with various progressive examples. The famous Lax–Milgram theorems are proved: they establish a correspondence between a coercive sesquilinear form and a bijective closed operator. These theorems allow defining explicit operators such as the Dirichlet Laplacian (whose abstract domain is characterized via a regularity theorem).

In Chapter 3, the spectrum of a closed operator is defined. The basic properties of the spectrum of bounded operators are established. In particular, the spectral radius is characterized in the case of normal operators. Then, thanks to the holomorphic function theory, the famous Riesz projections are defined. We explain that they are generalizations of the usual projections on the characteristic subspaces of matrices. This chapter ends with the basic definition of Fredholm operators and of two important subsets of the spectrum, which are related to the «Fredholmness»: the discrete and the essential spectrum. In some respects, the spectral theory of a Fredholm operator may be reduced to the spectral theory of a matrix.

Chapter 4 is about compact operators. Many elementary properties of this special class of bounded operators are recalled. Compact operators play an important role when trying to establish that an operator is Fredholm. In the perspective of dealing with examples, compact subsets of L^2 (giving rise to compact inclusion maps) are characterized via some L^p-version of the Ascoli theorem, and the famous Rellich theorems are established. At the end of this chapter, the reader has all the elements to understand why the Dirichlet Laplacian on a smooth bounded subset of $\mathbb{R}^d$ has compact resolvent.

Chapter 5 is devoted to the study of Fredholm operators. This theory is explained by means of a matrix formalism, sometimes called « Grushin formalism ». In the case of matrices, one can compute the difference between the number of rows and lines. When dealing with Fredholm operators, such a number can also be determined. Then, it is called the index of the Fredholm operator. Roughly speaking, Fredholm operators of index 0 are the ones whose spectra can be described, thanks to square matrices. Then, we explain the connection between Fredholm and compact operators: Fredholm operators can be inverted modulo compact operators (in the sense of Atkinson's theorem; see Proposition 5.5). By using the « Grushin method », we perform the spectral analysis of compact operators, and establish the famous Fredholm alternative. This chapter also contains various illustrations of the aforementioned concepts:

– the complex Airy operator is analyzed via a succession of small exercises;

– an elementary perturbation analysis, in the spirit of Kato's book, is performed via the Grushin method;

– the index of Toeplitz operators on the circle is studied (via Fourier series).

Chapter 6 deals with self-adjoint operators. It starts with the elementary analysis of compact normal operators. From this, we can infer the description of the spectrum of self-adjoint operators with compact resolvent. As a first explicit illustration of this point, the spectral analysis of the harmonic oscillator is performed. Then, we provide the reader with various characterizations (by means of the Weyl sequences) of the essential and discrete spectra of a self-adjoint operator. In particular, we show that the discrete and essential spectra form a partition of the spectrum. When a self-adjoint operator is bounded from below, we show the « min-max » characterization of the eigenvalues (they can be computed from the values of the quadratic form associated with the operator). Examples of applications are given like Sturm–Liouville theory and Weyl's law in one dimension. A section is also devoted to the analysis of the smallest eigenvalue of the Schrödinger operator describing the hydrogen atom.

In Chapter 7, we prove the Hille–Yosida and Stone theorems. Basically, they state that, under convenient assumptions, there is a correspondence between some semigroups and closed operators. The Stone theorem tackles the case when we have a (continuous) unitary group $(U_t)_{t \in \mathbb{R}}$. In this case, the group can be seen as the solution of a differential equation (somehow the

Schrödinger equation) associated with a self-adjoint operator $\mathscr{L}$, and written as $U_t = e^{it\mathscr{L}}$.

In Chapter 8, we return to the spectral analysis of self-adjoint operators. We construct a general functional calculus for self-adjoint operators. In other words, we explain how to define $f(\mathscr{L})$ for Borelian functions f, so that, *e.g.*, $(fg)(\mathscr{L}) = f(\mathscr{L})g(\mathscr{L})$. This calculus starts with the special case $f(t) = e^{it}$, and is based on the inverse Fourier transform. When constructing the functional calculus, a Borel measure appears. It is called the « spectral measure ». For all $u \in$ H, it is denoted by $d\mu_{u,u}$, and it satisfies

$$\langle f(\mathscr{L})u, u \rangle = \int_{\mathbb{R}} f(\lambda) d\mu_{u,u}(\lambda),$$

and the measure is given by $\mu_{u,u}(\Omega) = \langle \mathbb{1}_\Omega(\mathscr{L})u, u \rangle$. The functional calculus is used to characterize the discrete and essential spectra. Then, the spectral measure is analyzed, thanks to the Lebesgue decomposition theorem. This decomposition allows to split the ambient Hilbert space as an orthogonal sum of closed subspaces associated with different kinds of spectra (absolutely continuous, pure point, and singular continuous). We give an elementary criterion to check that a part of the spectrum is absolutely continuous.

Chapter 9 is devoted to particular classes of compact operators: the trace-class and Hilbert–Schmidt operators. As their name suggests, trace-class operators are operators for which a trace can be defined, say, by means of a Hilbert basis (a complete orthonormal system). We give examples of such operators and compute their traces. In particular, to make our general discussion on the properties of Hilbert–Schmidt operators more concrete, we analyze the « local traces » of the Laplacian on $\mathbb{R}^d$.

In Chapter 10, we present some important applications of the functional calculus. We prove a version of Lieb's Variational Principle, which allows estimating traces of trace-class operators. We illustrate this « principle » by coming back to the study of local traces of the Laplacian. Then, we use the functional calculus to prove the Stone formula, which relates the resolvent near an interval J of the real axis and spectral projections. We use it to deduce a criterion to check the absolute continuity of the spectrum in this interval. This final chapter ends with an elementary version of the Mourre theory. Assuming that a commutator is positive on the range of $\mathbb{1}_J(\mathscr{L})$, we show that the aforementioned criterion for absolute continuity of the spectrum is satisfied by establishing a « Limiting Absorption Principle ». This theory is exemplified for the Schrödinger operator in one dimension.

A list of errata and corrections will be maintained on our respective websites. Please let us know about any error you may find.

Have a good walk!

Christophe Cheverry
Nicolas Raymond

CONTENTS

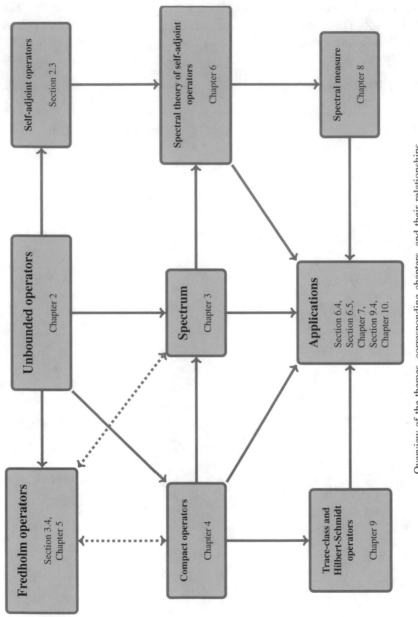

Overview of the themes, corresponding chapters, and their relationships

CHAPTER 1

A FIRST LOOK AT SPECTRAL THEORY

The aim of this opening chapter is to give a flavor of basic problems in spectral theory. This starts in Section 1.1 with some general comments on quantum physics and related mathematical issues. This is done more concretely in Section 1.2 where a practical question coming from the variational calculus is used to motivate the study of unbounded operators and their spectral properties.

1.1. From quantum physics to spectral considerations

A few years after the birth of spectral theory, one discovered that it could explain the emission spectra of the atoms. This is a good illustration of scientific serendipity. Indeed, Hilbert spaces, and even the word *spectrum*, were *independently* introduced for different motivations (see, for instance, the doctoral dissertation of E. Schmidt [**39**]). Nowadays, spectral theory has become an important area in mathematical physics because it links the energy levels of quantum mechanical systems to the spectrum of operators. Two main postulates (which are part of the Dirac–Von Neumann axioms) of quantum mechanics are the following:

   i. The *pure states* of a quantum system are described by rays $u \neq 0$ in a separable complex Hilbert space H.

   ii. The *observables* are represented by self-adjoint linear operators $T : H \longrightarrow H$.

© The Author(s), under exclusive license
to Springer Nature Switzerland AG 2021
C. Cheverry and N. Raymond, *A Guide to Spectral Theory*,
Birkhäuser Advanced Texts Basler Lehrbücher,
https://doi.org/10.1007/978-3-030-67462-5_1

A *state* $u \in \mathsf{H}$ provides a probability distribution for the outcome of each possible measurement. An *observable* $T : \mathsf{H} \longrightarrow \mathsf{H}$ represents a physical quantity that can be measured (like position and momentum). The two ingredients H and T play a fundamental role in what follows.

Examples to keep in mind. — We start the discussion by exhibiting classical examples of H and T, from the simplest to the most complex. Let us fix $\lambda \in \mathbb{C}$, $N \in \mathbb{N}$, a complex matrix $A = (a_{ij})_{1 \leqslant i,j \leqslant N}$ of size $N \times N$, and a bounded complex sequence $(\lambda_n)_{n \in \mathbb{N}}$. In the table below, we consider four linear applications:

		H	$T : \mathsf{H} \longrightarrow \mathsf{H}$
1)	Homothety	$\mathbb{C}$	$u \longmapsto \lambda u$
2)	Linear map in finite dimension	$\mathbb{C}^N$	$u \longmapsto Au$
3)	Linear (diagonal) map in infinite dimension	$\ell^2(\mathbb{N}; \mathbb{C})$	$u \longmapsto (\lambda_n u_n)_n$
4)	Shift operator	$\ell^2(\mathbb{N}; \mathbb{C})$	$u \longmapsto (u_{n+1})_n$

Their norms are respectively given by

		$\|T\|$
1)	Homothety	$|\lambda|$
2)	Linear map in finite dimension	$\left(\sum_{1 \leqslant i,j \leqslant N} |a_{ij}|^2 \right)^{\frac{1}{2}}$
3)	Linear (diagonal) map in infinite dimension	$\sup_{n \in \mathbb{N}} |\lambda_n|$
4)	Shift operator	1

The definition of Tu in cases of 1), 2), 3), and 4) is, for all $u \in H$, not a problem. Moreover, the action of T gives rise to a bounded operator with a (finite) norm $\|T\|$ controlled as indicated on the right-hand side. Let us now provide the reader with a slightly more problematic example, which is the following:

	H	$T : \mathsf{H} \longrightarrow \mathsf{H}$	$\|T\|$
5) Laplace operator	$L^2(\mathbb{R}; \mathbb{C})$	$u \longmapsto -\partial_x^2 u$?	?

A list of questions. — The situation 5) raises a number of issues. In case 5), we would like to answer the following questions:

— What is the *domain of definition* Dom (T) of T? For instance, the function $\mathbb{1}_{[0,1]}$ is in $L^2(\mathbb{R};\mathbb{C})$ while the distribution $\partial_x^2(\mathbb{1}_{[0,1]})$ is not. The choice of the target set H plays in this discussion a very important role.

— What do we mean when we say that T is *continuous*? When Dom $(T) = $ H, in a linear setting, this means that T is *bounded*. But otherwise?

— What is the *spectrum* sp(T) of the operator T? Basically, the spectrum of T is a subset of $\mathbb{C}$ which is a generalization of the spectrum of matrices. Thus,

  - in case 1), the spectrum is simply $\{\lambda\}$.
  - in case 2), the spectrum is the set of the eigenvalues of A. In particular, when the matrix A is self-adjoint meaning that A is Hermitian or that $A^* := \bar{A}^{\mathrm{T}} = A$, we know that all eigenvalues are real (and that there is an associated orthonormal basis of eigenvectors). This implies that the spectrum is made of a finite number $(\leqslant N)$ of real numbers.
  - in case 3), the vector $u_i = (\delta_{in})_n$ is an eigenvector with eigenvalue λ_i. Thus, we can guess that the spectrum contains, at least, the set $\{\lambda_n; n \in \mathbb{N}\}$. We will see that it also contains the limit points of the sequence $(\lambda_n)_{n\in\mathbb{N}}$.
  - What happens concerning the shift operator of case 4) ? A vector of the form

$$u_\lambda := (1, \lambda, \lambda^2, \cdots, \lambda^n, \cdots) \in \mathbb{C}^{\mathbb{N}}$$

satisfies $Tu_\lambda = \lambda u_\lambda$. When $|\lambda| < 1$, we find that $u_\lambda \in \ell^2(\mathbb{N};\mathbb{C})$, and u_λ is clearly an eigenvector. It turns out that sp(T) is the closure of such λ, that is, the unit disk $D = \{\lambda \in \mathbb{C}; |\lambda| \leqslant 1\}$. By the way, note that the vectors u_λ with $|\lambda| \geqslant 1$ are not eigenvectors because they do not belong to the space $\ell^2(\mathbb{N};\mathbb{C})$. We will describe the *nature* of the spectrum (for instance, we will define the *discrete* and *essential* spectra) and its topological properties. Remember that the spectrum of an operator may be empty (Sect. 5.4), with a non-empty interior, or even equal to the whole complex plane $\mathbb{C}$. There will be a few surprises!

— What about the *functional calculus*? What does it means to compute T^2, T^3, $\cdots$, e^T, e^{iT} or more generally $f(T)$ when f is a measurable function? For instance, in case 5), the domain of T is

(1.1.1.1)
$$\text{Dom}\,(T) = \mathsf{H}^2(\mathbb{R}, \mathbb{C}) = \{u \in \mathsf{L}^2(\mathbb{R}, \mathbb{C}) : \xi^2 \hat{u}(\xi) \in \mathsf{L}^2(\mathbb{R}, \mathbb{C})\}$$
$$\subsetneq \mathsf{L}^2(\mathbb{R}; \mathbb{C}),$$

where $\hat{u}$ is the Fourier transform of u. The domain of e^{iT} is the whole $\mathsf{L}^2(\mathbb{R}, \mathbb{C})$. How to explain this difference?

A more elaborate statement. — We want to briefly illustrate the preceding discussion by commenting the following important statement, which is called Stone's theorem (see Theorem 7.12 for a more precise formulation).

Theorem 1.1. — *Let* $T : \text{Dom}(T) \longrightarrow \mathsf{H}$ *be a self-adjoint operator. Then, the solution to the evolution equation*

(1.1.1.2)
$$\partial_t u = iTu, \qquad u(0, \cdot) = u_0 \in \mathsf{H}$$

is given by $t \longmapsto u(t) = U_t u_0$ *where* $(U_t)_{t \in \mathbb{R}}$ *is a unitary group defined by* $U_t \equiv e^{itT}$.

When $T \equiv -\partial_x^2$ and $\mathsf{H} = \mathsf{L}^2(\mathbb{R})$, the equation (1.1.1.2) is nothing but the *Schrödinger equation* of quantum mechanics. The notion of the *self-adjoint* operator evoked above will be explained later in this book. Since the solution may be expressed as $U_t \equiv e^{itT}$, we see here that the functional calculus can help to solve partial differential equations.

Proof. — Let us consider this theorem in two particular cases:

— In case 1), the condition « self-adjoint » implies that $\lambda \in \mathbb{R}$ since

$$\langle \lambda u, u \rangle = \langle u, \lambda u \rangle, \quad \forall u \in \mathbb{C} \implies (\lambda - \bar{\lambda})|u|^2 = 0, \forall u \in \mathbb{C}$$
$$\implies \lambda \in \mathbb{R}.$$

Then, the solution to (1.1.1.2) is given by $t \longmapsto u(t) = e^{i\lambda t} u_0$. The operator U_t is a rotation and preserves the norms of vectors (it is a unitary operator).

— In case 5), denoting by $\hat{u} \equiv \mathscr{F}u$ the Fourier transform of u, it may be checked that $Tu \in \mathsf{H}$ if and only if $u \in \mathrm{Dom}\,(T)$ with $\mathrm{Dom}\,(T)$ as in (1.1.1.1). In this setting, T is self-adjoint if and only if $T = T^*$ (see Definition 2.42). This implies that T must be symmetric (see Definition 2.55) in the sense that, for all $(u, v) \in \mathrm{Dom}\,(T)^2$, we have

$$\langle Tu, v \rangle = \langle u, T^*v \rangle = -\int_{\mathbb{R}} \partial_x^2 u(x)\bar{v}(x)\,\mathrm{d}x$$

$$= -\int_{\mathbb{R}} u(x)\partial_x^2 \bar{v}(x)\,\mathrm{d}x$$

$$= \langle u, Tv \rangle.$$

The Fourier multiplier $\mathbb{R} \ni \xi \mapsto \xi^2$ is not a bounded function. That is why the operator $-\partial_x^2 : \mathsf{L}^2(\mathbb{R}) \longrightarrow \mathsf{L}^2(\mathbb{R})$ is not well defined. We find that $\hat{u}(t, \xi) = e^{it\xi^2}\hat{u}_0(\xi)$. By contrast, the Fourier multiplier $e^{it\xi^2}$ is bounded, and it is even of modulus 1. That is why the action $U_t : \mathsf{L}^2(\mathbb{R}) \longrightarrow \mathsf{L}^2(\mathbb{R})$, which is defined by $U_t u = \mathscr{F}^{-1}(e^{it\xi^2}\hat{u})$, is a unitary operator defined on the *whole* L^2-space.

$\square$

1.2. A paradigmatic spectral problem

The aim of this section is to help the reader in revising some notions that she/he perhaps already encountered in the past, or to introduce her/him with elementary tools. In Sections 1.2.1 and 1.2.2 is discussed one of the most simple spectral problems involving a differential equation. The spirit is clearly inherited from [4, Chapter VI, p. 400].

1.2.1. A question. — Fix some interval $I \subset \mathbb{R}$. We endow the space $\mathsf{L}^2(I)$ with the usual scalar product

$$\langle u, v \rangle_{\mathsf{L}^2(I)} = \int_I u(x)\,\overline{v(x)}\,\mathrm{d}x.$$

We define (see Appendix A.4)

$$\mathsf{H}^1(I) = \{\psi \in \mathsf{L}^2(I) : \psi' \in \mathsf{L}^2(I)\},$$

and we endow it with the following Hermitian form

$$\langle u, v \rangle_{\mathsf{H}^1(I)} = \langle u, v \rangle_{\mathsf{L}^2(I)} + \langle u', v' \rangle_{\mathsf{L}^2(I)} .$$

Lemma 1.2. — $(\mathsf{H}^1(I), \langle \cdot, \cdot \rangle_{\mathsf{H}^1(I)})$ *is a Hilbert space.*

We define

$$\mathsf{H}_0^1(I) = \overline{\mathscr{C}_0^\infty(I)}^{\mathsf{H}^1} ,$$

where the bar denotes the closure.

Lemma 1.3. — $(\mathsf{H}_0^1(I), \langle \cdot, \cdot \rangle_{\mathsf{H}^1(I)})$ *is a Hilbert space.*

Select $a \in \mathbb{R}$ and $b \in \mathbb{R}$ such that $a < b$. We work on the bounded interval $J := (a, b)$. We can always define

$$(1.1.2.3) \qquad \lambda_1 = \inf_{\substack{\psi \in \mathsf{H}_0^1(J) \\ \psi \neq 0}} \frac{\int_J |\psi'|^2 \, \mathrm{d}x}{\int_J |\psi|^2 \, \mathrm{d}x} \geq 0 .$$

Question: | What is the value of λ_1 ?

1.2.2. An answer. —

Lemma 1.4 (Sobolev embedding). — *The following assertions hold.*

(i) *We have* $\mathsf{H}^1(\mathbb{R}) \subset \mathscr{C}_{\to 0}^0(\mathbb{R})$ *and for all* $\psi \in \mathsf{H}^1(\mathbb{R})$,

$$\forall x \in \mathbb{R}, \quad |\psi(x)| \leq \frac{1}{\sqrt{2}} \|\psi\|_{\mathsf{H}^1(\mathbb{R})} .$$

(ii) *We have* $\mathsf{H}_0^1(J) \subset \mathscr{C}^0(\overline{J})$ *and, for all* $\psi \in \mathsf{H}_0^1(J)$, $\psi(a) = \psi(b) = 0$ *and*

$$\forall x \in J, \quad |\psi(x)| \leq |J|^{\frac{1}{2}} \|\psi'\|_{\mathsf{L}^2(J)} .$$

(iii) *For all* $\psi \in \mathsf{H}_0^1(J)$, *we have, for all* $x, y \in J$,

$$|\psi(x) - \psi(y)| \leq \sqrt{|x - y|} \|\psi'\|_{\mathsf{L}^2(J)} .$$

Proof. — Let us deal with (i). We use the (unitary) Fourier transform, *i.e.*, defined by

$$\widehat{\psi}(\xi) = \frac{1}{\sqrt{2\pi}} \int_{\mathbb{R}} e^{-ix\xi} \psi(x) \, \mathrm{d}x , \quad \forall \psi \in \mathscr{S}(\mathbb{R}) ,$$

and extended to $L^2(\mathbb{R})$. We get[1]

$$\|\psi\|_{H^1(\mathbb{R})}^2 = \int_{\mathbb{R}} \langle \xi \rangle^2 |\widehat{\psi}(\xi)|^2 \, d\xi, \qquad \langle \xi \rangle := (1 + \xi^2)^{1/2}.$$

In particular, we deduce, by Cauchy–Schwarz, that $\widehat{\psi} \in L^1(\mathbb{R})$. By using the inverse Fourier transform, we get

$$\forall x \in \mathbb{R}, \quad \psi(x) = \frac{1}{\sqrt{2\pi}} \int_{\mathbb{R}} \widehat{\psi}(\xi) e^{ix\xi} \, d\xi.$$

By dominated convergence, we see that ψ is continuous. Moreover, it goes to 0 at infinity by the Riemann–Lebesgue lemma. In addition,

$$\forall x \in \mathbb{R}, \quad |\psi(x)| \leqslant (2\pi)^{-\frac{1}{2}} \|\widehat{\psi}\|_{L^1(\mathbb{R})}$$

$$\leqslant (2\pi)^{-\frac{1}{2}} \|\langle \xi \rangle^{-1}\|_{L^2(\mathbb{R})} \|\langle \xi \rangle \widehat{\psi}\|_{L^2(\mathbb{R})}.$$

Let us now consider (ii). Select some $\psi \in H_0^1(J)$. Let us extend ψ by zero outside J and denote by $\underline{\psi}$ this extension. We have $\underline{\psi} \in L^2(\mathbb{R})$. Since $\mathscr{C}_0^\infty(J)$ is dense in $H_0^1(J)$, we can consider a sequence $(\psi_n)_n$ with $\psi_n \in \mathscr{C}_0^\infty(J)$ converging to ψ in H^1-norm. Note that $(\underline{\psi_n})_n$ is a Cauchy sequence in $H^1(\mathbb{R})$ since (1.1.2.4)

$$\|\underline{\psi_n} - \underline{\psi_m}\|_{H^1(\mathbb{R})} = \|\psi_n - \psi_m\|_{H^1(J)}, \qquad \forall (m, n) \in \mathbb{N}^2.$$

Since $H^1(\mathbb{R})$ is a complete metric space, the sequence $(\underline{\psi_n})_n$ converges in $H^1(\mathbb{R})$ to some $v \in H^1(\mathbb{R})$. Since $(\underline{\psi_n})_n$ converges in $L^2(\mathbb{R})$ to $\underline{\psi}$, we get $v = \underline{\psi} \in H^1(\mathbb{R})$. By (i), we deduce that $\underline{\psi}$ is continuous on J. Coming back to (1.1.2.4) and using again (i), we get that $(\underline{\psi_n})_n$ uniformly converges to $\underline{\psi}$. In particular, $\psi(a) = \psi(b) = 0$. Then, we can write

$$\psi_n(x) - \psi_n(y) = \int_x^y \psi_n'(t) \, dt, \qquad \forall (x, y) \in J^2.$$

From the Cauchy–Schwarz inequality, we get

$$|\psi_n(x) - \psi_n(y)| = |y - x|^{1/2} \|\psi_n'\|_{L^2(J)}.$$

Passing to the limit ($n \to +\infty$), since $\psi_n(a) = 0$, we obtain both (ii) and (iii). □

From (ii), we can infer that

$$\int_J |\psi(x)|^2 dx \leqslant |J|^2 \|\psi'\|^2_{L^2(J)} \, .$$

This already implies that λ_1 is finite, and moreover that

(1.1.2.5) $$0 < |J|^{-2} = (b-a)^{-2} \leqslant \lambda_1.$$

Lemma 1.5. — *Consider E and F two normed vector spaces, and $T \in \mathcal{L}(E, F)$. Then, if (u_n) weakly converges to u in E, the sequence (Tu_n) weakly converges in F.*

Proof. — The adjoint of T, denoted by T', is defined by

$$\forall \ell \in F' \quad T'(\ell) = \ell \circ T \, .$$

The application T' is linear and continuous. Consider the sequence (u_n) and $\ell \in F'$. Since $T'(\ell) \in E'$, we get that $T'(\ell)(u_n) = \ell(Tu_n)$ converges to $T'(\ell)(u) = \ell(Tu)$.   □

Lemma 1.6. — *The infimum* (1.1.2.3) *is a minimum $\psi \in H^1_0(J)$ with $\psi \neq 0$.*

Proof. — Let (ψ_n) be a minimizing sequence such that we have $\|\psi_n\|_{L^2(J)} = 1$. In particular (ψ'_n) is bounded in $L^2(J)$. By Lemma 1.4, (ψ_n) is equicontinuous on the (compact) interval $[a, b]$ and pointwise bounded. We can apply the Ascoli theorem and (after extraction) we may assume that (ψ_n) uniformly converges to ψ on $[a, b]$ and therefore in $L^2(J)$. We get $\|\psi\|_{L^2(J)} = 1$. Since (ψ_n) is bounded in $H^1_0(J)$, we can assume that it is weakly convergent (to ϕ) in $H^1_0(J)$, and thus in $L^2(J)$ (by Lemma 1.5). Since $\mathscr{C}^\infty_0(J)$ is dense in $L^2(J)$ (as a consequence of Lemma 1.9), we must have $\phi = \psi$. Since (ψ'_n) weakly converges in $L^2(J)$ to ψ' (again by Lemma 1.5), we deduce that

$$\liminf_{n \to +\infty} \|\psi'_n\|_{L^2(J)} \geqslant \|\psi'\|_{L^2(J)} \, .$$

As a consequence, we have

$$\lambda_1 \geqslant \|\psi'\|^2_{L^2(J)} \, ,$$

where $\psi \in H^1_0(J)$ and $\|\psi\|_{L^2(J)} = 1$. It follows that $\|\psi'\|^2_{L^2(J)} = \lambda_1$.   □

The last proof can easily be adapted to show a Rellich lemma, saying that the embedding $\iota : \mathsf{H}_0^1(J) \hookrightarrow \mathsf{L}^2(J)$ is « compact ». Along these lines, we have implicitly dealt with a « compact operator ». All these notions are discussed in Chapter 4.

Lemma 1.7. — *Let $\psi \in \mathsf{H}_0^1(\overline{J})$ be a minimum. Then the function ψ is smooth, in $\mathscr{C}^\infty(\overline{J})$, and it satisfies (in a classical sense) the following differential equation:*

(1.1.2.6) $$-\psi'' = \lambda_1 \psi.$$

Proof. — Let $\varphi \in \mathscr{C}_0^\infty(J)$. Given $\epsilon \in \mathbb{R}$, we define

$$f(\epsilon) := \int_J |(\psi + \epsilon\varphi)'|^2 \, \mathrm{d}x - \lambda_1 \int_J |\psi + \epsilon\varphi|^2 \, \mathrm{d}x.$$

Let ψ be a minimum. By construction, we must have $f(0) = 0$ and $f(\epsilon) \geqslant 0$ for all $\epsilon \in \mathbb{R}$. It follows that $f'(0) = 0$, that is

$$\int_J (\psi'\overline{\varphi}' + \overline{\psi}'\varphi') \, \mathrm{d}x = \lambda_1 \int_J (\psi\overline{\varphi} + \overline{\psi}\varphi) \, \mathrm{d}x.$$

Test this identity for all $\varphi \in \mathscr{C}_0^\infty(J, \mathbb{R})$ to obtain that

$$\int_J (\psi + \overline{\psi})' \varphi' \, \mathrm{d}x = \lambda_1 \int_J (\psi + \overline{\psi})\varphi \, \mathrm{d}x.$$

Do the same with $i\varphi$ to get that

$$\int_J (\psi - \overline{\psi})' \varphi' \, \mathrm{d}x = \lambda_1 \int_J (\psi - \overline{\psi})\varphi \, \mathrm{d}x.$$

We deduce that

$$\int_J \psi'\varphi' \, \mathrm{d}x = \lambda_1 \int_J \psi\varphi \, \mathrm{d}x, \qquad \forall \varphi \in \mathscr{C}_0^\infty(J, \mathbb{R}).$$

This shows that $(\psi')'$ belongs to $\mathsf{L}^2(\overline{J})$ and equals $\lambda_1\psi$ (see Section A.4). Therefore, $\psi \in \mathsf{H}^2(\overline{J})$, and we have (1.1.2.6). Then, an iterative argument based on (1.1.2.6) shows that $\psi \in \mathsf{H}^n(\overline{J}) \subset \mathscr{C}^{n-1}(\overline{J})$ for all n. Thus, $\psi \in \mathscr{C}^\infty(\overline{J})$ and (1.1.2.6) is satisfied in a classical sense. $\qquad\square$

The discerning reader can remark that a more direct proof would be accessible by using derivatives in the sense of distributions. She/He could also observe that our initial question has led to the introduction of the « unbounded » operator $-\Delta = -\partial_x^2$ (whose definition involves questions about its domain of definition). The action of $-\Delta$ is indeed obvious on $H^2(J)$ but not on the space $H_0^1(J)$ (which was the a priori regularity needed for ψ in Lemma 1.7). These domain issues are explained in Chapter 2.

Lemma 1.8. — *The λ_1 for which there are nontrivial solutions to (1.1.2.6) being 0 at a and b are exactly the numbers $(b-a)^{-2}n^2\pi^2$ where $n \in \mathbb{N}^*$. The corresponding solutions are proportional to $\sin(n\pi(x-a)(b-a)^{-1})$.*

Proof. — Since ψ is smooth, by Cauchy–Lipschitz theorem, the non-zero solutions to (1.1.2.6) being 0 at a are given by $\psi(x) = \sin(\sqrt{\lambda_1}(x-a))$. The extra condition $\psi(b) = 0$ is satisfied if and only if $\sqrt{\lambda_1}(b-a) = n\pi$ for some $n \in \mathbb{N}^*$. $\square$

We have found the answer to our preliminary question. In accordance with the rough estimate (1.1.2.5), we have proved that

$$\boxed{\lambda_1 = (b-a)^{-2}\,\pi^2\,.}$$

In the language of spectral theory, this equality tells us that the smallest eigenvalue of the Dirichlet Laplacian on J (which is a self-adjoint operator) is $(b-a)^{-2}\,\pi^2$. The notions to fully understand this sentence are introduced in Chapters 3, 4, 5 and 6.

1.3. Some density results

This section is here to help the reader to check that everything is clear in her/his mind about basic density results that will be used very often in this book.

Let us consider a sequence of smooth non-negative functions $(\rho_n)_{n\in\mathbb{N}^*}$ with $\int_{\mathbb{R}^d} \rho_n(x)\,\mathrm{d}x = 1$ with $\mathrm{supp}\rho_n = B\left(0, \frac{1}{n}\right)$. Consider a smooth function with compact support $0 \leqslant \chi \leqslant 1$ equal to 1 in a neighborhood of 0, and define $\chi_n(\cdot) = \chi(n^{-1}\cdot)$.

Lemma 1.9. — *Let $p \in [1, +\infty)$. Let $f \in L^p(\mathbb{R}^d)$. Then, $\rho_n \star f$ and $\chi_n(\rho_n \star f)$ converges to f in $L^p(\mathbb{R}^d)$. In particular, $\mathscr{C}_0^\infty(\mathbb{R}^d)$ is dense in $(L^p(\mathbb{R}), \|\cdot\|_{L^p(\mathbb{R}^d)})$.*

Proof. — Let $\varepsilon > 0$ and $f \in \mathscr{C}_0^0(\mathbb{R}^d)$ such that $\|f - f_0\|_{L^p(\mathbb{R}^d)} \leqslant \varepsilon$. We have

$$\rho_n \star f_0(x) - f_0(x) = \int_{\mathbb{R}^d} \rho_n(y)(f_0(x - y) - f_0(x))\, \mathrm{d}y\,,$$

and, by the Hölder inequality (with measure $\rho_n\, \mathrm{d}y$),

$$\|\rho_n \star f_0 - f_0\|_{L^p(\mathbb{R}^d)}^p \leqslant \int_{\mathbb{R}^d} \int_{\mathbb{R}^d} \rho_n(y)|f_0(x - y) - f_0(x)|^p\, \mathrm{d}y\, \mathrm{d}x\,.$$

By using the uniform continuity of f_0 and the support of ρ_n, we see that $\rho_n \star f_0$ converges to f_0 in $L^p(\mathbb{R}^d)$. It remains to notice that

$$\|\rho_n \star (f - f_0)\|_{L^p(\mathbb{R}^d)} \leqslant \|f - f_0\|_{L^p(\mathbb{R}^d)}\,,$$

to see that $\rho_n \star f$ converges to f in $L^p(\mathbb{R}^d)$.

Then, we consider

$$\|(1 - \chi_n)\rho_n \star f\|_{L^p(\mathbb{R}^d)}^p$$

$$\leqslant \int_{\mathbb{R}^d} (1 - \chi_n(x))^p \int_{\mathbb{R}^d} \rho_n(y)|f(x - y)|^p\, \mathrm{d}y\, \mathrm{d}x\,,$$

and we get

$$\|(1 - \chi_n)\rho_n \star f\|_{L^p(\mathbb{R}^d)}^p$$

$$\leqslant \int_{|x|\geqslant n-1} \int_{\mathbb{R}^d} (1 - \chi_n(x + y))^p \rho_n(y)|f(x)|^p\, \mathrm{d}y\, \mathrm{d}x$$

$$\leqslant \int_{|x|\geqslant n-1} |f(x)|^p\, \mathrm{d}x\,,$$

and the conclusion follows since $f \in L^p(\mathbb{R}^d)$. $\qquad\qquad\square$

Lemma 1.10. — *Let $k \in \mathbb{N}$. $\mathscr{C}_0^\infty(\mathbb{R}^d)$ is dense in the Sobolev space $(H^k(\mathbb{R}^d), \|\cdot\|_{H^k(\mathbb{R}^d)})$.*

Proof. — The definition of $H^k(\mathbb{R}^d)$ is recalled in Section A.4. Let us only deal with the case $k = 1$. Let $f \in H^1(\mathbb{R}^d)$. We let $f_n = \chi_n(\rho_n \star f)$.

First, notice that (f_n) converges to f in $L^2(\mathbb{R}^d)$. Then, we have (first, in the sense of distributions, and then in the usual sense)

$$(1.1.3.7) \qquad f_n' = \chi_n' \rho_n \star f + \chi_n \rho_n \star f'\,.$$

This can be checked by considering $\langle f_n', \varphi \rangle$ with $\varphi \in \mathscr{C}_0^\infty(\mathbb{R}^d)$, and using the Fubini theorem. The first term in (1.1.3.7) converges to 0 in $L^2(\mathbb{R}^d)$ and the second one goes to f' in $L^2(\mathbb{R}^d)$.

$\square$

Consider

$$B^1(\mathbb{R}) = \{\psi \in H^1(\mathbb{R}) : x\psi \in L^2(\mathbb{R})\} \subset L^2(\mathbb{R}).$$

We let, for all $\varphi, \psi \in B^1(\mathbb{R})$,

$$Q(\varphi, \psi) = \langle \varphi, \psi \rangle_{H^1(\mathbb{R})} + \langle x\varphi, x\psi \rangle_{L^2(\mathbb{R})}.$$

Lemma 1.11. — $(B^1(\mathbb{R}), Q)$ *is a Hilbert space.*

The following lemma will be convenient.

Lemma 1.12. — $\mathscr{C}_0^\infty(\mathbb{R})$ *is dense in* $(B^1(\mathbb{R}), \|\cdot\|_{B^1(\mathbb{R})})$.

Proof. — Let us recall Lemma 1.10. Let $f \in B^1(\mathbb{R})$. As in Lemma 1.10, we introduce the sequence $f_n = \chi_n(\rho_n \star f)$. We have seen that f_n goes to f in $H^1(\mathbb{R})$. Let us prove that xf_n goes to xf in $L^2(\mathbb{R})$. Since $xf \in L^2(\mathbb{R})$, $\chi_n(\rho_n \star (xf))$ goes to $xf \in L^2(\mathbb{R})$. We write

$$xf_n(x) - xf(x) = x\chi_n\rho_n \star f(x) - xf(x)$$
$$= n^{-1}\chi_n\tilde{\rho}_n \star f(x) + \chi_n\rho_n \star (xf) - xf(x),$$

with $\tilde{\rho}_n(y) = n^2 y\rho(ny)$. Then, we get

$$\|\chi_n\tilde{\rho}_n \star f\|_{L^2(\mathbb{R})} \leqslant \|\tilde{\rho}_n\|_{L^1(\mathbb{R})}\|f\|_{L^2(\mathbb{R})} = \|(\cdot)\rho(\cdot)\|_{L^1(\mathbb{R})}\|f\|_{L^2(\mathbb{R})}.$$

The conclusion follows.

$\square$

Exercise 1.13. — We let, for all $\varphi, \psi \in \mathscr{S}(\mathbb{R})$,

$$Q_\pm(\varphi, \psi) = \langle \varphi, \psi \rangle_{L^2(\mathbb{R})} + \langle (\pm\partial_x + x)\varphi, (\pm\partial_x + x)\psi \rangle_{L^2(\mathbb{R})}.$$

i. Prove that $\mathscr{C}_0^\infty(\mathbb{R})$ is dense in $\mathscr{S}(\mathbb{R})$ for the $\mathcal{V}_\pm$-norm $\sqrt{Q_\pm(\cdot, \cdot)}$.

ii. Consider[2]
$$\mathcal{V}_\pm = \{\psi \in L^2(\mathbb{R}) : (\pm\partial_x + x)\psi \in L^2(\mathbb{R})\} \subset L^2(\mathbb{R}).$$
Show that $(\mathcal{V}_\pm, Q_\pm)$ is a Hilbert space.

iii. Let $f \in \mathcal{V}_\pm$. Show that the sequence $f_n = \chi_n(\rho_n \star f)$ converges to f for the $\mathcal{V}_\pm$-norm $\sqrt{Q_\pm(\cdot, \cdot)}$. *Hint: Prove first that, for all $\varphi \in \mathscr{S}(\mathbb{R})$, and all $n \geqslant 1$,*

$$|\langle (\pm\partial_x + x)(\rho_n \star f) - \rho_n \star [(\pm\partial_x + x)f], \varphi \rangle| \leqslant \frac{C}{n}\|\varphi\|_{L^2(\mathbb{R})}.$$

Then, check that

$$\|(\pm\partial_x + x)(\chi_n \rho_n \star f) - \chi_n(\pm\partial_x + x)(\rho_n \star f)\|_{L^2(\mathbb{R})}$$
$$\leqslant \frac{C}{n}\|\rho_n \star f\|_{L^2(\mathbb{R})} \leqslant \frac{C}{n}\|f\|_{L^2(\mathbb{R})}.$$

Remark 1.14. — Exercise 1.13 shows that $\mathscr{C}_0^\infty(\mathbb{R})$ is dense in $\mathcal{V}_\pm$ (equipped with the natural norm). This density result could also be proved by using spectral results established in this book. Consider $f \in \mathcal{V}_\pm$ such that, for all $\varphi \in \mathscr{S}(\mathbb{R})$,

$$\langle f, \varphi \rangle_{\mathcal{V}_\pm} := \langle f, \varphi \rangle + \langle (\pm\partial_x + x)f, (\pm\partial_x + x)\varphi \rangle = 0,$$

which can also be written as

$$\langle f, \varphi \rangle + \langle f, (\mp\partial_x + x)(\pm\partial_x + x)\varphi \rangle = 0,$$

or also

$$\langle f, \varphi \rangle + \langle f, (-\partial_x^2 + x^2 \mp 1)\varphi \rangle = 0.$$

We will see in Chapter 6 (cf. Proposition 6.10) that there exists a complete orthonormal system[3] (contained in $\mathscr{S}(\mathbb{R})$) denoted by $(g_n)_{n \in \mathbb{N}}$ such that

$$(-\partial_x^2 + x^2)g_n = (2n + 1)g_n.$$

We deduce that $\langle f, g_n \rangle = 0$ for all $n \in \mathbb{N}$. Since the system $(g_n)_{n \in \mathbb{N}}$ is complete, we get $f = 0$. This shows that $\mathscr{S}(\mathbb{R})$ is dense

---

[2]$\psi \in \mathcal{V}_\pm$ means that there exists $f \in L^2(\mathbb{R})$ such that, we have

(1.1.3.8) $\qquad \forall \varphi \in \mathscr{C}_0^\infty(\mathbb{R}), \quad \langle \psi, (\mp\partial_x + x)\varphi \rangle = \langle f, \varphi \rangle.$

This f is unique and is denoted by $f =: (\pm\partial_x + x)\psi$. By density, (1.1.3.8) also holds for all $\varphi \in \mathscr{S}(\mathbb{R})$.

[3]The corresponding functions are the famous Hermite functions and they diagonalize the « harmonic oscillator ».

in $\mathcal{V}_{\pm}$. This is then a small exercise (using smooth cutoff functions) to check that $\mathscr{C}_0^\infty(\mathbb{R})$ is itself dense in $\mathscr{S}(\mathbb{R})$ for the $\mathcal{V}_{\pm}$-norm.

In this book, we will meet Sobolev spaces on open subsets of $\mathbb{R}^d$. Let us discuss the case of $\mathsf{H}^1(\mathbb{R}_+)$. In particular, we need to be careful with the density of smooth functions. Behind the proof of the following proposition lies a general argument related to extension operators (which will appear later, see Section 4.2.2.2).

Proposition 1.15. — *We have* $\mathsf{H}^1(\mathbb{R}_+) \subset \mathscr{C}^0(\overline{\mathbb{R}_+})$. *Moreover,* $\mathscr{C}_0^\infty(\overline{\mathbb{R}_+})$ *is dense in* $\mathsf{H}^1(\mathbb{R}_+)$.

Proof. — Let $\psi \in \mathsf{H}^1(\mathbb{R}_+)$. We define $\underline{\psi}$, the function defined by $\underline{\psi}(x) = \psi(x)\mathbb{1}_{\mathbb{R}_+}(x) + \psi(-x)\mathbb{1}_{\mathbb{R}_-}(x)$. Let us prove that $\underline{\psi} \in \mathsf{H}^1(\mathbb{R})$ and $\|\underline{\psi}\|_{\mathsf{H}^1(\mathbb{R})}^2 = 2\|\psi\|_{\mathsf{H}^1(\mathbb{R}_+)}^2$. Obviously, we have $\underline{\psi} \in \mathsf{L}^2(\mathbb{R})$. Let $\varphi \in \mathscr{C}_0^\infty(\mathbb{R})$ and consider

$$
\begin{aligned}
\langle \underline{\psi}, \varphi' \rangle_{\mathsf{L}^2(\mathbb{R})} &= \int_0^{+\infty} \psi(t)\varphi'(t)\,\mathrm{d}t + \int_{-\infty}^0 \psi(-t)\varphi'(t)\,\mathrm{d}t \\
&= \int_0^{+\infty} \psi(t)(\varphi'(t) + \varphi'(-t))\,\mathrm{d}t \\
&= \int_0^{+\infty} \psi(t)\Phi'(t)\,\mathrm{d}t,
\end{aligned}
$$

with $\Phi(t) = \varphi(t) - \varphi(-t)$, for all $t \in \mathbb{R}$. Consider a smooth even function χ being 0 on $\left(-\frac{1}{2}, \frac{1}{2}\right)$ and 1 away from $(-1, 1)$. We let $\chi_n(t) = \chi(nt)$. We have $(\chi_n\Phi)_{|[0,+\infty)} \in \mathscr{C}_0^\infty(\mathbb{R}_+)$. Since $\psi \in \mathsf{H}^1(\mathbb{R}_+)$, there exists a function $f \in \mathsf{L}^2(\mathbb{R}_+)$ such that, for all $\phi \in \mathscr{C}_0^\infty(\mathbb{R}_+)$,

$$
\langle \psi, \phi' \rangle_{\mathsf{L}^2(\mathbb{R}_+)} = -\langle f, \phi \rangle_{\mathsf{L}^2(\mathbb{R}_+)}.
$$

By changing ψ into $\chi_n\psi$, we have

$$\langle \underline{\psi}, \chi_n\varphi' \rangle_{\mathsf{L}^2(\mathbb{R})}$$

$$= \int_0^{+\infty} \psi(t)\chi_n(t)\Phi'(t)\, \mathrm{d}t$$

$$= \int_0^{+\infty} \psi(t)(\chi_n\Phi)'(t)\, \mathrm{d}t - \int_0^{+\infty} \psi(t)\chi_n'(t)\Phi(t)\, \mathrm{d}t$$

$$= -\langle f, \chi_n\Phi \rangle_{\mathsf{L}^2(\mathbb{R}_+)} - \int_0^{+\infty} \psi(t)\chi_n'(t)\Phi(t)\, \mathrm{d}t.$$

By using the behavior of Φ at 0 and a support consideration, we get

$$\lim_{n\to+\infty} \int_0^{+\infty} \psi(t)\chi_n'(t)\Phi(t)\, \mathrm{d}t = 0\,,$$

and we deduce

$$\langle \underline{\psi}, \varphi' \rangle_{\mathsf{L}^2(\mathbb{R})} = -\langle f, \Phi \rangle_{\mathsf{L}^2(\mathbb{R}_+)}\,,$$

and thus

$$|\langle \underline{\psi}, \varphi' \rangle_{\mathsf{L}^2(\mathbb{R})}|^2 \leqslant 2\|f\|_{\mathsf{L}^2(\mathbb{R}_+)}^2 \|\varphi\|_{\mathsf{L}^2(\mathbb{R})}^2\,.$$

This proves that $\underline{\psi} \in \mathsf{H}^1(\mathbb{R})$ and the relation between the H^1-norms follows. Thus $\underline{\psi} \in \mathscr{C}^0(\mathbb{R})$. The conclusion about the density follows from Lemma 1.10. $\qquad\square$

CHAPTER 2

UNBOUNDED OPERATORS

The aim of this chapter is to describe what a *(closed) linear operator* is. It also aims at drawing the attention of the Reader to the *domain* of such an operator. Such domains will be explicitly described (such as the domain of the Dirichlet Laplacian). We will see that *closed* operators are natural generalizations of continuous operators. Then, we will define what the adjoint of an operator is, and explore the special case when the adjoint of an operator coincides with itself (*self-adjoint* operators). We will give classical criteria to determine if an operator is self-adjoint and illustrate these criteria by means of explicit examples. The Reader will be provided with a canonical way (the Lax–Milgram theorems) of defining an operator from a continuous and coercive sesquilinear form. Let us again underline here that the domain of the operator is as important as its action. Changing the domain can strongly change the spectrum.

2.1. Definitions

In this chapter, E and F are Banach spaces.

Definition 2.1 (**Unbounded operator**). — An *unbounded operator* $T : E \longrightarrow F$ is a pair $(\mathrm{Dom}\,(T), T)$ where

— $\mathrm{Dom}\,(T)$ is a linear subspace of E;
— T is a linear map from $\mathrm{Dom}\,(T)$ to F.

In contrast with bounded operators, unbounded operators on a given space do not form an algebra, nor even a linear space (because each one is defined on its own domain). Moreover, the term « unbounded operator » may be misleading. Note that

— *unbounded* does not mean *not bounded*. As a matter of fact, a bounded operator can be seen as an unbounded (closed) operator whose domain is the whole space.
— *unbounded* should be understood as "not necessarily bounded".
— *operator* should be understood as *linear operator*.

Let us also recall that $\mathcal{L}(E, F)$ denotes the set of bounded (*i.e.*, continuous) linear applications from E to F.

***Definition 2.2* (Domain).** — The set $\mathrm{Dom}\,(T)$ is called the *domain* of T.

The domain of an operator is a linear subspace, not necessarily the whole space. It is not necessarily closed. It will often (but not always) be assumed to be dense.

***Definition 2.3* (Range).** — The linear subspace

$$\mathrm{ran}\,T := \big\{Tx : x \in \mathrm{Dom}\,(T)\big\}$$

is called the *range* of T.

***Exercise 2.4.* —** Take $E = F = \mathsf{L}^2(\mathbb{R})$ and $(\mathrm{Dom}\,(T), T) = (\mathscr{C}_0^\infty(\mathbb{R}), -\partial_x^2)$. What is the range of T?

Solution: This is the linear subset of $\mathscr{C}_0^\infty(\mathbb{R})$ made of functions f satisfying

$$\int_{\mathbb{R}} f(y)\,\mathrm{d}y = 0\,, \qquad \int_{\mathbb{R}} \left(\int_{-\infty}^{x} f(y)dy\right)\mathrm{d}x = 0\,.$$

$\circ$

***Definition 2.5.* —** We say that T is *densely defined* when $\mathrm{Dom}\,(T)$ is dense in E.

***Definition 2.6* (Graph).** — The *graph* $\Gamma(T)$ of $(\mathrm{Dom}\,(T), T)$ is

$$\Gamma(T) = \big\{(x, Tx), x \in \mathrm{Dom}\,(T)\big\} \subset E \times F\,.$$

Definition 2.7 (Graph norm). — Let $(\mathrm{Dom}\,(T), T)$ be some unbounded operator. For all $x \in \mathrm{Dom}\,(T)$, we let

(2.2.1.1) $$\|x\|_T := \|x\|_E + \|Tx\|_F.$$

The pair $(\mathrm{Dom}\,(T), \|\cdot\|_T)$ is a normed vector space. The norm $\|\cdot\|_T$ is called the *graph norm*.

Definition 2.8 (Sesquilinear form associated with the graph norm)

Let $(\mathrm{Dom}\,(T), T)$ be some unbounded operator between two Hilbert spaces E and F. For all $(x, y) \in \mathrm{Dom}\,(T)^2$, we let

(2.2.1.2) $$\langle x, y \rangle_T = \langle x, y \rangle_E + \langle Tx, Ty \rangle_F.$$

The Hermitian inner product $\langle \cdot, \cdot \rangle_T$ is called the *sesquilinear form associated with the graph norm*.

Definition 2.9. — Let $(\mathrm{Dom}\,(T), T)$ and $(\mathrm{Dom}\,(S), S)$ be two operators. We say that S is an *extension* of T when $\Gamma(T) \subset \Gamma(S)$. In this case, we simply write $T \subset S$.

Proposition 2.10. — *We have* $T \subset S$ *if and only if* $\mathrm{Dom}\,(T) \subset \mathrm{Dom}\,(S)$ *and* $S_{|\mathrm{Dom}\,(T)} \equiv T$.

Proof. — By definition, the operator S is an extension of T when for all $x \in \mathrm{Dom}\,(T)$, we have $(x, Tx) \in \Gamma(S)$, that is, $(x, Tx) = (\tilde{x}, S\tilde{x})$ for some $\tilde{x} \in \mathrm{Dom}\,(S)$. Necessarily, we must have $x = \tilde{x} \in \mathrm{Dom}\,(S)$ and $Tx = S\tilde{x} = Sx$. The converse is obvious. $\square$

Definition 2.11 (Closed operator). — The unbounded operator $(\mathrm{Dom}\,(T), T)$ is said to be *closed* when $\Gamma(T)$ is a closed subset of $E \times F$ (equipped with the product norm $\|(u, v)\| = \|u\|_E + \|v\|_F$).

Proposition 2.12. — *The following assertions are equivalent.*

  (i) $(\mathrm{Dom}\,(T), T)$ *is closed.*
  (ii) *For all* $(u_n) \in \mathrm{Dom}\,(T)^{\mathbb{N}}$ *such that* $u_n \to u$ *in* E *and* $Tu_n \to v$ *in* F, *we have* $u \in \mathrm{Dom}\,(T)$ *and* $v = Tu$.
  (iii) $(\mathrm{Dom}\,(T), \|\cdot\|_T)$ *is a Banach space.*

Proof. — (i) $\implies$ (ii) The two conditions $u_n \to u$ and $v_n = Tu_n \to v$ mean that

$$(u_n, v_n) \in \Gamma(T), \qquad (u_n, v_n) \to (u, v) \text{ in } E \times F.$$

Since $\Gamma(T)$ is closed, we must have $(u, v) \in \Gamma(T)$, that is, $u \in \text{Dom}\,(T)$ and $v = Tu$.

(ii) $\implies$ (iii) Consider a Cauchy sequence $(u_n) \in \text{Dom}\,(T)^{\mathbb{N}}$ for the graph norm. This implies that (u_n) is a Cauchy sequence in the Banach space E, and that (Tu_n) is a Cauchy sequence in the Banach space F. Therefore (u_n, v_n) converges to some $(u, v) \in E \times F$. In view of (ii), we have $u \in \text{Dom}\,(T)$ and $u_n \to u$ for the graph norm.

(iii) $\implies$ (i) Consider a sequence $(u_n, Tu_n) \in \Gamma(T)^{\mathbb{N}}$ converging to some $(u, v) \in E \times F$ for the product norm. Then (u_n) is a Cauchy sequence for the graph norm, and therefore it converges to some $\tilde{u} \in \text{Dom}\,(T)$. Since Banach spaces are separated, we must have $u = \tilde{u}$ and $v = T\tilde{u}$, which means that $(u, v) \in \Gamma(T)$.

$\square$

Exercise 2.13. — Take $E = F = \mathsf{L}^2(\mathbb{R}^d)$. Prove through two methods that the operator $(\text{Dom}\,(T), T) = (\mathsf{H}^2(\mathbb{R}^d), -\Delta)$ is closed.

Solution: Recalling that $\|u\|_T = \|u\|_{\mathsf{L}^2} + \|\Delta u\|_{\mathsf{L}^2}$, the key point is that

$$(2.2.1.3) \qquad C_1 \|\langle \xi \rangle^2 \hat{u}\|_{\mathsf{L}^2(\mathbb{R}^d)} \leqslant \|u\|_T \leqslant C_2 \|\langle \xi \rangle^2 \hat{u}\|_{\mathsf{L}^2(\mathbb{R}^d)},$$

and that

$$(2.2.1.4) \qquad \tilde{C}_1 \|u\|_{\mathsf{H}^2(\mathbb{R}^d)} \leqslant \|\langle \xi \rangle^2 \hat{u}\|_{\mathsf{L}^2(\mathbb{R}^d)} \leqslant \tilde{C}_2 \|u\|_{\mathsf{H}^2(\mathbb{R}^d)}.$$

— First method. Let $(u_n, v_n) \in \Gamma(T)^{\mathbb{N}}$ be such that $(u_n, v_n) \to (u, v)$ in $\mathsf{L}^2(\mathbb{R}^d) \times \mathsf{L}^2(\mathbb{R}^d)$. We must have $v_n = -\Delta u_n$, and therefore $v_n \to -\Delta u$ in $\mathscr{D}'(\mathbb{R}^d)$. Since the limit is unique, this means that $v = -\Delta u \in \mathsf{L}^2(\mathbb{R}^d)$. From (2.2.1.3) and (2.2.1.4), we deduce that $u \in \mathsf{H}^2(\mathbb{R}^d)$, and therefore we have $(u, v) \in \Gamma(T)$.

— Second (more direct) method. Observe that (2.2.1.3) and (2.2.1.4) imply that $(\text{Dom}\,(T), \|\cdot\|_T)$ and $(\mathsf{H}^2(\mathbb{R}^d), \|\cdot\|_{\mathsf{H}^2})$

are two isomorphic normed spaces. The second one is a Banach space, so is the first one. Criterion (iii) is satisfied. ○

Proposition 2.14. — *Let* $(\mathrm{Dom}\,(T), T)$ *be a closed operator. There exists* $c > 0$ *such that*

$$(2.2.1.5) \qquad \forall u \in \mathrm{Dom}\,(T), \quad \|Tu\| \geqslant c\|u\|,$$

if and only if T *is injective with a closed range.*

Proof. — Assume that the inequality holds. The injectivity is obvious. Let us consider (v_n) in the range of T such that (v_n) converges to $v \in F$. For all $n \in \mathbb{N}$, there exists $u_n \in \mathrm{Dom}\,(T)$ such that $v_n = Tu_n$. We deduce from (2.2.1.5) that (u_n) is a Cauchy sequence so that it converges to some $u \in E$. Since T is closed, we find that $u \in \mathrm{Dom}\,(T)$ and $v = Tu \in \mathrm{ran}\,T$.

Conversely, assume that T is injective with a closed range. Then $(\mathrm{ran}\,T, \|\cdot\|_F)$ is a Banach space. Then T induces a continuous bijection from $(\mathrm{Dom}\,(T), \|\cdot\|_T)$ to $(\mathrm{ran}\,T, \|\cdot\|_F)$. The inverse is continuous by the Banach isomorphism theorem, or by the open mapping theorem (see Section A.2). □

Exercise 2.15. — Prove that there exists a constant $c > 0$ such that

$$(2.2.1.6) \quad \forall \varphi \in \mathsf{H}^2(\mathbb{R}^d), \quad \|(-\Delta + 1)\varphi\|_{\mathsf{L}^2(\mathbb{R}^d)} \geqslant c\|\varphi\|_{\mathsf{L}^2(\mathbb{R}^d)}.$$

Show that this holds for $c = 1$. What is the optimal c?

Solution: Take $E = F = \mathsf{L}^2(\mathbb{R}^d)$. First, as we did in Exercise 2.13, we can prove that the operator $(\mathsf{H}^2(\mathbb{R}^d), -\Delta + 1)$ is closed. Indeed, thanks to the Fourier transform,

$$\|u\|_T = \|u\|_{\mathsf{L}^2} + \|-\Delta u + u\|_{\mathsf{L}^2} = \|\hat{u}\|_{\mathsf{L}^2} + \|(1 + |\xi|^2)\hat{u}\|_{\mathsf{L}^2}.$$

In particular, there exist $C_1, C_2 > 0$ such that, for all $u \in \mathsf{H}^2(\mathbb{R}^d)$,

$$C_1\|u\|_{\mathsf{H}^2(\mathbb{R}^d)} \leqslant \|u\|_T \leqslant C_2\|u\|_{\mathsf{H}^2(\mathbb{R}^d)}.$$

To obtain (2.2.1.6), in view of Proposition 2.14, it suffices to show that $-\Delta + 1$ is injective with a closed range.

— Assume that $\tilde{u} \in \mathsf{H}^2(\mathbb{R}^d)$ is such that $(-\Delta + 1)\tilde{u} = 0$. Then

$$\|(-\Delta + 1)\tilde{u}\|_{\mathsf{L}^2} = \|(1 + |\xi|^2)\widehat{\tilde{u}}\|_{\mathsf{L}^2} = 0,$$

implying that $\widehat{\tilde{u}} = 0$, and therefore $\tilde{u} = 0$. Thus, $-\Delta + 1$ is injective.

— Fix $v \in L^2(\mathbb{R}^d)$. Define w as the inverse Fourier transform of $\hat{w} := (1 + |\xi|^2)^{-1}\hat{v}$. Thus we have $v = (-\Delta + 1)w$, as well as $w \in H^2(\mathbb{R}^d)$. In other words, $\mathrm{ran}\,(-\Delta + 1) = L^2(\mathbb{R}^d)$, which is closed for the L^2-norm. Then, remark that

$$\|(-\Delta + 1)\varphi\|_{L^2(\mathbb{R}^d)} = \|(1 + |\xi|^2)\hat{\varphi}\|_{L^2} \geqslant \|\hat{\varphi}\|_{L^2(\mathbb{R}^d)} = \|\varphi\|_{L^2(\mathbb{R}^d)}.$$

The optimal constant is $c = 1$ as can be seen by testing the above inequality with a sequence of functions φ_n such that $\hat{\varphi}_n$ concentrates near $\xi = 0$, like $\hat{\varphi}_n(\xi) = n^{d/2}\chi(n\xi)$ where $\chi \in \mathscr{C}_0^\infty(\mathbb{R}^d)$ is a non-zero function.   ○

The following proposition follows from the results in Section A.2.

Proposition 2.16. — *[Closed graph theorem] Let* $(\mathrm{Dom}\,(T), T)$ *be an operator. Assume that* $\mathrm{Dom}\,(T) = E$. *Then, the operator* $(\mathrm{Dom}\,(T), T)$ *is closed if and only if* T *is bounded.*

In other words, on condition that $\mathrm{Dom}\,(T) = E$, the *closed graph theorem* says that T is continuous if and only if $\Gamma(T)$ is a closed subset of $E \times F$. Thus, the concept of a closed operator can be viewed as a generalization of the notion of a bounded (or continuous) operator.

Exercise 2.17. — Prove Proposition 2.16 by using the open mapping theorem.

Solution:
$\Longrightarrow$ Assume that $\Gamma(T)$ is a closed. Then, $\Gamma(T)$ equipped with the product norm of $E \times F$ is a Banach space, and the application

$$U : \Gamma(T) \longrightarrow E$$
$$(x, Tx) \longmapsto x$$

is a linear bounded bijection. By the open mapping theorem, U^{-1} is bounded and therefore

$$\exists C \in \mathbb{R}, , \quad \|x\|_E + \|Tx\|_F \leqslant C\|x\|_E .$$

This means that T is bounded with a norm less than $C - 1$.

$\Longleftarrow$ Conversely, assume that T is bounded. Select any sequence $(u_n) \in E^{\mathbb{N}}$ such that $u_n \to u$ and $Tu_n \to v$. Since $\mathrm{Dom}\,(T) = E$, we have $u \in \mathrm{Dom}\,(T)$. On the other hand, since T is linear and bounded, it is continuous. We must have $Tu_n \to v = Tu$. We recover here the criterion (ii) of Proposition 2.12.   ○

Example 2.18. — Let $\Omega \subset \mathbb{R}^d$ and $K \in \mathsf{L}^2(\Omega \times \Omega)$. For all $\psi \in \mathsf{L}^2(\Omega)$, we let

$$T_K\psi(x) = \int_\Omega K(x,y)\psi(y)\,\mathrm{d}y\,.$$

$T_K : \mathsf{L}^2(\Omega) \to \mathsf{L}^2(\Omega)$ is well-defined and bounded. Moreover, $\|T_K\| \leqslant \|K\|_{\mathsf{L}^2(\Omega \times \Omega)}$.

Definition 2.19 (Closable operator). — $(\mathrm{Dom}\,(T), T)$ is said to be *closable* when it admits a closed extension.

Proposition 2.20. — *The following assertions are equivalent.*

(i) $\overline{(\mathrm{Dom}\,(T), T)}$ *is closable.*
(ii) $\overline{\Gamma(T)}$ *is the graph of an operator.*
(iii) *For all* $(u_n) \in \mathrm{Dom}\,(T)^{\mathbb{N}}$ *such that* $u_n \to 0$ *and* $Tu_n \to v$, *we have* $v = 0$.

Proof. — (i) $\Longrightarrow$ (ii). Let $(\mathrm{Dom}\,(S), S)$ be a closed extension of $(\mathrm{Dom}\,(T), T)$. Then $\Gamma(T) \subset \Gamma(S)$ and $\overline{\Gamma(T)} \subset \overline{\Gamma(S)} = \Gamma(S)$. Define

$$\mathrm{Dom}\,(R) = \big\{x \in E : \exists y \in F \text{ with } (x,y) \in \overline{\Gamma(T)}\big\}\,.$$

Given $x \in \mathrm{Dom}\,(R)$, the corresponding $y = Sx$ is uniquely defined. Note also that $\mathrm{Dom}\,(R)$ is a vector space and that $\mathrm{Dom}\,(R) \subset \mathrm{Dom}\,(S)$. For all $x \in \mathrm{Dom}\,(R)$, we let $Rx = Sx$. Thus we find $\Gamma(R) = \overline{\Gamma(T)}$.

(ii) $\Longrightarrow$ (i). Assume that $\overline{\Gamma(T)}$ is the graph of an operator R. Then R is a closed extension of T.

(ii) $\Longrightarrow$ (iii). Assume that $\overline{\Gamma(T)}$ is the graph $\Gamma(R)$ of an operator R. Let $(u_n) \in \mathrm{Dom}\,(T)^{\mathbb{N}}$ be such that $u_n \to 0$ and $Tu_n \to v$. Then $(0, v) \in \overline{\Gamma(T)} = \Gamma(R)$, and therefore $(0, v) \in \Gamma(R)$. It follows that $v = R0 = 0$.

(iii) $\Longrightarrow$ (ii). Consider $(x, y) \in \overline{\Gamma(T)}$ and $(x, \tilde{y}) \in \overline{\Gamma(T)}$. We may find sequences (x_n) and $(\tilde{x}_n)$ such that (x_n, Tx_n)

and $(\tilde{x}_n, T\tilde{x}_n)$ converge to (x, y) and $(x, \tilde{y})$, respectively. The sequence $u_n = x_n - \tilde{x}_n$ converges to 0 and Tu_n converges to $y - \tilde{y}$. Thus, $y = \tilde{y}$. This shows that $\overline{\Gamma(T)}$ is a graph. $\square$

All unbounded operators are not closable. We give below a counter-example.

Exercise 2.21. — Take $E = \mathsf{L}^2(\mathbb{R}^d)$ and $F = \mathbb{C}$. Consider the operator T defined on the domain $\mathrm{Dom}\,(T) = \mathscr{C}_0^\infty(\mathbb{R}^d)$ by $T\varphi = \varphi(0)$. Then T is not closable.

Solution: We use the criterion (iii) of Proposition 2.20. Given some function $\varphi \in \mathscr{C}_0^\infty(\mathbb{R}^d; \mathbb{R}_+)$ satisfying $\varphi(0) = 1$, we define $u_n = \varphi_n(x) := \varphi(nx)$. By construction, we have $Tu_n = 1 \neq 0$ for all n, whereas $\|u_n\| = n^{-d/2}\|u_1\|$ goes to zero. $\circ$

Assume that the operator $(\mathrm{Dom}\,(T), T)$ is closable. Then, by (ii) of Proposition 2.20, we can find $(\mathrm{Dom}\,(R), R)$ such that

$$(2.2.1.7) \qquad \overline{\Gamma(T)} = \Gamma(R) = \left\{ (u, Ru)\,;\, u \in \mathrm{Dom}\,(R) \right\}.$$

First, note that the operator R is uniquely determined by the characterization (2.2.1.7). Moreover, we have $\mathrm{Dom}\,(T) \subset \mathrm{Dom}\,(R)$ and $\Gamma(T) \subset \Gamma(R) = \overline{\Gamma(R)}$. Thus, the operator R is a closed extension of T. Let S be another closed extension of T. Then

$$\Gamma(T) \subset \Gamma(S) \quad \Longrightarrow \quad \overline{\Gamma(T)} = \Gamma(R) \subset \overline{\Gamma(S)} = \Gamma(S)$$

which means that S is an extension of R. By this way, the operator R appears as a closed extension of T, which (in the sense of the graph inclusion) is smaller than all the others.

Definition 2.22 **(Closure).** — Assume that $(\mathrm{Dom}\,(T), T)$ is closable. Then, the operator $(\mathrm{Dom}\,(R), R)$ defined by (2.2.1.7) is called the *closure* of $(\mathrm{Dom}\,(T), T)$.

We have a more explicit characterization of the closure.

Proposition 2.23. — *Assume that the operator* $(\mathrm{Dom}\,(T), T)$ *is closable. Then, the closure of* $(\mathrm{Dom}\,(T), T)$ *is the operator* $(\mathrm{Dom}\,(\overline{T}), \overline{T})$ *defined by*

— *the domain*

$$\mathrm{Dom}\,(\overline{T}) := \big\{ x \in E\,; \textit{there exists a sequence } (x_n) \in \mathrm{Dom}\,(T)^{\mathbb{N}}$$
$$\textit{satisfying } x_n \to x \textit{ and } (Tx_n) \textit{ converges in } F$$
$$\textit{and, we can find } y \in F \textit{ such that,}$$
$$\textit{for any such sequence, we have } Tx_n \to y \big\},$$

— *and the relation* $\overline{T}x = y$ *for any* $x \in \mathrm{Dom}\,(\overline{T})$.

The following two exercises give the proof of Proposition 2.23.

Exercise 2.24. — Prove that $\overline{T}$ is well-defined, and that $\overline{T}$ is indeed an extension of T.

Solution: By assumption, we can find a closed operator $(\mathrm{Dom}\,(S), S)$ such that $\Gamma(T) \subset \Gamma(S)$. Let $x \in \mathrm{Dom}\,(T)$. The stationary sequence $(x_n) = (x)$ is such that $Tx_n \to Tx$. Any other sequence $(x_n) \in \mathrm{Dom}\,(T)^{\mathbb{N}}$ satisfying $x_n \to x$ and $(Tx_n = Sx_n)_n$ converging (to some z) in F is such that $(x_n, Sx_n) \to (x, z) \in \Gamma(S)$, and therefore $z = Sx = Tx$. Since Tx is the only possible limit value of such a sequence (Tx_n), we have indeed $x \in \mathrm{Dom}\,(\overline{T})$. Moreover, we have $\overline{T}x = Tx$ when $x \in \mathrm{Dom}\,(T) \subset \mathrm{Dom}\,(\overline{T})$.

For more general $x \in \mathrm{Dom}\,(\overline{T})$, note that y is uniquely identified by x. It follows that $\mathrm{Dom}\,(\overline{T})$ is a linear subspace of E, and that $\overline{T}$ defines a linear operator which is an extension of T. ○

Exercise 2.25. — Assume that $(\mathrm{Dom}\,(T), T)$ is closable, and prove that $\overline{T}$ is the smallest closed extension of T in the sense of the graph inclusion.

Solution: The fact that $\Gamma(\overline{T})$ is closed is a consequence of Cantor's diagonal argument (check the details). Fix $x \in \mathrm{Dom}\,(\overline{T})$. We can find a sequence $(x_n) \in \mathrm{Dom}\,(T)^{\mathbb{N}}$ satisfying $x_n \to x$ and $(Tx_n = Sx_n)_n \to \overline{T}x$ in F. Since $\Gamma(S)$ is closed, we must have $(x, \overline{T}x) \in \Gamma(S)$, and therefore $\Gamma(\overline{T}) \subset \Gamma(S)$. ○

The operator $(\mathrm{Dom}\,(R), R)$ defined by (2.2.1.7) is the same as $(\mathrm{Dom}\,(\overline{T}), \overline{T})$. From now on, it is denoted by $\overline{T}$. Remember the following important result.

Proposition 2.26. — *Assume that* $(\mathrm{Dom}\,(T), T)$ *is closable. Then, we have* $\Gamma(\overline{T}) = \overline{\Gamma(T)}$.

Exercise 2.27. — The closure of $(\mathscr{C}_0^\infty(\mathbb{R}^d), -\Delta)$ is the operator $(\mathrm{H}^2(\mathbb{R}^d), -\Delta)$.

Solution: Take $u \in \mathrm{H}^2(\mathbb{R}^d)$. Since $\mathscr{C}_0^\infty(\mathbb{R}^d)$ is dense in $\mathrm{H}^2(\mathbb{R}^d)$, we can find a sequence $(u_n) \in \mathscr{C}_0^\infty(\mathbb{R}^d)^{\mathbb{N}}$ such that $u_n \to u$ for the norm of $\mathrm{H}^2(\mathbb{R}^d)$. Then, $-\Delta u_n \to -\Delta u$ for the norm of $\mathrm{L}^2(\mathbb{R}^d)$. This means that the graph of $(\mathrm{H}^2(\mathbb{R}^d), -\Delta)$ is contained in the closure of the graph of $(\mathscr{C}_0^\infty(\mathbb{R}^d), -\Delta)$. But, as seen in Exercise 2.13, the operator $(\mathrm{H}^2(\mathbb{R}^d), -\Delta)$ is closed. Thus, it is the smallest closed extension. ○

2.2. Adjoint and closedness

2.2.1. About duality and orthogonality. — In this section, E and F are vector spaces.

Definition 2.28. — Let $T \in \mathcal{L}(E, F)$. For all $\varphi \in F' = \mathcal{L}(F, \mathbb{C})$, we let $T'(\varphi) = \varphi \circ T \in E'$.

Proposition 2.29. — *Let* $T \in \mathcal{L}(E, F)$. *Then* $T' \in \mathcal{L}(F', E')$ *and* $\|T\|_{\mathcal{L}(E,F)} = \|T'\|_{\mathcal{L}(F',E')}$.

Proof. — T' is clearly linear. Let us show that it is continuous. We have

$$\|T'\|_{\mathcal{L}(F',E')} = \sup_{\varphi \in F' \setminus \{0\}} \frac{\|T'\varphi\|_{E'}}{\|\varphi\|_{F'}}$$

$$= \sup_{\varphi \in F' \setminus \{0\}} \sup_{x \in E \setminus \{0\}} \frac{\|T'\varphi(x)\|_F}{\|x\|_E \|\varphi\|_{F'}}$$

$$\leqslant \|T\|_{\mathcal{L}(E,F)}.$$

For the converse inequality, we write, thanks to a corollary of the Hahn–Banach theorem (see Corollary A.2),

$$\|T\|_{\mathcal{L}(E,F)} = \sup_{x \in E \setminus \{0\}} \frac{\|Tx\|_F}{\|x\|_E} = \sup_{x \in E \setminus \{0\}} \sup_{\varphi \in F' \setminus \{0\}} \frac{\|\varphi(Tx)\|}{\|\varphi\|_{F'} \|x\|_E}$$

$$\leqslant \|T'\|_{\mathcal{L}(F',E')}.$$

$\square$

Definition 2.30. — If $A \subset E$, we let

$$A^\perp = \{\varphi \in E' : \varphi_{|A} = 0\} \subset E',$$

and, for all $B \subset E'$, we let

$$B^\circ = \{x \in E : \forall \varphi \in B, \varphi(x) = 0\} \subset E.$$

By construction, both $A^\perp$ and B° are closed sets. There is a deep connection between these notions of orthogonality and the adjoint.

Proposition 2.31. — *Let* $T \in \mathcal{L}(E, F)$. *We have*

$$\ker T' = (\operatorname{ran} T)^\perp \subset F', \qquad \ker T = (\operatorname{ran} T')^\circ \subset E.$$

Proof. — Concerning the first equality, just remark that

$$\begin{aligned}
\ker T' &= \{\varphi \in F' : \forall x \in F : \varphi(Tx) = 0\} \\
&= \{\varphi \in F' : \varphi_{|\operatorname{ran} T} = 0\}.
\end{aligned}$$

Concerning the second one, we have (thanks to the Hahn–Banach theorem; see Section A.1)

$$\begin{aligned}
(\operatorname{ran} T')^\circ &= \{x \in E : \forall \varphi \in \operatorname{ran} T' : \varphi(x) = 0\} \\
&= \{x \in E : \forall \psi \in F' : T'(\psi)(x) = 0\} \\
&= \{x \in E : \forall \psi \in F' : \psi \circ T(x) = 0\} \\
&= \{x \in E : T(x) = 0\} = \ker T.
\end{aligned}$$

$\square$

Lemma 2.32. — *Assume that* $(E, \|\cdot\|)$ *is a Banach space. Let us write* $E = E_1 \oplus E_2$ *with* E_1 *and* E_2 *closed. Then, the projections* Π_{E_1} *and* Π_{E_2} *are bounded.*

Proof. — For all $x \in E$, there exists a unique $(x_1, x_2) \in E_1 \times E_2$ such that $x = x_1 + x_2$. We introduce the norm defined for all $x \in E$ by

$$\|x\|' = \|x_1\| + \|x_2\|.$$

Since E_1 and E_2 are closed, $(E, \|\cdot\|')$ is a Banach space. We have

$$\forall x \in E, \quad \|x\| \leqslant \|x\|'.$$

By the Banach theorem (see Theorem A.6), $\|\cdot\|$ and $\|\cdot\|'$ are equivalent, and thus there exists $C > 0$ such that

$$\forall x \in E, \quad \|x\|' \leqslant C\|x\|.$$

□

Let us recall the notion of codimension.

Definition 2.33. — Let E be a vector space. Let E_1 and E_2 be two subspaces such that $E = E_1 \oplus E_2$. Assume that $\dim E_2 < +\infty$. Then, all the supplements of E_1 are finite dimensional and have the same dimension. This dimension is called the codimension of E_1. It is denoted by $\operatorname{codim} E_1$.

The notion of orthogonality is convenient to estimate the codimension.

Proposition 2.34. — *Assume that E is a Banach space. Let us write $E = E_1 \oplus E_2$ with E_1 closed and E_2 finite dimensional. Then, we have $\dim E_1^\perp = \dim E_2 = \operatorname{codim} E_1$.*

Proof. — Consider $N \in \mathbb{N} \setminus \{0\}$. Let $(e_n)_{1 \leqslant n \leqslant N}$ be a basis of E_2. We can consider $(e_n^*)_{1 \leqslant n \leqslant N}$ the dual basis, which satisfies

$$e_i^* \in E_2', \qquad \forall (i,j) \in \{1, \cdots, n\}^2, \quad e_i^*(e_j) = \delta_{ij}.$$

We consider $(e_n^* \Pi_{E_2})_{1 \leqslant n \leqslant N}$. By Lemma 2.32, this is a free family in E' being 0 on E_1. Thus $\dim E_1^\perp \geqslant N$. If $\varphi \in E_1^\perp$ and $x \in E$, we write $x = x_1 + x_2$, with $(x_1, x_2) \in E_1 \times E_2$, and thus

$$\forall x \in E, \quad \varphi(x) = \varphi(x_2) = \sum_{n=1}^{N} e_n^*(x_2) \varphi(e_n)$$

$$= \sum_{n=1}^{N} e_n^*(\Pi_{E_2} x) \varphi(e_n).$$

In other words

$$\varphi = \sum_{n=1}^{N} \varphi(e_n) e_n^* \Pi_{E_2}, \quad e_n^* \Pi_{E_2} \in E'$$

so that $\dim E_1^\perp \leqslant N$.

□

2.2.2. Adjoint of bounded operators in Hilbert spaces. — In this section, we assume that $E = F = \mathsf{H}$ is a separable Hilbert space with norm $\| \cdot \|$. The notion of the *adjoint* of an operator is first introduced in the bounded case.

Proposition 2.35 (Adjoint of a bounded operator)

    Let $T \in \mathcal{L}(\mathsf{H})$ be a bounded operator. For all $x \in \mathsf{H}$, there exists a unique $T^*x \in \mathsf{H}$ such that

$$\forall y \in \mathsf{H}, \quad \langle Ty, x \rangle = \langle y, T^*x \rangle, \quad \langle x, Ty \rangle = \langle T^*x, y \rangle.$$

The application $T^* : \mathsf{H} \to \mathsf{H}$ is a bounded operator called the adjoint of T.

Proof. — The linear application

$$
\begin{array}{rcl}
L : \mathsf{H} & \longrightarrow & \mathbb{C} \\
y & \longmapsto & \langle Ty, x \rangle, \quad |\langle Ty, x \rangle| \leqslant (\|T\|\|x\|)\|y\|
\end{array}
$$

is continuous on H. The Riesz representation theorem guarantees the existence of $T^*x \in \mathsf{H}$ such that $Ly = \langle y, T^*x \rangle$. $\qquad \square$

There is, of course, a relation between T^* and T'.

Definition 2.36. — Let us denote by $\mathscr{J} : \mathsf{H} \to \mathsf{H}'$ the canonical application defined by

$$\forall u \in \mathsf{H}, \quad \forall \varphi \in \mathsf{H}, \quad \mathscr{J}(u)(\varphi) = \langle \varphi, u \rangle.$$

We recall that $\mathscr{J}$ is a bijective isometry by the Riesz representation theorem. In particular, given $v \in \mathsf{H}'$, the element $u = \mathscr{J}^{-1}(v) \in \mathsf{H}$ can be recovered through the relation

$$\forall \varphi \in \mathsf{H}, \quad v(\varphi) = \langle \varphi, \mathscr{J}^{-1}(v) \rangle.$$

Proposition 2.37. — *Let $T \in \mathcal{L}(\mathsf{H})$. We have $T^* = \mathscr{J}^{-1}T'\,\mathscr{J}$.*

Proof. — Consider $(x, y) \in \mathsf{H}^2$ and

$$
\begin{aligned}
\langle x, \mathscr{J}^{-1}T'\,\mathscr{J}y \rangle = T'\,\mathscr{J}y(x) = (\mathscr{J}y)(Tx) &= \langle Tx, y \rangle \\
&= \langle x, T^*y \rangle.
\end{aligned}
$$

$\qquad \square$

Exercise 2.38. — We let $\mathsf{H} = \ell^2(\mathbb{Z}, \mathbb{C})$, equipped with the usual Hermitian scalar product. For all $u \in \mathsf{H}$, we let, for all $n \in \mathbb{Z}$, $(S_- u)_n = u_{n-1}$ and $(S_+ u)_n = u_{n+1}$.

i. Show that S_- and S_+ are bijective isometries.

ii. Prove that $S_\pm^* = S_\mp$.

Solution:

i. Obvious with $S_- = S_+^{-1}$ and $S_+ = S_-^{-1}$.

ii. Note that

$$\langle S_\pm u, v \rangle = \sum_{n \in \mathbb{Z}} u_{n\pm 1} \bar{v}_n = \sum_{n \in \mathbb{Z}} u_n \bar{v}_{n \mp 1} = \langle u, S_\mp v \rangle.$$

○

2.2.3. Adjoint of unbounded operators in Hilbert spaces. —
In this section, we assume again that $E = F = \mathsf{H}$ is a separable Hilbert space with norm $\| \cdot \|$. Let $(\mathrm{Dom}\,(T), T)$ be an operator. Consider the application

$$\Phi_x : \begin{array}{rcl} \big(\mathrm{Dom}\,(T), \| \cdot \|\big) & \longrightarrow & (\mathbb{C}, |\cdot|) \\ y & \longmapsto & \langle Ty, x \rangle. \end{array}$$

***Definition 2.39* (Domain of T^*).** — The domain of T^* is the linear subspace

$$(2.2.2.8) \qquad \mathrm{Dom}\,(T^*) := \big\{ x \in \mathsf{H} \; : \; \Phi_x \text{ is continuous} \big\}.$$

In other words, an element $x \in \mathsf{H}$ is in $\mathrm{Dom}\,(T^*)$ if and only if we can find a constant $C_x \in \mathbb{R}_+$ such that

$$\forall y \in \mathrm{Dom}\,(T), \quad |\langle Ty, x \rangle| \leqslant C_x \|y\|.$$

Proposition 2.40. — *Let $(\mathrm{Dom}\,(T), T)$ be an operator with a dense domain. Then, for all $x \in \mathrm{Dom}\,(T^*)$, there exists a unique $T^*x \in \mathsf{H}$ such that*

$$\forall y \in \mathrm{Dom}\,(T), \quad \langle Ty, x \rangle = \langle y, T^*x \rangle.$$

Proof. — Given $x \in \mathrm{Dom}\,(T^*)$, the application Φ_x is by definition a continuous linear form on $\mathrm{Dom}\,(T)$. When $\mathrm{Dom}\,(T)$ is dense in H, it can be uniquely extended as a continuous linear form on the whole space H. Then, the existence and uniqueness of T^*x is a consequence of the Riesz representation theorem. □

***Definition 2.41* (Adjoint of an unbounded operator)**
Let $(\mathrm{Dom}\,(T), T)$ be an operator with dense domain. Then $(\mathrm{Dom}\,(T^*), T^*)$ is a linear operator called the *adjoint* of $(\mathrm{Dom}\,(T), T)$.

Remember that we can deal with T^* only when $\mathrm{Dom}\,(T)$ is dense. That is why this condition will be often implicitly assumed, as in the definition below.

Definition 2.42 (Self-adjoint operator). — We say that $(\mathrm{Dom}\,(T), T)$ is *self-adjoint* when the following conditions are both satisfied:

(2.2.2.9a) $$\mathrm{Dom}\,(T) = \mathrm{Dom}\,(T^*),$$

(2.2.2.9b) $$\forall u \in \mathrm{Dom}\,(T), \quad Tu = T^*u.$$

To express that $(\mathrm{Dom}\,(T), T)$ is self-adjoint, we will sometimes simply write $T = T^*$. But keep in mind that this includes the two conditions (2.2.2.9a) and (2.2.2.9b).

Exercise 2.43. — Work in the context of Exercise 2.38, and show that $S_+ + S_-$ is self-adjoint.

Solution: Recall that $\mathrm{Dom}\,(S_+) = \mathrm{Dom}\,(S_-) = \mathrm{Dom}\,(S_+ + S_-) = \ell^2(\mathbb{Z}, \mathbb{C})$. Fix $v \in \ell^2(\mathbb{Z}, \mathbb{C})$. By the Cauchy–Scwarz inequality, we have

$$|\langle (S_+ + S_-)u, v \rangle| = \left| \sum_{n \in \mathbb{Z}} (u_{n+1} + u_{n-1}) \bar{v}_n \right| \leqslant 2 \|u\|_{\ell^2} \|v\|_{\ell^2},$$

which ensures that $v \in \mathrm{Dom}\,(S_+ + S_-)^*$, and therefore

$$\mathrm{Dom}\,(S_+ + S_-) = \mathrm{Dom}\,(S_+ + S_-)^* = \ell^2(\mathbb{Z}, \mathbb{C}).$$

On the other hand,

$$\begin{aligned}
\langle (S_+ + S_-)u, v \rangle &= \sum_{n \in \mathbb{Z}} (u_{n+1} + u_{n-1}) \bar{v}_n \\
&= \sum_{n \in \mathbb{Z}} u_n (\bar{v}_{n+1} + \bar{v}_{n-1}) \\
&= \langle u, (S_+ + S_-)v \rangle,
\end{aligned}$$

which implies (2.2.2.9b). ○

Example 2.44. — Let us consider $(X, \mathcal{A}, \mu)$ as a measure space, with a σ-finite measure μ. We let $\mathsf{H} = \mathsf{L}^2(X, \mathcal{A}, \mu)$ and consider a $\mathbb{C}$-valued measurable function f that is finite almost everywhere. We define

$$\mathrm{Dom}\,(T_f) = \{ \psi \in \mathsf{H} : f\psi \in \mathsf{H} \}, \forall \psi \in \mathrm{Dom}\,(T_f), T_f\psi = f\psi.$$

    i. If $f \in L^\infty(X, \mathcal{A}, \mu)$, we have $\mathrm{Dom}\,(T_f) = H$ and T_f is bounded.

    ii. The domain of the adjoint of T_f is given by $\mathrm{Dom}\,(T_f)$ and $T_f^* = T_{\bar{f}}$. In particular, when f is real valued, T_f is self-adjoint.

Exercise 2.45. — Take $H = L^2(\mathbb{R})$. Consider $\mathrm{Dom}\,(T) = H^1(\mathbb{R})$ and $T = -i\partial_x$. What is $(\mathrm{Dom}\,(T^*), T^*)$? And if we choose $\mathrm{Dom}\,(T) = \mathscr{C}_0^\infty(\mathbb{R})$?

Solution: Let $f \in L^2(\mathbb{R})$. We find that

$$\Phi_f : \big(H^1(\mathbb{R}), \|\cdot\|_{L^2}\big) \longrightarrow (\mathbb{C}, |\cdot|)$$
$$g \longmapsto -i\int_{\mathbb{R}} g'(x)\bar{f}(x)\,\mathrm{d}x$$

is continuous if and only if $f \in H^1(\mathbb{R})$. It follows that $\mathrm{Dom}\,(T^*) = H^1(\mathbb{R})$. Moreover, knowing that $f \in H^1(\mathbb{R})$, an integration by parts gives

$$\Phi_f(g) = -i\int_{\mathbb{R}} g'(x)\bar{f}(x)\,\mathrm{d}x = \int_{\mathbb{R}} g(x)\overline{-if'}(x)\,\mathrm{d}x$$

which means that $T^* = T$. Now, if we choose $\mathrm{Dom}\,(T) = \mathscr{C}_0^\infty(\mathbb{R})$, for the same reasons as before, we still have $\mathrm{Dom}\,(T^*) = H^1(\mathbb{R})$. But T is not self-adjoint (we do not have $T = T^*$) since the condition (2.2.2.9a) is not satisfied.   ○

Proposition 2.46. — *If T is an operator with a dense domain, we have*

$$\Gamma(T^*) = \big\{(x, y) \in H \times H \,;\, \langle z, y\rangle - \langle Tz, x\rangle = 0, \forall z \in \mathrm{Dom}\,(T)\big\}.$$

Proof. — By definition

$$\Gamma(T^*) = \big\{(x, y) \in \mathrm{Dom}\,(T^*) \times H \,;\, y - T^*x = 0\big\}.$$

Since $\mathrm{Dom}\,(T)$ is dense, this means that $\Gamma(T^*)$ is equal to

$$\big\{(x, y) \in \mathrm{Dom}\,(T^*) \times H \,;\, \langle z, y - T^*x\rangle = 0, \forall z \in \mathrm{Dom}\,(T)\big\}$$

or, equivalently, to

$$\big\{(x, y) \in \mathrm{Dom}\,(T^*) \times H \,;\, \langle z, y\rangle - \langle Tz, x\rangle = 0, \forall z \in \mathrm{Dom}\,(T)\big\}$$
$$\subset \big\{(x, y) \in H \times H \,;\, \langle z, y\rangle - \langle Tz, x\rangle = 0, \forall z \in \mathrm{Dom}\,(T)\big\}.$$

Conversely, assume that $(x, y) \in H \times H$ satisfies the above condition. Then, the application $\Phi_x : \mathrm{Dom}\,(T) \longrightarrow \mathbb{C}$ given by $\Phi_x(z) = \langle Tz, x \rangle = \langle z, y \rangle$ is continuous with norm less than $\|y\|$. This implies that $x \in \mathrm{Dom}\,(T^*)$, and therefore we have the opposite inclusion. $\qquad \square$

We can equip $H \times H$ with the natural scalar product of $H \times H$ denoted by

$$(x, y) \cdot (\tilde{x}, \tilde{y}) = \langle x, \tilde{x} \rangle + \langle y, \tilde{y} \rangle.$$

Proposition 2.47. — *Let us define* $J : H \times H \ni (x, y) \mapsto (-y, x) \in H \times H$. *If* T *is an operator with a dense domain, we have*

$$\Gamma(T^*) = J(\Gamma(T))^{\perp}, \quad \overline{\Gamma(T)} = J(\Gamma(T^*))^{\perp}.$$

In particular, T^* *is closed.*

Proof. — From Proposition 2.46, we know that

$$\Gamma(T^*) = \big\{ (x, y); \langle y, z \rangle - \langle x, Tz \rangle = 0, \forall z \in \mathrm{Dom}\,(T) \big\}$$

$$(2.2.2.10) \quad = \big\{ (x, y); (x, y) \cdot J(z, Tz) = 0, \forall z \in \mathrm{Dom}\,(T) \big\}$$

$$= J(\Gamma(T))^{\perp}.$$

It follows that $\Gamma(T^*)$ is closed as the orthogonal of the set $J(\Gamma(T))$. Moreover

$$\Gamma(T^*)^{\perp} = \big(J(\Gamma(T))^{\perp} \big)^{\perp} = \overline{J(\Gamma(T))} = J(\overline{\Gamma(T)}).$$

Since $J \circ J = -\mathrm{Id}$, we have

$$\overline{\Gamma(T)} = -J(\Gamma(T^*)^{\perp}) = J(\Gamma(T^*)^{\perp}) = J(\Gamma(T^*))^{\perp}.$$

$$\square$$

Exercise 2.48. — Let T be an operator with a dense domain. Show directly that T^* is closed.

Solution: Fix any $(x, z) \in \overline{\Gamma(T^*)}$. We can find a sequence $(x_n, z_n) \in \Gamma(T^*)^{\mathbb{N}}$ with $z_n = T^* x_n$ converging to (x, z). Then, for all $y \in \mathrm{Dom}\,(T)$, we have

$$\langle y, z \rangle = \lim_{n \to \infty} \langle y, T^* x_n \rangle = \lim_{n \to \infty} \langle Ty, x_n \rangle = \langle Ty, x \rangle.$$

We deduce that the application $\Phi_x : \mathrm{Dom}\,(T) \longrightarrow \mathbb{C}$ given by $\Phi_x(y) = \langle Ty, x \rangle = \langle y, z \rangle$ is continuous with norm less than $\|z\|$.

This implies that $x \in \mathrm{Dom}\,(T^*)$ and $T^*x = z$. In other words, we still have $(x, z) \in \Gamma(T^*)$. ○

Proposition 2.49. — *Consider two operators* $(\mathrm{Dom}\,(T), T)$ *and* $(\mathrm{Dom}\,(S), S)$ *with dense domains. If* $T \subset S$, *we have* $S^* \subset T^*$.

Proof. — It is a consequence of Proposition 2.47. □

Proposition 2.50. — *Let* T *be an operator with a dense domain. Then,* T *is closable if and only if* $\mathrm{Dom}\,(T^*)$ *is dense. In this case,* $(T^*)^* = T^{**} = \overline{T}$.

Proof. — Assume that $\mathrm{Dom}\,(T^*)$ is dense. Then, by Proposition 2.47 applied to T^*, we have

$$\Gamma((T^*)^*) = J(\Gamma(T^*))^\perp = J\big(J(\Gamma(T))^\perp\big)^\perp = \overline{\Gamma(T)} = \overline{\Gamma((T^*)^*)}.$$

In other words, $(T^*)^* = T^{**}$ is closed, with graph $\overline{\Gamma(T)}$. It follows that T is closable, and more precisely $\overline{T} = T^{**}$.

Now, assume that T is closable. Take $v \in \mathrm{Dom}\,(T^*)^\perp$. We have

$$\forall z \in \mathrm{Dom}\,(T^*), \quad 0 = \langle v, z \rangle = (0, v) \cdot (-T^*z, z)$$

which means that

$$(0, v) \in J(\Gamma(T^*))^\perp = \overline{\Gamma(T)} = \Gamma(\overline{T})$$

and therefore $v = \overline{T}0 = 0$. Thus, we get

$$\big(\mathrm{Dom}\,(T^*)^\perp\big)^\perp = \{0\}^\perp = \mathrm{H} = \overline{\mathrm{Dom}\,(T^*)}$$

as expected. □

Proposition 2.51. — *If* T *is closable with a dense domain, then* $T^* = \overline{T}^*$ *where* $\overline{T}^* = (\overline{T})^*$.

Proof. — By Proposition 2.50, we know that $\mathrm{Dom}\,(T^*)$ is dense, and that $T^{**} = \overline{T}$. It follows that $\overline{T}^* = (T^{**})^* = (T^*)^{**}$. By Proposition 2.47, we recall that T^* is closed, and therefore closable. Applying Proposition 2.50 with T^* in place of T, we get that $(T^*)^{**} = \overline{T^*}$. Since $\overline{T^*} = T^*$, we can conclude that $\overline{T}^* = T^*$. □

Proposition 2.52. — *Let us consider a densely defined operator* T. *We have*

$$\ker(T^*) = \mathrm{ran}\,(T)^\perp, \quad \ker(T^*)^\perp = \overline{\mathrm{ran}\,(T)}.$$

In particular, T^* *is injective if and only if* T *has a dense range.*

Proof. — Let $x \in \ker T^*$ and $y \in \operatorname{ran}(T)$. We can write $y = Tz$ with $z \in \operatorname{Dom}(T)$. Then

$$\forall y \in \operatorname{ran}(T), \quad \langle x, y \rangle = \langle x, Tz \rangle = \langle T^*x, z \rangle = \langle 0, z \rangle = 0$$

which implies that $x \in \operatorname{ran}(T)^\perp$.

Conversely, let $y \in \mathsf{H}$ be such that $\langle y, Tx \rangle = 0$ for all $x \in \operatorname{Dom}(T)$ so that $y \in \operatorname{Dom}(T^*)$, and we have $\langle T^*y, x \rangle = 0$ for all $x \in \operatorname{Dom}(T)$. Since the domain $\operatorname{Dom}(T)$ is dense, this implies that $T^*y = 0$. $\qquad\square$

2.2.4. Creation and annihilation operators. — To illustrate the preceding (rather abstract) propositions, we discuss here two important examples . Take $\mathsf{H} = \mathsf{L}^2(\mathbb{R})$. Let us introduce the following differential operators, acting on $\operatorname{Dom}(a) = \operatorname{Dom}(c) = \mathscr{S}(\mathbb{R})$,

$$a := \frac{1}{\sqrt{2}}(\partial_x + x), \quad c := \frac{1}{\sqrt{2}}(-\partial_x + x).$$

The domains of their adjoints are

$$\operatorname{Dom}(a^*) := \{\psi \in \mathsf{L}^2(\mathbb{R}); (-\partial_x + x)\psi \in \mathsf{L}^2(\mathbb{R})\},$$
$$\operatorname{Dom}(c^*) := \{\psi \in \mathsf{L}^2(\mathbb{R}); (\partial_x + x)\psi \in \mathsf{L}^2(\mathbb{R})\}.$$

Observe that $\mathscr{S}(\mathbb{R}) \subset \operatorname{Dom}(u^*)$ and that $\mathscr{S}(\mathbb{R}) \subset \operatorname{Dom}(c^*)$. It follows that $\operatorname{Dom}(a^*)$ and $\operatorname{Dom}(c^*)$ are dense in $\mathsf{L}^2(\mathbb{R})$. By Proposition 2.50, the two operators a and c are closable with $\bar{a} = a^{**}$ and $\bar{c} = c^{**}$. On the other hand

$$\forall \psi \in \operatorname{Dom}(a^*), \quad a^*\psi = \frac{1}{\sqrt{2}}(-\partial_x + x)\psi = c\psi,$$

$$\forall \psi \in \operatorname{Dom}(c^*), \quad c^*\psi = \frac{1}{\sqrt{2}}(\partial_x + x)\psi = a\psi.$$

We have $a \subset c^*$ and $c \subset a^*$. Moreover, by Proposition 2.47, $\Gamma(a^*)$ and $\Gamma(c^*)$ are closed, so that $\bar{a} \subset c^*$ and $\bar{c} \subset a^*$.

Lemma 2.53. — *We have*

(2.2.2.11)
$$\operatorname{Dom}(\bar{a}) = \operatorname{Dom}(\bar{c}) = \mathsf{B}^1(\mathbb{R}) = \{\psi \in \mathsf{H}^1(\mathbb{R}) : x\psi \in \mathsf{L}^2(\mathbb{R})\}.$$

Proof. — For all $u \in \mathscr{S}(\mathbb{R})$, we have

$$(2.2.2.12) \qquad 2\|au\|^2 = \|u'\|^2 + \|xu\|^2 - \|u\|^2$$
$$(2.2.2.13) \qquad 2\|cu\|^2 = \|u'\|^2 + \|xu\|^2 + \|u\|^2 .$$

Now, take $u \in \mathrm{Dom}\,(\bar{a})$. By definition, we have $(u, \bar{a}u) \in \overline{\Gamma(a)}$. There exists $(u_n) \in \mathrm{Dom}\,(a)^{\mathbb{N}}$ such that (u_n) converges to u and (au_n) converges to $\bar{a}u$. We deduce that (u'_n) and (xu_n) are Cauchy sequences in $\mathsf{L}^2(\mathbb{R})$. Thus, we get that $u' \in \mathsf{L}^2(\mathbb{R})$ and $xu \in \mathsf{L}^2(\mathbb{R})$, and therefore $\mathrm{Dom}\,(\bar{a}) \subset \mathsf{B}^1(\mathbb{R})$. We can proceed in the same way for $\bar{c}$.

We deal now with the reversed inclusion. Take $u \in \mathsf{B}^1(\mathbb{R})$. By Lemma 1.12, there exists a sequence (u_n) of smooth functions with compact support such that u_n converges to u in $\mathsf{B}^1(\mathbb{R})$. In particular, (au_n) and (cu_n) converge in $\mathsf{L}^2(\mathbb{R})$, so that $(u, au) \in \overline{\Gamma(a)} = \Gamma(\bar{a})$ as well as $(u, cu) \in \overline{\Gamma(c)} = \Gamma(\bar{c})$. This implies $u \in \mathrm{Dom}\,(\bar{a})$ and $u \in \mathrm{Dom}\,(\bar{c})$. $\qquad\square$

Lemma 2.54. — *The closures $\bar{a}$ and $\bar{c}$ of a and c are adjoints of each other, and they share the same domain $\mathsf{B}^1(\mathbb{R})$.*

Proof. — We use the results of Exercise 1.13. For example, if $\psi \in \mathrm{Dom}\,(c^*)$, we have $\psi \in \mathsf{L}^2(\mathbb{R})$ and $(\partial_x + x)\psi \in \mathsf{L}^2(\mathbb{R})$. There exists $(\psi_n) \in \mathscr{S}(\mathbb{R})^{\mathbb{N}}$ such that ψ_n converges to ψ and $(\partial_x + x)\psi_n$ converges to $(\partial_x + x)\psi \in \mathsf{L}^2(\mathbb{R})$. Using (2.2.2.13), we get that (ψ'_n) and $(x\psi_n)$ are Cauchy sequences with $\mathsf{L}^2(\mathbb{R})$ with limits ψ' and $x\psi$. Thus $\psi \in \mathsf{B}^1(\mathbb{R})$. From the inclusion $\bar{a} \subset c^*$ and from (2.2.2.11), we deduce that

$$\mathrm{Dom}\,(c^*) \subset \mathsf{B}^1(\mathbb{R}) = \mathrm{Dom}\,(\bar{a}) \subset \mathrm{Dom}\,(c^*) .$$

We can deal with a^* in the same way to obtain

$$\mathrm{Dom}\,(c^*) = \mathrm{Dom}\,(\bar{a}) = \mathsf{B}^1(\mathbb{R}) = \mathrm{Dom}\,(\bar{c}) = \mathrm{Dom}\,(a^*) .$$

We deduce that

$$\bar{a} = c^*, \quad \bar{c} = a^* .$$

By Proposition 2.51, we get

$$\bar{a}^* = a^* = \bar{c}, \quad \bar{c}^* = c^* = \bar{a} .$$

$\qquad\square$

2.3. Self-adjoint operators and essentially self-adjoint operators

In Quantum Mechanics, the states of a system are represented by normalized vectors u in a Hilbert space H. In other words, a *state* is an element $u \in H$ such that $\|u\| = 1$. Then, each dynamical variable (e.g., position, momentum, orbital angular momentum, spin, and energy) is associated with an operator $(\text{Dom}\,(T), T)$, called an *observable*, that acts on H.

Assume that $u \equiv u_\lambda$ is an eigenvector of T with eigenvalue λ. The eigenvalue equation

$$T u_\lambda = \lambda u_\lambda$$

means that if a measurement of the observable T is made while the system of interest belongs to the state u_λ, then the observed value of that particular measurement returns the eigenvalue λ with certainty. However, if the system of interest is in a general state $u \in H$, the Born rule stipulates that the eigenvalue λ is returned with probability $|\langle u, u_\lambda \rangle|^2$. For physical consistency[1], the mean value of a dynamical variable T must be a real number, that is,

(2.2.3.14)
$$\forall u \in \text{Dom}\,(T), \quad \langle Tu, u \rangle = \overline{\langle Tu, u \rangle} = \langle u, Tu \rangle \in \mathbb{R}.$$

2.3.1. Symmetric and self-adjoint operators. — The starting point is (2.2.3.14).

Definition 2.55. — An operator T is said to be symmetric if (2.2.3.14) is satisfied.

Proposition 2.56. — *An operator T is symmetric if and only if*

(2.2.3.15) $$\forall u, v \in \text{Dom}\,(T), \quad \langle Tu, v \rangle = \langle u, Tv \rangle.$$

Proof. — From (2.2.3.14), we deduce that

$$\forall (u, v) \in \text{Dom}\,(T)^2, \quad \langle T(u+v), u+v \rangle \in \mathbb{R},$$

or, equivalently,

(2.2.3.16) $$\forall (u, v) \in \text{Dom}\,(T)^2, \quad \langle Tu, v \rangle + \overline{\langle u, Tv \rangle} \in \mathbb{R}.$$

---

[1]Note, however, that there is a non-self-adjoint Quantum Mechanics, related to dissipative systems.

This implies that

(2.2.3.17) $\forall (u, v) \in \mathrm{Dom}\,(T)^2, \quad \mathrm{Im}\,\langle Tu, v \rangle = \mathrm{Im}\,\langle u, Tv \rangle \,.$

By testing (2.2.3.16) with iv, we obtain

(2.2.3.18) $\forall (u, v) \in \mathrm{Dom}\,(T)^2, \quad i\langle Tu, v \rangle - i\overline{\langle u, Tv \rangle} \in \mathbb{R} \,.$

This means that

(2.2.3.19) $\forall (u, v) \in \mathrm{Dom}\,(T)^2, \quad \mathrm{Re}\,\langle Tu, v \rangle = \mathrm{Re}\,\langle u, Tv \rangle \,.$

Combine (2.2.3.17) and (2.2.3.19) to get (2.2.3.15). $\square$

Proposition 2.57. — *A densely defined operator T is symmetric if and only if $T \subset T^*$.*

In other words, a densely defined operator T is symmetric if and only if

(2.2.3.20)
$$\mathrm{Dom}\,(T) \subset \mathrm{Dom}\,(T^*) \quad , \quad \forall u \in \mathrm{Dom}\,(T), \quad Tu = T^*u \,.$$

In view of (2.2.2.9), any self-adjoint operator is symmetric, and a densely defined symmetric operator is self-adjoint if and only if $\mathrm{Dom}\,(T) = \mathrm{Dom}\,(T^*)$.

Proof. — Let T be a symmetric densely defined operator. Let $u \in \mathrm{Dom}\,(T)$. From (2.2.3.15), we have

$$\forall v \in \mathrm{Dom}\,(T), \quad |\langle u, Tv \rangle| = |\langle Tu, v \rangle| \leqslant \|Tu\|\|v\| \,,$$

and therefore $u \in \mathrm{Dom}\,(T^*)$. We have also

$$\forall (u, v) \in \mathrm{Dom}\,(T)^2, \quad \langle u, Tv \rangle = \langle T^*u, v \rangle = \langle Tu, v \rangle \,.$$

Since the domain $\mathrm{Dom}\,(T)$ is assumed to be dense in H, by the Riesz representation theorem, we must have $T^*u = Tu$, and therefore $(u, Tu) \in \Gamma(T^*)$ so that $\Gamma(T) \subset \Gamma(T^*)$.

Conversely, assume that $T \subset T^*$ or equivalently that $\Gamma(T) \subset \Gamma(T^*)$. Thus, given $u \in \mathrm{Dom}\,(T)$, we must have $(u, Tu) \in \Gamma(T^*)$, and thus $Tu = T^*u$. Then

$$\forall u \in \mathrm{Dom}\,(T), \quad \overline{\langle Tu, u \rangle} = \langle u, Tu \rangle = \langle u, T^*u \rangle = \langle Tu, u \rangle \,,$$

which is exactly (2.2.3.14). $\square$

Proposition 2.58. — *Let us consider a densely defined symmetric operator T. Then, T is closable and $T \subset \overline{T} \subset T^*$. Moreover, T is self-adjoint if and only if $T = \overline{T} = T^*$.*

Proof. — A densely defined symmetric operator satisfies

$$\mathsf{H} \subset \overline{\mathrm{Dom}\,(T)} \subset \overline{\mathrm{Dom}\,(T^*)} \subset \mathsf{H} \quad \Longrightarrow \quad \overline{\mathrm{Dom}\,(T^*)} = \mathsf{H}\,.$$

By Proposition 2.50, the operator T is closable. By Proposition 2.51, the operator T^* is closed, so that $T \subset \overline{T} \subset T^*$. Thus, it is self-adjoint if and only if $T = \overline{T} = T^*$. $\qquad\qquad\square$

Exercise 2.59. — Take $\mathsf{H} = \mathsf{L}^2(\mathbb{R})$. Show that $(\mathscr{C}_0^\infty(\mathbb{R}, \mathbb{C}), -i\partial_x)$ is symmetric.

Solution: Just perform an integration by parts to see that

$$\forall f \in \mathscr{C}_0^\infty(\mathbb{R}), \qquad \int_{\mathbb{R}} (-i\partial_x f)\,\bar f \,\mathrm{d}x = \int_{\mathbb{R}} f\,\overline{(-i\partial_x f)}\,\mathrm{d}x\,.$$

○

Exercise 2.60. — Let $P \in \mathbb{R}[X]$ be a polynomial of degree n. Work with $\mathsf{H} = \mathsf{L}^2(\mathbb{R})$. Show that the differential operator $(\mathsf{H}^n(\mathbb{R}), P(D))$ is symmetric. Here $D = -i\partial_x$. Use the Fourier transform and Example 2.44.

Solution: For all $u \in \mathsf{H}^n(\mathbb{R})$, since $P(\cdot)$ is real valued, we have

$$\langle P(D)u, u \rangle = \langle P(\xi)\hat u, \hat u \rangle = \langle \hat u, \overline{P(\xi)}\hat u \rangle = \langle \hat u, P(\xi)\hat u \rangle$$
$$= \langle u, P(D)u \rangle\,.$$

○

Exercise 2.61. — Give an example of some unbounded operator on $\mathsf{H} = \mathsf{L}^2(\mathbb{T})$ that is densely defined and not symmetric. Here $\mathbb{T} = \mathbb{R}/(2\pi\mathbb{Z})$.

Solution: Just take $(\mathscr{C}^\infty(\mathbb{T}), \partial_x)$, and remark

$$\langle \partial_x(e^{ix}), e^{ix} \rangle = \int_{\mathbb{T}} ie^{ix}e^{-ix}\,\mathrm{d}x = i \notin \mathbb{R}.$$

○

Proposition 2.62. — *Consider a symmetric operator T. Let $z = \alpha + i\beta$ with $(\alpha, \beta) \in \mathbb{R} \times (\mathbb{R} \setminus \{0\})$. Then, $T - z\mathrm{Id}$ is injective and*

$$(2.2.3.21) \qquad \forall u \in \mathrm{Dom}\,(T), \quad \|(T - z\mathrm{Id})u\| \geqslant |\beta| \|u\|.$$

Moreover, if T is closed, the operator $T - z\mathrm{Id}$ has a closed range.

Proof. — Let $u \in \mathrm{Dom}\,(T)$. We have

$$\begin{aligned}
&\|(T - z\mathrm{Id})u\|^2 \\
&= \|(T - \alpha)u - i\beta u\|^2 \\
&= \|(T - \alpha)u\|^2 + \beta^2 \|u\|^2 + 2\mathrm{Re}\,\langle (T - \alpha)u, (-i\beta)u \rangle.
\end{aligned}$$

On the other hand

$$\langle (T - \alpha)u, (-i\beta)u \rangle = i\beta \big[\langle Tu, u \rangle - \alpha \|u\|^2\big] \in i\mathbb{R}.$$

It follows that

$$\|(T - zI)u\|^2 = \|(T - \alpha)u\|^2 + \beta^2 \|u\|^2 \geqslant \beta^2 \|u\|^2.$$

It remains to apply Proposition 2.14. $\qquad\square$

Proposition 2.63. — *Let T be a densely defined, closed, and symmetric operator. Then, T is self-adjoint if and only if T^* is symmetric.*

Proof. — If T is self-adjoint, we have $T = T^* = (T^*)^*$ and $\Gamma(T^*) \equiv \Gamma\big((T^*)^*\big) \subset \Gamma\big((T^*)^*\big)$, which means that T^* is symmetric.

Now, assume that T is densely defined with $T \equiv \overline{T} \subset T^*$. By Proposition 2.50, we already know that $(T^*)^* = \overline{T}$. Moreover, if T^* is symmetric, we have $T^* \subset (T^*)^*$. Thus,

$$T \equiv \overline{T} \subset T^* \subset (T^*)^* \equiv \overline{T} \equiv T,$$

and therefore $T = T^*$. $\qquad\square$

Proposition 2.64. — *Let T be a densely defined symmetric operator. The three following assertions are equivalent.*

  (i) *T is self-adjoint.*
  (ii) *T is closed and $\ker(T^* \pm i) = \{0\}$.*
  (iii) $\mathrm{ran}\,(T \pm i) = \mathsf{H}$.

Proof. — (i) $\implies$ (ii). If T is self-adjoint, we have $T = \overline{T} = T^*$ and T is closed. Let $x \in \ker(T^* \pm iI)$. Then

$$\mathbb{R} \ni \langle x, Tx \rangle = \langle x, T^*x \rangle = \langle Tx, x \rangle = \langle x, \mp ix \rangle = \pm i\|x\|^2 \in \mathbb{R}.$$

This is possible only if $x = 0$.

(ii) $\implies$ (iii). By Proposition 2.52,

$$\ker(T^* \pm i) = \operatorname{ran}(T \mp i)^\perp = \{0\}.$$

Thus, $T \mp i$ has a dense range. Since T is closed, by Proposition 2.62, the range $\operatorname{ran}(T \mp i)$ is closed. In other words, we have $\operatorname{ran}(T \mp i) = \overline{\operatorname{ran}(T \mp i)} = \mathsf{H}$.

(iii) $\implies$ (i). Assume that $\operatorname{ran}(T \pm i) = \mathsf{H}$. First, let us prove that $\operatorname{Dom}(T^*) \subset \operatorname{Dom}(T)$. To this end, take $u \in \operatorname{Dom}(T^*)$ and consider $(T^* - i)u$. There exists $v \in \operatorname{Dom}(T)$ such that $(T^* - i)u = (T - i)v$. Since T is symmetric, we have $Tv = T^*v$ and $(T^* - i)u = (T^* - i)v$. It follows that

$$u - v \in \ker(T^* - i) = \operatorname{ran}(T + i)^\perp = \mathsf{H}^\perp = \{0\},$$

and therefore $u = v \in \operatorname{Dom}(T)$. Thus $\operatorname{Dom}(T^*) \subset \operatorname{Dom}(T)$. Since we have (2.2.3.20), it follows that $\operatorname{Dom}(T^*) = \operatorname{Dom}(T)$, so that $T = T^*$. $\qquad\square$

Exercise 2.65. — Take $\mathsf{H} = L^2(\mathbb{R}^d, \mathbb{C})$. Select some potential $V \in L^\infty(\mathbb{R}^d, \mathbb{R})$. Then, consider the corresponding operator $T = -\Delta + V$ with domain $\mathsf{H}^2(\mathbb{R}^d)$. Is it self-adjoint? Conclusion?

Solution: Since $\mathsf{H}^2(\mathbb{R}^d)$ is dense in $L^2(\mathbb{R}^d)$ for the L^2-norm, the operator T is densely defined. Since $\Delta = \operatorname{div}(\nabla)$ and because $V(\cdot)$ is real valued, two integration by parts show that T is symmetric. By definition, an element $u \in L^2(\mathbb{R}^d)$ is in $\operatorname{Dom}(T)$ if and only if the application

$$\begin{aligned} \Phi_u : \mathsf{H}^2(\mathbb{R}^d) &\longrightarrow \mathbb{C} \\ v &\longmapsto \int_{\mathbb{R}} (-\Delta v + Vv)\bar{u}\,dx \end{aligned}$$

is continuous for the L^2-norm. Since $V(\cdot)$ is bounded, we have

$$\left| \int_{\mathbb{R}} Vv\bar{u}\,dx \right| \leqslant (\|V\|_{L^\infty}\|u\|_{L^2})\|v\|_{L^2}.$$

This is equivalent (after Fourier transform) to the continuity of

$$\tilde{\Phi}_u : H^2(\mathbb{R}^d) \longrightarrow \mathbb{C}$$
$$v \longmapsto \int_{\mathbb{R}} |\xi|^2 \hat{v} \, \overline{\hat{u}} \, d\xi,$$

which is satisfied if and only if $v \in H^2(\mathbb{R}^d)$. This means that $\mathrm{Dom}\,(T^*) = H^2(\mathbb{R}^d) = \mathrm{Dom}\,(T)$.

Conclusion: knowing that T is self-adjoint, we can use the criteria (ii) and (iii). The condition (iii) says that, for any $g \in L^2(\mathbb{R}^d)$, the equation $(-\Delta u + V \pm i)u = g$ has a solution in $H^2(\mathbb{R}^d)$, while the condition (ii) guarantees the uniqueness of such solution.   ○

Exercise 2.66. — Take $H = L^2(\mathbb{R}_+)$.

   i. Is the operator $(H^1(\mathbb{R}_+), -i\partial_x)$ symmetric?
  ii. Is the operator $(H^1_0(\mathbb{R}_+), -i\partial_x)$ symmetric?
 iii. Show that the domain of the adjoint of $(H^1_0(\mathbb{R}_+), -i\partial_x)$ is $H^1(\mathbb{R}_+)$.
 iv. By using Proposition 2.64, prove that $(H^1_0(\mathbb{R}_+), -i\partial_x)$ is not self-adjoint.

Exercise 2.67. — Take $H = L^2(\mathbb{R}_+)$. We let

$$\mathrm{Dom}\,(T) = \{\psi \in H^2(\mathbb{R}_+) : u'(0) = -u(0)\}$$

and $T = -\partial_x^2$. Is this operator symmetric? We recall that $H^2(\mathbb{R}_+)$ is continuously embedded in $\mathscr{C}^1(\overline{\mathbb{R}_+})$ (see Proposition 1.15). Is T self-adjoint? In order to answer this question, the reader is invited to consider Section 2.5, and to introduce an appropriate Hermitian sesquilinear form on $H^1(\mathbb{R}_+)$.

2.3.2. Essentially self-adjoint operators. —

Definition 2.68. — A (densely defined) symmetric operator is essentially self-adjoint if its closure is self-adjoint.

Proposition 2.69. — *Let T be a (densely defined) symmetric operator. Then, T is essentially self-adjoint if and only if $\overline{T} = T^*$.*

Proof. — First, by Proposition 3.40, the operator T is closable.

$\implies$ If T is essentially self-adjoint, we have $\overline{T}^* = \overline{T}$. Then, by Proposition 2.51, we have $\overline{T}^* = T^*$, and therefore $\overline{T} = T^*$.

$\impliedby$ Assume that $\overline{T} = T^*$. By Proposition 2.50, we have $T^{**} = \overline{T}$, so that $\overline{T}^* = T^{**} = \overline{T}$.   □

Exercise 2.70. — Take $\mathsf{H} = \mathsf{L}^2(\mathbb{R}^d)$, and consider the operator $(\mathscr{C}_0^\infty(\mathbb{R}^d), -\Delta)$. Show that this operator is essentially self-adjoint. What is the adjoint?

Solution: As seen before, the closure is $(\mathsf{H}^2(\mathbb{R}^d), -\Delta)$, which is self-adjoint. This is T^*.

                                                                        ○

Proposition 2.71. — *If T is essentially self-adjoint, the operator T has a unique self-adjoint extension, which is $\overline{T}$.*

Proof. — Let us consider a self-adjoint extension S of T. We have $T \subset S$. By Proposition 2.49 , we get

$$\overline{T} \subset \overline{S} = S = S^* \subset T^* = \overline{T}.$$

Necessarily, we must have $\overline{T} = S$.   □

Proposition 2.72. — *Let T be a (densely defined) symmetric operator. The three following assertions are equivalent.*

  (i) T *is essentially self-adjoint.*
  (ii) $\overline{\ker(T^* \pm i)} = \{0\}$.
  (iii) $\overline{\mathrm{ran}\,(T \pm i)} = \mathsf{H}$.

Proof. — Assume (i). Then, $\overline{T}$ is self-adjoint. By Propositions 2.51 and 2.64, we have

$$\ker(\overline{T}^* \pm i) = \ker(T^* \pm i) = \{0\}.$$

Assume (ii) . Then, by Proposition 2.64, $\mathrm{ran}\,(\overline{T} \pm i) = \mathsf{H}$ and it follows that the operator $\overline{T}$ is self-adjoint. Obviously, (ii) and (iii) are equivalent by Proposition 2.52.   □

Exercise 2.73. — Take $\mathsf{H} = \mathsf{L}^2(I)$ with $I = (0,1)$. Consider $(\mathscr{C}_0^\infty(I), -\partial_x^2)$. Prove in two different ways, using respectively (ii) and (iii), that this operator is not essentially self-adjoint. What is the closure of this operator? Explain why this closure is not self-adjoint.

Solution: The proof is by contradiction and, each time, it relies on the function $f_\lambda \in L^2(I)$ which is given by $0 \neq f_\lambda(x) = e^{\lambda x}$ with $\lambda^2 = \pm i$.

— Through the criterion (ii), first, observe that $H^2(I) \subset \text{Dom}\,(T^*)$. Indeed, an integration by parts gives

$$\forall u \in \mathscr{C}_0^\infty(I), \quad |\langle -\partial_x^2 u, v \rangle| = \left| \int_I u\, \partial_x^2 \bar{v}\, dx \right|$$
$$\leqslant \|v\|_{H^2(I)} \|u\|_{L^2(I)},$$

as well as $T^* v = -\partial_x^2 v$. But $f_\lambda \in H^2(I)$ is such that

$$-\partial_x^2 f_\lambda \pm i f_\lambda = 0,$$

that is $f_\lambda \in \ker(T^* \pm i)$, which is a contradiction.
— Through the criterion (iii), notice that

$$\forall u \in \mathscr{C}_0^\infty(I), \quad \langle e^{i\lambda x}, -\partial_x^2 u \pm iu \rangle = (\lambda^2 \mp i)\langle e^{i\lambda x}, u \rangle = 0,$$

which implies that $0 \neq e^{i\lambda x} \in \overline{\text{ran}\,(T \pm i)}^\perp$, and therefore (iii) does not hold.

The closure is $(H_0^1(I) \cap H^2(I), -\partial_x^2)$ which is symmetric. However, the domain of its adjoint is $H^2(I)$ which is strictly bigger than $H_0^1(I) \cap H^2(I)$.   ∘

2.3.3. Essential self-adjointness of Schrödinger operators. —

Lemma 2.74. — *Let $f \in L_{loc}^2(\mathbb{R}^d)$ such that $\Delta f \in L_{loc}^2(\mathbb{R}^d)$. Then, there exists a sequence $(f_n) \in \mathscr{C}_0^\infty(\mathbb{R}^d)^{\mathbb{N}}$ such that (f_n) converges to f and (Δf_n) converges to Δf in $L_{loc}^2(\mathbb{R}^d)$.*

Proof. — It is sufficient to adapt the proof of Lemma 1.10.   □

Lemma 2.75. — *Let φ and χ be two smooth functions with compact supports, with χ real valued. We have*

$$\int_{\mathbb{R}^d} \chi^2 |\nabla \varphi|^2\, dx \leqslant 2\|\chi \Delta \varphi\| \|\chi \varphi\| + 4\|(\nabla \chi)\varphi\|^2.$$

Proof. — We write

$$\langle \Delta\varphi, \chi^2\varphi \rangle = \langle \nabla\varphi, \nabla(\chi^2\varphi) \rangle = \|\chi\nabla\varphi\|^2 + 2\langle \chi\nabla\varphi, (\nabla\chi)\,\varphi \rangle \,.$$

By using the Cauchy–Schwarz inequality, we have

$$2|\langle \chi\nabla\varphi, (\nabla\chi)\,\varphi \rangle| \leqslant \frac{1}{2}\|\chi\nabla\varphi\|^2 + 2\|(\nabla\chi)\varphi\|^2 \,.$$

We deduce the desired estimate.   □

Lemma 2.76. — *Let $f \in L^2_{loc}(\mathbb{R}^d)$ such that $\Delta f \in L^2_{loc}(\mathbb{R}^d)$. Then $f \in H^1_{loc}(\mathbb{R}^d)$.*

Proof. — We consider the sequence (f_n) given in Lemma 2.74 and we use Lemma 2.75 with $\varphi = f_n - f_p$. We easily deduce that (∇f_n) is convergent in $L^2_{loc}(\mathbb{R}^d)$ and that the limit is ∇f in the sense of distributions.   □

Lemma 2.77. — *Let $f \in L^2_{loc}(\mathbb{R}^d)$ such that $\Delta f \in L^2_{loc}(\mathbb{R}^d)$. Then $f \in H^2_{loc}(\mathbb{R}^d)$.*

Proof. — Let χ be a smooth function with compact support. We just have to show that $\chi f \in H^2(\mathbb{R}^d)$. We have

$$\Delta(\chi f) = \chi\Delta f + 2\nabla\chi \cdot \nabla f + f\Delta\chi \in L^2(\mathbb{R}^d)$$

by Lemma 2.76. Thus, by considering the Fourier transform of χf, we easily find that $\langle \xi \rangle^2 \widehat{\chi f} \in L^2(\mathbb{R}^d)$ and we deduce that $\chi f \in H^2(\mathbb{R}^d)$.   □

Proposition 2.78. — *Let us consider $V \in \mathscr{C}^\infty(\mathbb{R}^d, \mathbb{R})$ and the operator T with domain $\mathscr{C}^\infty_0(\mathbb{R}^d)$ acting as $-\Delta + V$. We assume that T is semi-bounded from below, i.e., there exists $C \in \mathbb{R}$ such that*

$$\forall u \in \mathscr{C}^\infty_0(\mathbb{R}^d), \quad \langle Tu, u \rangle \geqslant C\|u\|^2 \,.$$

Then, T is essentially self-adjoint.

Proof. — Up to a translation of V, we can assume that $C = 1$. Let us prove that the range of $T \pm i$ is dense. Let us consider $f \in L^2(\mathbb{R}^d)$ such that, for all $u \in \mathscr{C}^\infty_0(\mathbb{R}^d)$,

$$\langle f, (T \pm i)u \rangle = 0 \,.$$

We get, in the sense of distributions, that

$$(-\Delta + V \mp i)f = 0 \,.$$

With Lemma 2.77, we get that $f \in H^2_{loc}(\mathbb{R}^d)$. By induction, we get that $f \in H^\infty_{loc}(\mathbb{R}^d)$. From this and the Sobolev embedding $H^s(\mathbb{R}^d) \hookrightarrow \mathscr{C}^0(\mathbb{R}^d)$ when $s > \frac{d}{2}$, we deduce that $f \in \mathscr{C}^\infty(\mathbb{R}^d)$.

Now, take $u \in \mathscr{C}^\infty(\mathbb{R}^d)$ and consider $\chi \in \mathscr{C}^\infty_0(\mathbb{R}^d, \mathbb{R})$ supported in $B(0, 2)$ and equal to 1 on $B(0, 1)$. For all $n \geqslant 1$, we let, for all $x \in \mathbb{R}^d$,

$$\chi_n(x) = \chi(n^{-1}x) .$$

We write

$$\langle f, (T \pm i)(\chi_n^2 u) \rangle = 0 ,$$

and we have

$$\langle f, (T \pm i)(\chi_n^2 u) \rangle = \int_{\mathbb{R}^d} \left(\nabla f \nabla (\chi_n^2 \overline{u}) + (V \mp i)\chi_n^2 f \overline{u} \right) \, \mathrm{d}x .$$

We get

$$\int_{\mathbb{R}^d} \nabla f \nabla (\chi_n^2 \overline{u}) \, \mathrm{d}x = \int_{\mathbb{R}^d} \chi_n \nabla f \nabla (\chi_n \overline{u}) \, \mathrm{d}x$$
$$+ \int_{\mathbb{R}^d} \nabla f \cdot (\nabla \chi_n) \chi_n \overline{u} \, \mathrm{d}x .$$

Thus,

$$\int_{\mathbb{R}^d} \nabla f \nabla (\chi_n^2 \overline{u}) \, \mathrm{d}x = \int_{\mathbb{R}^d} \nabla (\chi_n f) \nabla (\chi_n \overline{u}) \, \mathrm{d}x$$
$$- \int_{\mathbb{R}^d} f \nabla \chi_n \cdot \nabla (\chi_n \overline{u}) \, \mathrm{d}x + \int_{\mathbb{R}^d} \nabla f \cdot (\nabla \chi_n) \chi_n \overline{u} \, \mathrm{d}x ,$$

and

$$\int_{\mathbb{R}^d} \nabla f \nabla (\chi_n^2 \overline{u}) \, \mathrm{d}x$$
$$= \int_{\mathbb{R}^d} \nabla (\chi_n f) \nabla (\chi_n \overline{u}) \, \mathrm{d}x - \int_{\mathbb{R}^d} f |\nabla \chi_n|^2 \overline{u}) \, \mathrm{d}x$$
$$+ \int_{\mathbb{R}^d} \nabla f \cdot (\nabla \chi_n) \chi_n \overline{u}) \, \mathrm{d}x - \int_{\mathbb{R}^d} f \chi_n \nabla \chi_n \cdot \nabla \overline{u} \, \mathrm{d}x .$$

We can choose $u = f$, take the real part to get

$$\int_{\mathbb{R}^d} |\nabla (\chi_n f)|^2 + V |\chi_n f|^2 \, \mathrm{d}x = \int_{\mathbb{R}^d} |f \nabla \chi_n|^2 \, \mathrm{d}x .$$

The r.h.s. goes to zero when n goes to $+\infty$. By assumption, this implies that

$$\liminf_{n \to +\infty} \|\chi_n f\|^2 = 0.$$

The conclusion follows from the Fatou lemma. $\square$

Example 2.79. — The operator with domain $\mathscr{C}_0^\infty(\mathbb{R})$ acting as $-\partial_x^2 + x^2$ is essentially self-adjoint. Show that, in fact, this operator is bounded from below by 1.

Exercise 2.80. — Take $\mathsf{H} = \mathsf{L}^2(\mathbb{R}^2)$. We take $\mathrm{Dom}\,(T) = \mathscr{C}_0^\infty(\mathbb{R}^2)$. For $\psi \in \mathrm{Dom}\,(T)$, we let

$$T\psi = (-\partial_{x_1}^2 + (-i\partial_{x_2} - x_1)^2)\psi.$$

Is this operator essentially self-adjoint?

2.4. Polar decomposition

Proposition 2.81. — *[Square root of a non-negative operator] Let $T \in \mathcal{L}(\mathsf{H})$ be a non-negative operator. There exists a unique non-negative operator $S \in \mathcal{L}(\mathsf{H})$ such that $S^2 = T$. The operator S commutes with T.*

Proof. — Let us prove the existence. Multiplying T by a small factor, we can always assume that $\|T\| < 1$. Let us write $T = \mathrm{Id} - R$, with $R = \mathrm{Id} - T$. Let us notice that $\|R\| \leqslant 1$. Indeed, for all $u \in \mathsf{H}$, we have

$$0 \leqslant \|u\|^2 - \|u\|\|Tu\| \leqslant \langle Ru, u \rangle = \|u\|^2 - \langle Tu, u \rangle \leqslant \|u\|^2.$$

By using the Cauchy–Schwarz, we get, for all $(u, v) \in \mathsf{H}^2$,

$$|\langle Ru, v \rangle| \leqslant |\langle Ru, u \rangle|^{\frac{1}{2}} |\langle Rv, v \rangle|^{\frac{1}{2}} \leqslant \|u\|\|v\|.$$

Thus, $\|R\| \leqslant 1$. Let $\mathbb{D} := \{z; |z| < 1\}$ be the open unit disk. Now, $\mathbb{D} \ni z \mapsto s(z) = (1-z)^{\frac{1}{2}}$ has a power series expansion at 0 which is

$$s(z) = 1 + \sum_{n \geqslant 1} c_n z^n, \quad 0 \leqslant c_n = \frac{(2n)!}{(2n-1)(n!)^2 4^n}.$$

Moreover, this power series is absolutely convergent on $\overline{\mathbb{D}}$. We let $S = s(R)$ and notice that $S \geqslant 0$, and, by Cauchy product, $S^2 = s^2(R) = T$.

Let us now show the uniqueness. Let S' be a non-negative operator such that $S'^2 = T$. Then, S' commutes with T and thus with S. We have

$$(S - S')(S + S')(S - S') = 0.$$

Since $(S - S')S(S - S') \geqslant 0$ and $(S - S')S'(S - S') \geqslant 0$, both equal 0, and $(S - S')^3 = 0$. Then, $(S - S')^4 = 0$ so that $(S - S')^2 = 0$ and then $S = S'$.

$\square$

Definition 2.82. — Let $T \in \mathcal{L}(\mathsf{H})$. We define $|T| = (T^*T)^{\frac{1}{2}}$.

Proposition 2.83. — *All $T \in \mathcal{L}(\mathsf{H})$ can be written as a linear combination of four unitary operators.*

Proof. — First, we notice that

$$T = \frac{T + T^*}{2} + i\frac{T - T^*}{2i},$$

and observe that we have to show that all bounded self-adjoint operator can be written as a linear combination of two unitary operators. Thus, consider $T \in \mathcal{L}(\mathsf{H})$ a self-adjoint operator. We may assume that $\|T\| \leqslant 1$. Then, we have

$$T = \frac{1}{2}\left(T + i(\mathrm{Id} - T^2)^{\frac{1}{2}}\right) + \frac{1}{2}\left(T - i(\mathrm{Id} - T^2)^{\frac{1}{2}}\right).$$

$\square$

Definition 2.84. — We say that $U \in \mathcal{L}(\mathsf{H})$ is a partial isometry when, for all $\psi \in \ker(U)^{\perp}$, $\|U\psi\| = \|\psi\|$.

Proposition 2.85 (Polar decomposition). — *There exists a unique partial isometry U such that*

$$T = U|T|, \quad \ker U = \ker T(= \ker |T|).$$

Proof. — Let us prove the uniqueness. Consider U_1 and U_2 two such isometries. We have

$$U_1|T| = U_2|T|.$$

Thus, $U_1 = U_2$ on $\mathrm{Im}|T|$ and then on $\overline{\mathrm{Im}|T|}$. On $\overline{\mathrm{Im}|T|}^{\perp} = \ker U_j$, we have $U_1 = U_2 = 0$. Therefore, $U_1 = U_2$.

Let us now establish the existence of the decomposition. We have, for all $x \in \mathsf{H}$, $\|Tx\| = \||T|x\|$. In particular, we have

$$\forall (x_1, x_2) \in \mathsf{H}^2, \quad |T|x_1 = |T|x_2 \Longrightarrow Tx_1 = Tx_2.$$

Thus, there exists an application $U : \operatorname{ran} |T| \to \operatorname{ran} T$ such that, for all $x \in \mathsf{H}$, $U|T|x = Tx$. This application U is a linear isometry. In particular, it can be extended as a linear isometry $U : \overline{\operatorname{ran} |T|} \to \overline{\operatorname{ran} T}$. On $\overline{\operatorname{ran} |T|}^{\perp} = \ker |T|$, we extend U by 0. In particular, we have $\ker |T| \subset \ker U$. The reverse inclusion is also true. Indeed, consider $y \in \mathsf{H}$ such that $Uy = 0$. Writing $y = y_1 + y_2$ with $y_1 \in \overline{\operatorname{ran} |T|}$ and $y_2 \in \overline{\operatorname{ran} |T|}^{\perp}$, we have $Uy = Uy_1$ and $0 = \|Uy\| = \|y_1\|$. Thus, $y \in \overline{\operatorname{ran} |T|}^{\perp} = \ker |T|$. This shows that $\ker |T| = \ker U$ and that U is a partial isometry. $\square$

2.5. Lax–Milgram theorems

Let $\mathcal{V}$ be a Hilbert space.

***Definition 2.86* (Coercive form).** — A continuous sesquilinear form Q on $\mathcal{V} \times \mathcal{V}$ is said to be coercive when

$$(2.2.5.22) \qquad \exists \alpha > 0, \quad \forall u \in \mathcal{V}, \quad |Q(u, u)| \geqslant \alpha \|u\|_{\mathcal{V}}^2.$$

Theorem 2.87. — *Let Q be a continuous coercive sesquilinear form on $\mathcal{V} \times \mathcal{V}$. Then, the operator $\mathscr{A} : \mathcal{V} \to \mathcal{V}$ defined by*

$$(2.2.5.23) \qquad \forall u, v \in \mathcal{V}, \quad Q(u, v) = \langle \mathscr{A} u, v \rangle_{\mathcal{V}}$$

is a continuous isomorphism of $\mathcal{V}$ onto $\mathcal{V}$ with bounded inverse. The same applies to the adjoint operator $\mathscr{A}^*$ of $\mathscr{A}$.

Proof. — Fix $u \in \mathcal{V}$. Since Q is continuous, we have

$$\forall v \in \mathcal{V}, \quad |Q(u, v)| \leqslant (C\|u\|_{\mathcal{V}})\|v\|_{\mathcal{V}}.$$

By the Riesz representation theorem, we can find some $\mathscr{A} u \in \mathcal{V}$ such that

$$\forall v \in \mathcal{V}, \quad Q(u, v) = \langle \mathscr{A} u, v \rangle_{\mathcal{V}}.$$

The operator $\mathscr{A}$ is a linear continuous map, with norm bounded by
the above constant C. By construction and by Cauchy–Schwarz,
we have

$$\forall u \in \mathcal{V}, \quad \alpha \|u\|_{\mathcal{V}}^2 \leqslant |Q(u,u)| = |\langle \mathscr{A}u, u\rangle_{\mathcal{V}}| \leqslant \|\mathscr{A}u\|_{\mathcal{V}}\|u\|_{\mathcal{V}}.$$

This shows that $(\operatorname{ran} \mathscr{A})^\perp = \{0\}$, or equivalently that the range of
$\mathscr{A}$ is dense. This also implies that

$$\alpha \|u\|_{\mathcal{V}} \leqslant \|\mathscr{A}u\|_{\mathcal{V}}$$

which means that $\mathscr{A}$ is injective and has a closed range. Thus,
$\operatorname{ran} \mathscr{A} = \mathcal{V}$. The operator $\mathscr{A}$ is a continuous isomorphism of $\mathcal{V}$
onto $\mathcal{V}$ with the inverse bounded by α^{-1}.

The continuous sesquilinear form $\tilde{Q}$ defined on $\mathcal{V} \times \mathcal{V}$ through

$$\forall (u,v) \in \mathcal{V}^2, \quad \tilde{Q}(u,v) = \overline{Q(v,u)}$$

is coercive. The corresponding operator $\tilde{\mathscr{A}}$ satisfies

(2.2.5.24)
$$\forall u, v \in \mathcal{V}, \quad \tilde{Q}(u,v) = \langle \tilde{\mathscr{A}}u, v\rangle_{\mathcal{V}} = \overline{\langle \mathscr{A}v, u\rangle_{\mathcal{V}}} = \langle \mathscr{A}^*u, v\rangle_{\mathcal{V}}$$

and therefore $\tilde{\mathscr{A}} \equiv \mathscr{A}^*$ is a continuous isomorphism of $\mathcal{V}$ onto $\mathcal{V}$
with bounded inverse. $\square$

Example 2.88. — Take $\mathcal{V} = \mathsf{H}_0^1(I)$ with $I = (0,1)$ and

$$\forall (u,v) \in \mathcal{V}^2, \quad Q(u,v) = \int_0^1 u'\overline{v'}\,\mathrm{d}x\,.$$

This sesquilinear form Q is continuous on $\mathcal{V} \times \mathcal{V}$. We have the
Poincaré inequality[1]

$$\forall u \in \mathsf{H}_0^1(I), \quad \|u\|_{\mathsf{L}^2(I)} \leqslant \|u'\|_{\mathsf{L}^2(I)}\,.$$

It follows that

$$\forall u \in \mathsf{H}_0^1(I), \quad \|u\|_{\mathsf{H}_0^1(I)}^2 = \|u\|_{\mathsf{L}^2(I)}^2 + \|u'\|_{\mathsf{L}^2(I)}^2 \leqslant 2\|u'\|_{\mathsf{L}^2(I)}^2$$
$$= 2Q(u,u)\,.$$

---

[1] See Section 1.2, or prove this inequality first for $u \in \mathscr{C}_0^\infty(I)$ and extend it by
density.

We find (2.2.5.22) with $\alpha = 1/2$. According to Theorem 2.87, we can define an operator $\mathscr{A} : \mathcal{V} \to \mathcal{V}$ satisfying

(2.2.5.25)

$$\forall u, v \in \mathcal{V}, \quad \int_0^1 u'\overline{v'}\, dx = \langle \mathscr{A} u, v \rangle_{\mathsf{H}_0^1(I)}$$

$$= \int_0^1 (\mathscr{A} u)'\overline{v'}\, dx + \int_0^1 (\mathscr{A} u)\overline{v}\, dx.$$

○

In the preceding example, we cannot replace $\mathsf{H}_0^1(I)$ by $\mathsf{L}^2(I)$ because the form Q would not be well-defined. Still, we can assert that

$$\forall (u, v) \in \mathsf{H}_0^2(I) \times \mathsf{H}_0^1(I), \quad \int_0^1 u'\overline{v'}\, dx = -\int_0^1 \partial_x^2 u\, \overline{v'}\, dx$$

$$= \langle \mathscr{L} u, v \rangle_{\mathsf{L}^2(I)}$$

with $\mathscr{L} := -\partial_x^2$. Compare (2.2.5.25) with (2.5). This means that the action of Q can also be interpreted through the L^2-inner product. But this requires to introduce an auxiliary operator $\mathscr{L}$ which is defined only on a subspace of $\mathsf{H}_0^1(I)$, namely $\mathsf{H}^2(I) \cap \mathsf{H}_0^1(I)$.

Theorem 2.89. — *In addition to the hypotheses of Theorem 2.87, assume that* H *is a Hilbert space such that* $\mathcal{V}$ *is continuously embedded and dense in* H. *Introduce*

$$\mathsf{Dom}(\mathscr{L}) = \big\{ u \in \mathcal{V} : \text{the map } v \mapsto Q(u, v)$$

$$\text{is continuous on } \mathcal{V} \text{ for the norm of } \mathsf{H} \big\}.$$

Then the operator $\mathscr{L}$ *defined by*

$$\forall u \in \mathsf{Dom}(\mathscr{L}), \quad \forall v \in \mathcal{V}, \quad Q(u, v) = \langle \mathscr{L} u, v \rangle_{\mathsf{H}}$$

satisfies the following properties:

(i) $\mathscr{L}$ *is bijective from* $\mathsf{Dom}(\mathscr{L})$ *onto* H.
(ii) $\mathscr{L}$ *is closed.*
(iii) $\mathsf{Dom}(\mathscr{L})$ *is dense in* $\mathcal{V}$ *for* $\|\cdot\|_{\mathcal{V}}$, *and it is dense in* H *for* $\|\cdot\|_{\mathsf{H}}$.

Proof. — The sesquilinear product and the norm on H will be simply denoted by

$$\langle u, v \rangle_{\mathsf{H}} \equiv \langle u, v \rangle \quad , \quad \|u\|_{\mathsf{H}} \equiv \|u\| \, .$$

By density and the Riesz theorem, $\mathscr{L}$ is well defined on $\mathrm{Dom}(\mathscr{L})$. Let us deal with (i). For all $u \in \mathrm{Dom}\,(\mathscr{L})$, by Cauchy–Schwarz and due to the continuous embedding of $\mathcal{V}$ in H, we have

$$(2.2.5.26) \qquad \|\mathscr{L}u\|\|u\| \geqslant |\langle \mathscr{L}u, u \rangle| \geqslant \alpha \|u\|_{\mathcal{V}}^2 \geqslant \alpha c \|u\|^2 \, ,$$

where $c > 0$ is such that

$$\forall u \in \mathcal{V}, \quad c\|u\| \leqslant \|u\|_{\mathcal{V}} \, .$$

We deduce that $\mathscr{L}$ is injective. Let us prove the surjectivity. Fix some $w \in \mathsf{H}$. We look for an element $u \in \mathrm{Dom}\,(\mathscr{L})$ such that $\mathscr{L}u = w$. This is equivalent to

$$\forall \varphi \in \mathsf{H}, \quad \langle \mathscr{L}u, \varphi \rangle = \langle w, \varphi \rangle \, .$$

We notice that the application $\varphi \mapsto \langle w, \varphi \rangle$ is a continuous linear map on $(\mathcal{V}, \| \cdot \|_{\mathcal{V}})$. Thus, we can find some $v \in \mathcal{V}$ such that

$$\forall \varphi \in \mathcal{V}, \quad \langle w, \varphi \rangle = \langle v, \varphi \rangle_{\mathcal{V}}.$$

We let $u = \mathscr{A}^{-1}v \in \mathcal{V}$ so that

$$\forall \varphi \in \mathcal{V}, \quad \langle w, \varphi \rangle = Q(u, \varphi) \, .$$

We deduce that $u \in \mathrm{Dom}\,(\mathscr{L})$ and

$$\forall \varphi \in \mathcal{V}, \ \langle w, \varphi \rangle = \langle \mathscr{L}u, \varphi \rangle \, .$$

By density, we get $\mathscr{L}u = w$. Thus, $\mathscr{L}$ is surjective, and hence bijective.

Consider (ii). From (2.2.5.26), we get that $\mathscr{L}^{-1}$ is continuous, and that $\|\mathscr{L}^{-1}\| \leqslant (\alpha c)^{-1}$. It follows that $\mathscr{L}$ is closed as expected.

Now, we prove (iii). Let u be an element of $\mathcal{V}$ which is orthogonal to the domain $\mathrm{Dom}\,(\mathscr{L})$ for the sesquilinear form $\langle \cdot, \cdot \rangle_{\mathcal{V}}$. In other words

$$\forall v \in \mathrm{Dom}\,(\mathscr{L}), \quad \langle u, v \rangle_{\mathcal{V}} = 0 \, .$$

The operator $\mathscr{A} \in \mathcal{L}(\mathcal{V})$ is bijective. Thus,

$$\forall v \in \mathrm{Dom}\,(\mathscr{L}), \quad \langle u, \mathscr{A}v \rangle_{\mathcal{V}} = 0 \, ,$$

so that

$$\forall v \in \mathrm{Dom}\,(\mathscr{L}), \quad Q(v, u) = 0 \, ,$$

and therefore

$$\forall v \in \text{Dom}\,(\mathscr{L})\,, \quad \langle \mathscr{L}v, u \rangle = 0\,.$$

By surjectivity of $\mathscr{L}$, we get $u = 0$. This means that the domain $\text{Dom}(\mathscr{L})$ is dense in $\mathcal{V}$ for $\|\cdot\|_{\mathcal{V}}$, and therefore in $\mathcal{V}$ for $\|\cdot\|_{\mathsf{H}}$, and then in H for $\|\cdot\|_{\mathsf{H}}$. □

Let $\tilde{Q}$ be the adjoint sesquilinear form which is defined by

$$\forall u, v \in \mathcal{V}\,, \quad \tilde{Q}(u,v) = \overline{Q(v,u)}\,.$$

As above, we can introduce

$$\text{Dom}(\widetilde{\mathscr{L}}) = \big\{ u \in \mathcal{V} : \text{the map } v \mapsto \tilde{Q}(u,v)$$
$$\text{is continuous on } \mathcal{V} \text{ for the norm of } \mathsf{H} \big\}$$

and we can define the operator $\widetilde{\mathscr{L}}$ by

$$\forall u \in \text{Dom}(\widetilde{\mathscr{L}})\,, \quad \forall v \in \mathcal{V}\,, \quad \tilde{Q}(u,v) = \langle \widetilde{\mathscr{L}}u, v \rangle_{\mathsf{H}}\,.$$

Theorem 2.90. — *We have* $\widetilde{\mathscr{L}} = \mathscr{L}^*$.

Proof. — We first prove that $\mathscr{L}^* \subset \widetilde{\mathscr{L}}$. Let $u \in \text{Dom}\,(\mathscr{L}^*)$. Then

$$\forall \varphi \in \text{Dom}\,(\mathscr{L})\,, \quad \langle \mathscr{L}\varphi, u \rangle = \langle \varphi, \mathscr{L}^*u \rangle\,.$$

We notice that $\mathcal{V} \ni \varphi \mapsto \langle \varphi, \mathscr{L}^*u \rangle$ is continuous for $\|\cdot\|_{\mathcal{V}}$. Thus, there exists $v \in \mathcal{V}$ such that

$$\forall \varphi \in \mathcal{V}\,, \quad \langle \varphi, \mathscr{L}^*u \rangle = \langle \varphi, v \rangle_{\mathcal{V}}\,.$$

In particular, we have

$$\forall \varphi \in \text{Dom}\,(\mathscr{L})\,, \quad \langle \mathscr{L}\varphi, u \rangle = \langle \varphi, \mathscr{L}^*u \rangle = \langle \varphi, v \rangle_{\mathcal{V}}\,.$$

There exists $w \in \mathcal{V}$ such that $v = \mathscr{A}^*w$ and thus
$$\forall \varphi \in \text{Dom}\,(\mathscr{L})\,, \quad \langle \mathscr{L}\varphi, u \rangle = \langle \varphi, \mathscr{L}^*u \rangle = \langle \varphi, v \rangle_{\mathcal{V}} = Q(\varphi, w)$$
$$= \langle \mathscr{L}\varphi, w \rangle\,.$$

By surjectivity of $\mathscr{L}$, we get $u = w \in \mathcal{V}$. Then

$$\forall \varphi \in \text{Dom}\,(\mathscr{L})\,, \quad \overline{\tilde{Q}(u,\varphi)} = Q(\varphi, u) = \langle \varphi, \mathscr{L}^*u \rangle\,.$$

Since $\text{Dom}(\mathscr{L})$ is dense in $\mathcal{V}$, this gives rise to

$$\forall \varphi \in \mathcal{V}\,, \quad |\tilde{Q}(u,\varphi)| \leqslant \|\mathscr{L}^*u\| \|\varphi\|\,,$$

and therefore $u \in \mathrm{Dom}\,(\widetilde{\mathscr{L}})$, with

$$\forall \varphi \in \mathrm{Dom}\,(\mathscr{L})\,, \quad \overline{\langle \widetilde{\mathscr{L}}u, \varphi \rangle} = \langle \varphi, \widetilde{\mathscr{L}}u \rangle = \langle \varphi, \mathscr{L}^*u \rangle\,.$$

By density of $\mathrm{Dom}\,(\mathscr{L})$, this is possible only if $\widetilde{\mathscr{L}}u = \mathscr{L}^*u$. This means that $\mathscr{L}^* \subset \widetilde{\mathscr{L}}$. Let us now prove the converse inclusion. Let $u \in \mathrm{Dom}\,(\widetilde{\mathscr{L}})$. We have

$$\forall \varphi \in \mathrm{Dom}\,(\mathscr{L})\,, \quad \langle \mathscr{L}\varphi, u \rangle = Q(\varphi, u) = \overline{\widetilde{Q}(u, \varphi)} = \overline{\langle \widetilde{\mathscr{L}}u, \varphi \rangle}$$
$$= \langle \varphi, \widetilde{\mathscr{L}}u \rangle\,.$$

This gives rise to

$$\forall \varphi \in \mathrm{Dom}(\mathscr{L})\,, \quad |\langle \mathscr{L}\varphi, u \rangle| \leqslant \|\widetilde{\mathscr{L}}u\|\|\varphi\|\,.$$

It follows that $u \in \mathrm{Dom}\,(\mathscr{L}^*)$ and, since $\mathrm{Dom}(\mathscr{L})$ is dense in $\mathcal{V}$,

$$\forall \varphi \in \mathcal{V}\,, \quad \langle \varphi, \mathscr{L}^*u \rangle = \langle \varphi, \widetilde{\mathscr{L}}u \rangle\,.$$

Since $\mathcal{V}$ is dense in H, this gives $\mathscr{L}^*u = \widetilde{\mathscr{L}}u$. □

2.6. Examples

2.6.1. Dirichlet Laplacian. — Let $\Omega \subset \mathbb{R}^d$ be an open set. Here, we consider $\mathcal{V} = \mathsf{H}_0^1(\Omega)$ and we define the sesquilinear form

$$Q^{\mathrm{Dir}}(u, v) = \int_\Omega \nabla u \cdot \overline{\nabla v} + u\overline{v}\,\mathrm{d}x\,.$$

The form Q^{Dir} is Hermitian, continuous, and coercive on $\mathcal{V}$, with $\alpha = 1$. We find that $\mathscr{A} = \mathrm{Id}_\mathcal{V}$. The self-adjoint operator $\mathscr{L}^{\mathrm{Dir}} - \mathrm{Id}$ given by Theorem 2.89 is called the Dirichlet Laplacian on Ω. The domain of $\mathscr{L}^{\mathrm{Dir}}$ is

$$\mathrm{Dom}\,(\mathscr{L}^{\mathrm{Dir}}) = \{\psi \in \mathsf{H}_0^1(\Omega) : -\Delta\psi \in \mathsf{L}^2(\Omega)\}\,.$$

If the boundary of Ω is smooth (see Section 2.7 for more details), we have

$$\mathrm{Dom}\,(\mathscr{L}^{\mathrm{Dir}}) = \mathsf{H}_0^1(\Omega) \cap \mathsf{H}^2(\Omega)\,.$$

Remark 2.91. — This characterization of the domain is not true if the boundary is not smooth. To see this, the Reader can consider a sector Ω of opening $\alpha \in (\pi, 2\pi)$. In that case, the function $\psi = r^{\frac{\pi}{\alpha}} \sin(\alpha^{-1}\theta\pi)$ satisfies $\Delta\psi = 0$ but ψ is not H^2 near 0. Then, by using a convenient cutoff function (to get a function in $H_0^1(\Omega)$), we get a counter-example.

2.6.2. Neumann Laplacian. — Let $\Omega \subset \mathbb{R}^d$ be an open set. Here, we consider $\mathcal{V} = H^1(\Omega)$ and we define the sesquilinear form

$$Q^{\text{Neu}}(u, v) = \int_\Omega \nabla u \cdot \overline{\nabla v} + u\overline{v}\, dx\,.$$

The form Q is Hermitian, continuous, and coercive on $\mathcal{V}$. In Theorem 2.87, we have $\mathscr{A} = \text{Id}_\mathcal{V}$. The self-adjoint operator $\mathscr{L}^{\text{Neu}} - \text{Id}$ given by Theorem 2.89 is called the Neumann Laplacian on Ω. If the boundary of Ω is smooth, the domain of $\mathscr{L}^{\text{Neu}}$ is

$$\text{Dom}\,(\mathscr{L}^{\text{Neu}})$$
$$= \{\psi \in H^1(\Omega) : -\Delta\psi \in L^2(\Omega)\,, \quad \nabla\psi \cdot \mathbf{n} = 0 \text{ on } \partial\Omega\}\,,$$

and we have (and we admit that)

$$\text{Dom}\,(\mathscr{L}^{\text{Neu}}) = \{\psi \in H^1(\Omega) \cap H^2(\Omega) : \nabla\psi \cdot \mathbf{n} = 0 \text{ on } \partial\Omega\}\,.$$

Remark 2.92. — This characterization of the domain is not true if the boundary is not smooth.

2.6.3. Harmonic oscillator. — Let us consider the operator

$$\mathcal{H}_0 = (\mathscr{C}_0^\infty(\mathbb{R}), -\partial_x^2 + x^2)\,.$$

This operator is essentially self-adjoint as we have seen in Example 2.79. Let us denote by $\mathcal{H}$ its closure. The operator $\mathcal{H}$ is called the harmonic oscillator. We have

$$\text{Dom}\,(\mathcal{H}) = \text{Dom}\,(\mathcal{H}_0^*) = \{\psi \in L^2(\mathbb{R}) : (-\partial_x^2 + x^2)\psi \in L^2(\mathbb{R})\}\,.$$

We recall Lemma 1.11. Theorem 2.87 can be applied and $\mathscr{A} = \text{Id}$. Consider Theorem 2.89 with $H = L^2(\mathbb{R})$. The assumptions are satisfied since $B^1(\mathbb{R})$ is continuously embedded and dense in $L^2(\mathbb{R})$. The operator $\mathscr{L}$ associated with Q is self-adjoint, and its domain is

$$\text{Dom}\,(\mathscr{L}) = \{\psi \in B^1(\mathbb{R}) : (-\partial_x^2 + x^2)\psi \in L^2(\mathbb{R})\}\,.$$

The operator $\mathscr{L}$ satisfies in particular

$$\langle (-\partial_x^2 + x^2)u, v\rangle = Q(u, v) = \langle \mathscr{L}u, v\rangle\,,$$

for all $u, v \in \mathscr{C}_0^\infty(\mathbb{R})$. This shows that $\mathscr{L}$ is a self-adjoint extension of $\mathcal{H}_0$. Thus, $\mathscr{L} = \mathcal{H}$.

2.6.4. Exercise on the magnetic Dirichlet Laplacian. — Consider a bounded open set $\Omega \subset \mathbb{R}^2$ and a function $\phi \in \mathscr{C}^\infty(\overline{\Omega}, \mathbb{R})$. We let, for all $x \in \overline{\Omega}$,

$$B(x) = \Delta\phi(x),$$

and we assume that $B(x) \geqslant B_0 > 0$. We let

$$\mathbf{A} = (A_1, A_2) = (-\partial_{x_2}\phi, \partial_{x_1}\phi).$$

2.6.4.1. *Coercivity of the magnetic Laplacian.* —

i. Prove that, for all $\psi \in \mathscr{C}_0^\infty(\Omega)$,

$$\int_\Omega |(-i\nabla - \mathbf{A})\psi|^2 \mathrm{d}x$$

$$:= \int_\Omega |(-i\partial_1 - A_1)\psi|^2 + |(-i\partial_2 - A_2)\psi|^2 \mathrm{d}x$$

$$\geqslant \int_\Omega B(x)|\psi|^2 \mathrm{d}x.$$

We will note that

$$[-i\partial_1 - A_1, -i\partial_2 - A_2] = iB.$$

ii. Prove that the inequality given in i can be extended to $\psi \in \mathsf{H}_0^1(\Omega)$.

iii. For all $\varphi, \psi, \in \mathsf{H}_0^1(\Omega)$, we let

$$Q_{\mathbf{A}}(\varphi, \psi) = \int_\Omega (-i\nabla - \mathbf{A})\varphi \cdot \overline{(-i\nabla - \mathbf{A})\psi} \, \mathrm{d}x.$$

Show that $Q_{\mathbf{A}}$ is a continuous and coercive sesquilinear form on $\mathsf{H}_0^1(\Omega)$. *Hint: Show that, for all $\varepsilon > 0$ and all $\varphi \in \mathsf{H}_0^1(\Omega)$,*

$$Q_{\mathbf{A}}(\varphi, \varphi) \geqslant (1 - \varepsilon) \int_\Omega |\nabla\varphi|^2 \mathrm{d}x - \varepsilon^{-1} \int_\Omega |\mathbf{A}|^2 |\varphi|^2 \mathrm{d}x.$$

Let $\mathscr{L}_{\mathbf{A}}$ be the operator associated with this form via the Lax–Milgram theorem.

iv. Explain why $\mathscr{L}_{\mathbf{A}}$ is self-adjoint.

v. (a) Prove that

$$\mathrm{Dom}(\mathscr{L}_{\mathbf{A}}) = \{\psi \in \mathsf{H}_0^1(\Omega) : -\Delta\psi \in \mathsf{L}^2(\Omega)\}.$$

(b) Show that for all $\psi \in \mathrm{Dom}(\mathscr{L}_{\mathbf{A}})$,

$$\mathscr{L}_{\mathbf{A}}\psi = -\Delta\psi + 2i\mathbf{A} \cdot \nabla\psi + |\mathbf{A}|^2\psi.$$

2.6.4.2. *Magnetic Cauchy–Riemann operators.* — Consider the following differential operators

$$\partial_z = \frac{1}{2}(\partial_{x_1} - i\partial_{x_2}), \qquad \partial_{\bar{z}} = \frac{1}{2}(\partial_{x_1} + i\partial_{x_2}),$$

and

$$d_{\mathbf{A}} = -2i\partial_z - A_1 + iA_2, \qquad \widetilde{d_{\mathbf{A}}} = -2i\partial_{\bar{z}} - A_1 - iA_2.$$

More precisely, for all $\psi \in \mathscr{D}'(\Omega)$,

$$d_{\mathbf{A}}\psi = -2i\partial_z\psi - A_1\psi + iA_2\psi, \qquad \widetilde{d_{\mathbf{A}}}\psi = -2i\partial_{\bar{z}}\psi - A_1\psi - iA_2\psi.$$

We consider $(\mathsf{H}_0^1(\Omega), d_{\mathbf{A}})$.

i.   (a) Compute $\partial_{\bar{z}}(-A_1 + iA_2)$.
      (b) For all $\varphi \in \mathscr{C}_0^\infty(\Omega)$, give a simplified expression of $\widetilde{d_{\mathbf{A}}}(d_{\mathbf{A}}\varphi)$ by using only $\mathscr{L}_{\mathbf{A}}$ and B.

ii. Prove that

$$\forall \varphi \in \mathscr{C}_0^\infty(\Omega), \quad \|d_{\mathbf{A}}\varphi\|^2 = Q_{\mathbf{A}}(\varphi, \varphi) + \int_\Omega B|\varphi|^2 \mathrm{d}x.$$

iii.   (a) Is the operator $(\mathsf{H}_0^1(\Omega), d_{\mathbf{A}})$ closed?
      (b) Prove that $(\mathsf{H}_0^1(\Omega), d_{\mathbf{A}})$ is injective with a closed range.

iv. What is the adjoint $d_{\mathbf{A}}^*$ of $d_{\mathbf{A}}$?

v. Show that $\ker(d_{\mathbf{A}}^*) = [e^{-\phi}\mathcal{O}(\Omega)] \cap \mathsf{L}^2(\Omega)$. *We will admit that if $\psi \in \mathsf{L}^2(\Omega)$ satisfies $\partial_{\bar{z}}\psi = 0$, then $\psi \in \mathcal{O}(\Omega)$. Here, $\mathcal{O}(\Omega)$ denotes the set of the holomorphic functions on Ω.*

vi. Is $(\mathsf{H}_0^1(\Omega), d_{\mathbf{A}})$ surjective?

2.7. Regularity theorem for the Dirichlet Laplacian

Theorem 2.93. — *Let Ω be a bounded open set of class $\mathscr{C}^2$. Let $u \in \mathsf{H}_0^1(\Omega)$ and $f \in \mathsf{L}^2(\Omega)$ such that $-\Delta u = f$. Then, $u \in \mathsf{H}^2(\Omega)$.*

2.7.1. Difference quotients. —

Proposition 2.94. — *Let $p \in (1, +\infty]$ and $u \in L^p(\Omega)$. Then $u \in W^{1,p}(\Omega)$ if and only if there exists $C > 0$ such that, for all $\omega \subset\subset \Omega^{(1)}$ and $h \in (0, \operatorname{dist}(\omega, \complement\Omega))$, we have*
(2.2.7.27)
$$\|D_h u\|_{L^p(\omega)} \leqslant C, \quad D_h u = \frac{\tau_h u - u}{|h|}, \quad \tau_h u(\cdot) = u(\cdot + h).$$

In this case, we can take $C = \|\nabla u\|_{L^p(\Omega)}$. If $p = 1$ and $u \in W^{1,1}(\Omega)$, we still have

(2.2.7.28)
$$\|D_h u\|_{L^1(\omega)} \leqslant \|\nabla u\|_{L^1(\Omega)}.$$

Proof. — Consider $p \in [1, +\infty)$. For all $u \in \mathscr{C}_0^\infty(\mathbb{R}^d)$, the Taylor formula gives

$$\tau_h u(x) - u(x) = \int_0^1 \nabla u(x + th) \cdot h \, \mathrm{d}t.$$

With the Hölder inequality,

$$|\tau_h u(x) - u(x)|^p \leqslant |h|^p \int_0^1 |\nabla u(x + th)|^p \, \mathrm{d}t.$$

For $\omega \subset\subset \Omega$, we get

$$\|\tau_h u - u\|_{L^p(\omega)}^p \leqslant |h|^p \int_0^1 \int_{\omega + th} |\nabla u(y)|^p \, \mathrm{d}y \, \mathrm{d}t.$$

We can find $\omega' \subset\subset \Omega$ such that $\omega + th \subset \omega'$ for all $t \in [0, 1]$ and all $h \in (0, \operatorname{dist}(\omega, \complement\Omega))$. Then,

(2.2.7.29)
$$\|\tau_h u - u\|_{L^p(\omega)} \leqslant |h| \left(\int_{\omega'} |\nabla u(y)|^p \, \mathrm{d}y \right)^{\frac{1}{p}}.$$

For $u \in W^{1,p}(\Omega)$, we can find a sequence $(u_n) \subset \mathscr{C}_0^\infty(\mathbb{R}^d)$ such that $u_n \xrightarrow[n \to +\infty]{} u$ in $W^{1,p}(\omega)$ since $\mathscr{C}_0^\infty(\mathbb{R}^d)$ is dense in $W^{1,p}(\mathbb{R}^d)$. Thus, (2.2.7.29) is true for $u \in W^{1,p}(\Omega)$ (and also with $p = +\infty$). Then, (2.2.7.27) and (2.2.7.28) follow.

---

[1]This means that $\overline{\omega}$ is compact and $\overline{\omega} \subset \Omega$.

Conversely, for $p \in (1, +\infty]$, we consider $\varphi \in \mathscr{C}_0^\infty(\Omega)$. There exists $\omega \subset\subset \Omega$ such that $\operatorname{supp} \varphi \subset \omega$. We take $h \in (0, \operatorname{dist}(\omega, \complement\Omega))$, and we write

$$\left| \int_\Omega u \, D_{-h}\varphi \, \mathrm{d}x \right| = \left| \int_\Omega D_h u \, \varphi \, \mathrm{d}x \right| \leqslant \|D_h u\|_{\mathsf{L}^p(\omega)} \|\varphi\|_{\mathsf{L}^{p'}(\Omega)}$$
$$\leqslant C \|\varphi\|_{\mathsf{L}^{p'}(\Omega)}.$$

By using the dominated convergence theorem, we deduce that, for all $j \in \{1, \ldots, d\}$,

$$\left| \int_\Omega u \partial_j \varphi \, \mathrm{d}x \right| \leqslant C \|\varphi\|_{\mathsf{L}^{p'}(\Omega)}.$$

This shows that the distribution $\partial_j u$ belongs to $\mathsf{L}^p(\Omega)$ since $\mathsf{L}^p(\Omega) = (\mathsf{L}^{p'}(\Omega))'$ (only when $p > 1$). □

2.7.2. Partition of the unity. —

Lemma 2.95. — *Let Ω be a non-empty open set of $\mathbb{R}^d$ and $K \subset \Omega$ be a compact set. There exists $\chi \in \mathscr{C}_0^\infty(\mathbb{R}^d)$ such that*

$$0 \leqslant \chi \leqslant 1, \ and \quad \chi = 1 \quad in \ a \ neighborhood \ of \ K.$$

Proof. — There exists a non-negative function $\rho \in \mathscr{C}_0^\infty(\mathbb{R}^d)$ such that $\operatorname{supp}(\rho) \subset B(0,1]$ and $\int_{\mathbb{R}^d} \rho(x) \, \mathrm{d}x = 1$. Let $\varepsilon > 0$ and

$$K_\varepsilon = \{x \in \mathbb{R}^d : \operatorname{dist}(x, K) \leqslant \varepsilon\}.$$

Clearly, K_ε is compact, and $K \subset K_\varepsilon$. When $\Omega \neq \mathbb{R}^d$, we also let

$$\delta = \operatorname{dist}(K, \complement\Omega) > 0.$$

For all $\varepsilon \in (0, \delta)$, we have $K_\varepsilon \subset \Omega$.

Consider the smooth function defined by

$$\chi_\varepsilon(x) = \int_{\mathbb{R}^d} \mathbb{1}_{K_{2\varepsilon}}(y) \rho_\varepsilon(x - y) \, \mathrm{d}y, \quad \rho_\varepsilon(x) = \varepsilon^{-d} \rho(\varepsilon^{-1}x).$$

We have, for ε small enough,

$$\operatorname{supp}(\chi_\varepsilon) \subset \operatorname{supp}(\mathbb{1}_{K_{2\varepsilon}}) + B(0, \varepsilon] \subset K_{3\varepsilon} \subset \Omega.$$

Then, consider $x \in K_\varepsilon$. Then, $\operatorname{supp}(\rho_\varepsilon(x - \cdot)) \subset K_{3\varepsilon}$ and thus $\chi_\varepsilon(x) = 1$. □

Lemma 2.96. — *Let $K \subset \mathbb{R}^d$ be a compact set. Assume that*

$$K \subset \bigcup_{j=1}^{p} U_j \,,$$

where each U_j is an open set which cannot be removed. Then, there exist a family of non-empty open sets $(V_j)_{1 \leqslant j \leqslant p}$ such that

$$\forall j \in \{1, \ldots, p\}, \quad V_j \subset\subset U_j \,,$$

and

$$K \subset \bigcup_{j=1}^{p} V_j \,.$$

Proof. — Consider the non-empty compact set

$$K_1 = K \setminus \bigcup_{j=2}^{p} U_j \subset K \cap U_1 \,.$$

We let, for any $\varepsilon \in (0, \operatorname{dist}(K_1, \complement U_1))$,

$$V_1 = \{x \in \mathbb{R}^d : \operatorname{dist}(x, K_1) < \varepsilon\} \subset U_1 \,.$$

The set $\overline{V_1}$ is compact and $K_1 \subset V_1$. We have

$$K \subset V_1 \cup \bigcup_{j=2}^{p} U_j \,.$$

The result follows by induction. $\qquad\qquad\square$

Lemma 2.97. — *[Partition of the unity] Consider K and a family of open sets $(U_j)_{1 \leqslant j \leqslant p}$ as in Lemma 2.96. There exists a family of smooth functions $(\theta_j)_{1 \leqslant j \leqslant p}$ with compact supports such that*

$$\forall j \in \{1, \ldots, p\}, \quad \operatorname{supp}(\theta_j) \subset U_j \,,$$

and, in a neighborhood of K,

$$\sum_{j=1}^{p} \theta_j = 1 \,.$$

Proof. — We use Lemma 2.96 and then Lemma 2.95 to get the existence of $\chi_j \in \mathscr{C}_0^{\infty}(U_j)$ such that $\chi_j = 1$ on a neighborhood of $\overline{V_j}$. Then, we let

$$\theta_1 = \chi_1 , \quad \theta_2 = \chi_2(1-\chi_1) , \ldots , \theta_p = \chi_p(1-\chi_{p-1})\ldots(1-\chi_1) .$$

$\square$

2.7.3. Local charts. — Let $C = \{x \in \mathbb{R}^d : |x'| < 1 , |x_d| < 1\}$. Since $\partial\Omega$ is of class $\mathscr{C}^2$ and compact, there exist a family of open sets $(U_j)_{1\leqslant j\leqslant p}$ such that

$$\partial\Omega \subset \bigcup_{j=1}^p U_j ,$$

and $\mathscr{C}^2$-diffeomorphisms $\varphi_j : Q \to U_j$ with $\varphi_j \in \mathscr{C}^2(\overline{Q})$ and $\kappa_j := \varphi_j^{-1} \in \mathscr{C}^2(\overline{U_j})$ and $\varphi_j(C_0) = \partial\Omega \cap U_j$. There exists also an open set $U_0 \subset\subset \Omega$ such that

$$\overline{\Omega} \subset U_0 \cup \bigcup_{j=1}^p U_j .$$

We can apply Lemma 2.97 to get a family of smooth functions with compact supports $(\theta_j)_{0\leqslant j\leqslant p}$ such that

$$\theta_0 + \sum_{j=1}^p \theta_j = 1 .$$

2.7.4. Proof. — Let us write

$$u = \theta_0 u + \sum_{j=1}^r \theta_j u .$$

Note that $\theta_0 u \in H^1(\mathbb{R}^d)$. Moreover,

$$-\Delta(\theta_0 u) = -\Delta\theta_0 u - 2\nabla\theta_0\nabla u + \theta_0 f \in L^2(\mathbb{R}^d) .$$

By using the Fourier transform, we get $\theta_0 u \in H^2(\mathbb{R}^d)$.

Let us now prove that $\theta_j u \in H^2(\Omega)$ for all $j \in \{1,\ldots,p\}$. We let $v = \theta_j u$. We have

$$-\Delta(\theta_j u) = -\Delta\theta_j u - 2\nabla\theta_j\nabla u + \theta_j f = g \in L^2(\Omega) .$$

For all $\varphi \in H_0^1(\Omega)$, we have

$$\int_{\Omega\cap U_j} \nabla_x v\nabla_x\varphi \,dx = \int_{\Omega\cap U_j} g\varphi \,dx .$$

We let, for all $y \in Q_+$, $w(y) = v(\varphi_j(y))$. Note that, for all $x \in \Omega \cap U_j$,

$$v(x) = w(\varphi_j^{-1}(x)) \, .$$

By using the change of variable $x = \varphi_j(y)$, we get

$$\nabla_x = (d\kappa_j)^{\mathrm{T}} \nabla_y \, .$$

Letting $G_j = (d\varphi_j)^{\mathrm{T}} d\varphi_j$, we get

$$(2.2.7.30) \quad \int_{Q_+} \langle G_j^{-1} \nabla_y w, \nabla_y \tilde{\varphi} \rangle |G_j|^{\frac{1}{2}} \, \mathrm{d}y = \int_{Q_+} \tilde{g} \tilde{\varphi} |G_j|^{\frac{1}{2}} \, \mathrm{d}y \, ,$$

where $\tilde{f}(y) = f(\varphi_j(y))$. This holds in fact for all $\tilde{\varphi} \in \mathsf{H}_0^1(Q_+)$. Let us prove that $w \in \mathsf{H}^2(Q_+)$. Let us introduce the difference quotient

$$D_h u = \frac{\tau_h u - u}{|h|} \, .$$

Let us assume that h is parallel to the boundary $y_d = 0$ and that $|h|$ is small enough to have $D_{-h} D_h w \in \mathsf{H}_0^1(Q_+)$. Then, we can take $\tilde{\varphi} = D_{-h} D_h w$. By Proposition 2.94,

$$(2.2.7.31) \quad \int_{Q_+} \tilde{g} \tilde{\varphi} |G_j|^{\frac{1}{2}} \, \mathrm{d}y \leqslant C_j \|\tilde{g}\| \|\nabla D_h w\| \, .$$

Moreover,

$$\int_{Q_+} \langle G_j^{-1} \nabla_y w, \nabla_y \tilde{\varphi} \rangle |G_j|^{\frac{1}{2}} \, \mathrm{d}y$$

$$= \int_{Q_|} \langle D_h (G_j^{-1} \nabla_y w), \nabla_y D_h w \rangle |G_j|^{\frac{1}{2}} \, \mathrm{d}y \, .$$

Commuting D_h with G_j^{-1}, we find that

$$\int_{Q_+} \langle G_j^{-1} \nabla_y w, \nabla_y \tilde{\varphi} \rangle |G_j|^{\frac{1}{2}} \, \mathrm{d}y$$

$$\geqslant \int_{Q_+} \langle (G_j^{-1} \nabla_y D_h w), \nabla_y D_h w \rangle |G_j|^{\frac{1}{2}} \, \mathrm{d}y$$

$$- C \|w\|_{\mathsf{H}^1} \|\nabla D_h w\|_{\mathsf{L}^2} \, .$$

Since G_j is a positive definite matrix, we get, for some $\alpha > 0$,

$$\int_{Q_+} \langle G_j^{-1}\nabla_y w, \nabla_y \tilde{\varphi}\rangle |G_j|^{\frac{1}{2}}\, dy \geqslant \alpha \|\nabla_y D_h w\|^2$$
$$- C\|w\|_{\mathsf{H}^1}\|\nabla D_h w\|_{\mathsf{L}^2}.$$

Then, using the Young inequality,

$$\int_{Q_+} \langle G_j^{-1}\nabla_y w, \nabla_y \tilde{\varphi}\rangle |G_j|^{\frac{1}{2}}\, dy \geqslant \frac{\alpha}{2}\|\nabla_y D_h w\|^2 - C\|w\|_{\mathsf{H}^1}^2.$$

Note that we can prove with (2.2.7.30) (with $\tilde{\varphi} = w$), and the Poincaré inequality, that $\|w\|_{\mathsf{H}^1} \leqslant C\|\tilde{g}\|$. With (2.2.7.31), and again the Young inequality, we deduce that

$$\|\nabla_y D_h w\| \leqslant C\|\tilde{g}\|.$$

Proposition 2.94 implies that, for all $\ell \in \{0, \ldots, y_{d-1}\}$,

$$\|\nabla_y \partial_\ell w\| \leqslant C\|\tilde{g}\|.$$

It remains to control the normal derivative. Consider 2.2.7.30 with $\tilde{\varphi} \in \mathscr{C}_0^\infty(Q_+)$. The term in the left-hand side involving only the normal derivative is in the form $\alpha_{dd}(y)\partial_{y_d} w \partial_{y_d}\tilde{\varphi}$ with $\alpha_{dd} \geqslant \alpha > 0$. Thus, let us replace $\tilde{\varphi}$ by $\alpha_{dd}^{-1}\tilde{\varphi}$. Then, since all the other second order derivatives are controlled, we get

$$\left|\int_{Q_+} \partial_{y_d} w \partial_{y_d}\tilde{\varphi}\, dy\right| \leqslant C\|\tilde{g}\|\|\tilde{\varphi}\|.$$

This shows that $\partial_{y_d}^2 w$ belongs to $\mathsf{L}^2(Q_+)$.

Therefore, $w \in \mathsf{H}^2(Q_+)$ and then $u \in \mathsf{H}^2(\Omega)$.

2.8. Notes

i. The theorems in Section 2.5 were proved by Lax and Milgram in [**29**, Theorems 2.1 & 2.2]. Our presentation follows the book [**20**, Chapter 3], but Theorem 2.90 is added.

ii. Section 2.7 is essentially taken from the book [3, Section IX.6]. The (difference quotient) method (due to Nirenberg [33, p. 147]) has an interest of its own to establish elliptic estimates and *characterize* the domain of many operators. Indeed, the abstract Lax–Milgram characterization is not always very useful in practice. This addendum was suggested by L. Le Treust.

CHAPTER 3

SPECTRUM

This chapter describes the various elementary properties of the *spectrum*. We will first discuss the important case of bounded operators, and especially the remarkable resolvent bound for normal operators. Then, we will progressively consider more general closed operators and discuss the famous *Riesz projections*. Finally, we will say a few words about the Fredholm operators (and their indices). The main reason to do that is to define the *discrete* and the *essential* spectrum of a closed operator. Somehow, we will see that the Fredholm operators of index 0 are very close to being square matrices, at least from the spectral point of view.

3.1. Definitions and basic properties

3.1.1. Holomorphic functions valued in a Banach space. —
Let E be a Banach space.

Definition 3.1. — Let Ω be a non-empty open set in $\mathbb{C}$. We say that $f : \Omega \to E$ is holomorphic when, for all $z_0 \in \Omega$, the limit

$$\lim_{z \to z_0} \frac{f(z) - f(z_0)}{z - z_0}$$

exists. It is denoted by $f'(z_0)$.

Lemma 3.2. — *Let $A \subset E$ such that $\ell(A)$ is bounded for all $\ell \in E'$. Then A is bounded.*

© The Author(s), under exclusive license
to Springer Nature Switzerland AG 2021
C. Cheverry and N. Raymond, *A Guide to Spectral Theory*,
Birkhäuser Advanced Texts Basler Lehrbücher,
https://doi.org/10.1007/978-3-030-67462-5_3

Proof. — This is a consequence of the *uniform boundedness principle*, and of the Hahn–Banach theorem. □

Proposition 3.3. — *Let $f : \Omega \to E$. f be holomorphic if and only if it is weakly holomorphic, i.e., $\ell \circ f$ is holomorphic on Ω for all $\ell \in E'$.*

Proof. — Let us assume that $\ell \circ f$ is holomorphic on Ω for all $\ell \in E'$. Let us first prove that f is continuous. Take $z_0 \in \Omega$ and define for $r > 0$ such that $D(z_0, r) \subset \Omega$,

$$A = \left\{ \frac{f(z) - f(z_0)}{z - z_0} , z \in D(z_0, r) \setminus \{z_0\} \right\} \subset E .$$

We observe that $\ell(A)$ is bounded for all $\ell \in E'$. We deduce that A is bounded. This proves the continuity of f at z_0. Take $z_0 \in \Omega$ and Γ a circle with center z_0 and radius r such that $\overline{D(z_0, r)} \subset \Omega$. Since f is continuous, we can define, for $z \in D(z_0, r)$,

(3.3.1.1)
$$F(z) = \frac{1}{2i\pi} \int_\Gamma \frac{f(\zeta)}{\zeta - z} \, d\zeta$$

$$= \frac{1}{2i\pi} \sum_{n=0}^{+\infty} \left(\int_\Gamma \frac{f(\zeta)}{(\zeta - z_0)^{n+1}} \, d\zeta \right) (z - z_0)^n .$$

By the Cauchy formula, we get, for all $\ell \in E'$ and $z \in D(z_0, r)$,

$$\ell \circ f(z) = \frac{1}{2i\pi} \int_\Gamma \frac{\ell \circ f(\zeta)}{\zeta - z} \, d\zeta .$$

Using the Riemannian sums, we find

$$\ell (f(z) - F(z)) = 0 .$$

By the Hahn–Banach theorem, we deduce that $F(z) = f(z)$. From (3.3.1.1), it is easy to show that F (and therefore f) has a power series expansion on $D(z_0, r)$, and thus it is holomorphic. □

By using the classical Liouville theorem, we get the following.

Corollary 3.4. — *Let $f : \mathbb{C} \to E$ be holomorphic. If f is bounded, then it is constant.*

Proof. — Assume that we can find $z_0 \in \mathbb{C}$ and $z_1 \in \mathbb{C}$ such that $f(z_0) \neq f(z_1)$. Then, by the Hahn–Banach theorem, there exists some $\ell \in E'$ such that $\ell \circ f(z_0) \neq \ell \circ f(z_1)$. But the function

$\mathbb{C} \ni z \mapsto \ell \circ f(z)$ is holomorphic and bounded. By the classical Liouville theorem, it must be constant. This is a contradiction. $\quad\square$

3.1.2. Basic definitions and properties. — In this section, T is a bounded operator on a Banach space E (and $\mathrm{Dom}\,(T) = E$), or T is a closed (unbounded) operator $\big(\mathrm{Dom}\,(T), T\big)$ on $E = \mathsf{H}$.

Definition 3.5. — The *resolvent set* $\rho(T)$ of T is the set of all $z \in \mathbb{C}$ such that $T - z : \mathrm{Dom}\,(T) \to E$ is bijective.

Note that, by the closed graph theorem, if $z \in \rho(T)$,

$$R_T(z) := (T - z)^{-1} : (\mathsf{H}, \|\cdot\|) \to (\mathrm{Dom}\,(T), \|\cdot\|_T)$$

is bounded.

Definition 3.6. — The *spectrum* of T is the set $\mathrm{sp}(T) = \mathbb{C} \setminus \rho(T)$.

Definition 3.7. — An eigenvalue of T is a number $\lambda \in \mathbb{C}$ such that $\ker(T - \lambda) \neq \{0\}$. The set formed by the eigenvalues is called point spectrum. It is denoted by $\mathrm{sp}_p(T)$.

We have $\mathrm{sp}_p(T) \subset \mathrm{sp}(T)$.

Proposition 3.8. — *In finite dimension, the spectrum coincides with the point spectrum.*

Proof. — In finite dimension, the operator $T - z$ is injective if and only if $T - z$ is surjective, whereas the continuity is always guaranteed. $\quad\square$

Exercise 3.9. — Here $\mathsf{H} = \mathbb{C}^n$. Fix $\varepsilon > 0$, and define the matrix $M_n(\varepsilon) = (m_{i,j})_{\substack{1 \leqslant i \leqslant n \\ 1 \leqslant j \leqslant n}}$ with $m_{n,1} = \varepsilon$, $m_{i,i+1} = 1$ for all $i \in \{1, \ldots, n-1\}$, and 0 otherwise.

   i. What is the spectrum of $M_n(\varepsilon)$?
   ii. What is the behavior of the spectrum when n goes to $+\infty$?

Solution:

i. The eigenvalues λ_j^n of $M_n(\varepsilon)$ are distinct. They can be obtained by looking at the roots of the characteristic equation $X^n - \varepsilon = 0$. We find $\lambda_j^n = \sqrt[n]{\varepsilon}\, e^{2ij\pi/n}$ with $j \in \{0, \cdots, n-1\}$.

ii. A point $z \in \mathbb{C}$ is the limit of eigenvalues λ_j^n of $M_n(\varepsilon)$ when n goes to $+\infty$ if and only if $|z| = 1$. The spectrum tends to the unit circle.   ○

Exercise 3.10. — What are the spectra of $\overline{a}$ and $\overline{c}$ defined in Section 2.2.4?

Proposition 3.11. — *Assume that T is a bounded operator on E. Then, for $z \in \rho(T)$, the inverse operator $(T - z)^{-1}$ is bounded, and we have $R_T(z) = (T - z)^{-1}$. Moreover*

$$\mathsf{sp}(T) \subset \{z\,;\, |z| \leqslant \|T\|\}.$$

Proof. — The first assertion is a consequence of the *open mapping theorem.* Let $z \in \mathbb{C}$ be such that $\|T\| < |z|$. Then the operator $T - z$ is invertible with an inverse given by the (absolutely convergent) series

$$(T - z)^{-1} = -\sum_{n=0}^{+\infty} \frac{T^n}{z^{n+1}}.$$

$\square$

Proposition 3.12. — *$\rho(T)$ is an open set and $\rho(T) \ni z \mapsto R_T(z)$ is holomorphic.*

Proof. — Take $z_0 \in \rho(T)$. We let $M = \|(T - z_0)^{-1}\| > 0$. Consider $r \in (0, M^{-1})$. We have, for all $z \in D(z_0, r)$,

$$T - z = T - z_0 + z_0 - z = \left(\mathrm{Id} + (z_0 - z)(T - z_0)^{-1}\right)(T - z_0).$$

Since $|z - z_0|\|(T - z_0)^{-1}\| < 1$, we get that the operator

$$\mathrm{Id} + (z_0 - z)(T - z_0)^{-1} : E \to E$$

is bijective. It follows that $T - z$ is bijective for all $z \in D(0, r)$. A Neumann series gives the holomorphy of R_T. $\square$

Lemma 3.13 (Weyl sequences). — *Let us consider an unbounded closed operator $(T, \mathsf{Dom}(T))$. Assume that there exists a sequence $(u_n) \in \mathsf{Dom}(T)$ such that $\|u_n\|_{\mathsf{H}} = 1$ and*

$$\lim_{n \to +\infty} (T - \lambda)u_n = 0$$

in H. Then $\lambda \in \mathsf{sp}(T)$.

A sequence (u_n) as in Lemma 3.13 is called a Weyl sequence.

Proof. — Assume that $\lambda \in \rho(T)$. Since $(T - \lambda)^{-1}$ is bounded, we find

$$\lim_{n \to +\infty} (T - \lambda)^{-1}(T - \lambda)u_n = \lim_{n \to +\infty} u_n = 0$$

which is a contradiction. $\square$

Example 3.14. — We let $\mathsf{H} = \mathsf{L}^2(I)$, with $I = (0, 1)$. Take $f \in \mathscr{C}^0([0, 1], \mathbb{C})$. We consider the operator $T : \mathsf{L}^2(I) \ni \psi \mapsto f\psi \in \mathsf{L}^2(I)$. Note that T is bounded and $\|T\| \leqslant \|f\|_\infty$.

i. If $\lambda \notin \mathrm{ran}\,(f)$, then, the multiplication operator by $(f - \lambda)^{-1}$ is bounded and it is the inverse of $T - \lambda$. In particular, this shows that $\mathsf{sp}(T) \subset \mathrm{ran}\,(f)$.

ii. Select some $x_0 \in (0, 1)$ and let $\lambda = f(x_0)$. Let us consider $\chi \in \mathscr{C}_0^\infty(]-1, 1[)$ satisfying $\|\chi\|_{\mathsf{L}^2(\mathbb{R})} = 1$. Given $n \in \mathbb{N}$, we consider the sequence

$$u_n(x) = \sqrt{n}\chi(n(x - x_0)).$$

For n large enough, the support of u_n is included in $[0, 1]$. Moreover, we have

$$\|u_n\|_{\mathsf{H}} = 1, \quad \|(T - \lambda)u_n\|_{\mathsf{H}} \leqslant \sup_{|x - x_0| \leqslant 1/n} |f(x) - f(x_0)|,$$

which implies that

$$\lim_{n \to +\infty} (T - \lambda)u_n = 0.$$

By Lemma 3.13, this shows that $\lambda \in \mathsf{sp}(T)$. We get $f(I) \subset \mathsf{sp}(T)$. Since the spectrum is closed and f continuous, we get $f([0, 1]) \subset \mathsf{sp}(T)$.

iii. If λ is an eigenvalue of T, there exists $\psi \in \mathsf{L}^2(I)$ such that $\|\psi\|_{\mathsf{H}} = 1$ and $(f - \lambda)\psi = 0$. Thus the measure of $\{f = \lambda\}$ is positive. Conversely, if $A = \{f = \lambda\}$ has a non-zero measure, $\mathbb{1}_A$ is not zero and satisfies $T\mathbb{1}_A = \lambda\mathbb{1}_A$.

Actually, we can generalize this last example.

Exercise 3.15. — Use the notations of Example 2.44. We define the essential range of f as

$$\mathrm{ran}\,_{\mathsf{ess}}(f) = \{\lambda \in \mathbb{C} : \forall \varepsilon > 0\,, \mu(\{|f - \lambda| < \varepsilon\}) > 0\}.$$

i. Prove that if $\lambda \notin \mathrm{ran}\,_{\mathsf{ess}}(f)$, then $\lambda \in \rho(T_f)$.

ii. Let $\lambda \in \text{ran}_{\text{ess}}(f)$ and $\varepsilon > 0$. By using $A_\varepsilon = \{|f - \lambda| > \varepsilon\}$, find a function $\psi_\varepsilon \in \text{Dom}\,(T_f)$ such that

$$\|(T_f - \lambda)\psi_\varepsilon\|_{\mathsf{H}} \leqslant \varepsilon \|\psi_\varepsilon\|_{\mathsf{H}}\,.$$

iii. Conclude that $\text{ran}_{\text{ess}}(f) = \text{sp}(T_f)$.

Exercise 3.16. — Consider on $\ell^1(\mathbb{N})$ the shift operator T defined by $(Tu)_n = u_{n+1}$.

i. Show that $\text{sp}(T) \subset D(0,1]$.

ii. Show that $\text{sp}_p(T) = D(0,1[$. Conclusion ?

Solution:

i. Note that $\|T\| = 1$.

ii. Let $\lambda \in \mathbb{C}$ with $|\lambda| < 1$. Then $u_\lambda := (\lambda^n)_n \in \ell^1(\mathbb{N})$ is an eigenvector of T associated with the eigenvalue λ. This means that $D(0,1[\subset \text{sp}_p(T)$. Thus,

$$D(0,1[\subset \text{sp}_p(T) \subset \text{sp}(T) \subset D(0,1]\,.$$

Now, let $\lambda \in \mathbb{C}$ with $|\lambda| = 1$. Then λ is an eigenvalue of T if and only if we can find some non-zero vector $= u \in \ell^1(\mathbb{N})$ such that $u_{n+1} = \lambda u_n$ for all $n \in \mathbb{N}$. But this implies that $u = u_0 u_\lambda$ with $u_0 \neq 0$, whereas such u is not in $\ell^1(\mathbb{N})$. It follows that $\lambda \notin \text{sp}_p(T)$. Since $\text{sp}(T)$ is closed, we have $\text{sp}(T) = D(0,1]$. As a consequence, we can find spectral values that are not eigenvalues.

○

Exercise 3.17. — Here $\mathsf{H} = \ell^2(\mathbb{Z})$. We recall that $\mathsf{L}^2(\mathbb{S}^1, \mathbb{C})$ is isometric to $\ell^2(\mathbb{Z})$ via the Fourier series and the Parseval formula.

i. For all $u \in \mathsf{H}$, we let, for all $n \in \mathbb{Z}$, $(S_- u)_n = u_{n-1}$. By using the result of Exercise 3.15 (or Exercise 3.14) and the Fourier series, find the spectrum of S_-. What is the point spectrum of S_-?

ii. For all $u \in \mathsf{H}$, we let, for all $n \in \mathbb{Z}$, $(Tu)_n = u_{n-1} + u_{n-1}$. Find the spectrum of T.

Proposition 3.18 (Resolvent formula). — *For all $z_1, z_2 \in \rho(T)$, we have*

$$R_T(z_1)R_T(z_2) = R_T(z_2)R_T(z_1)\,,$$

and

(3.3.1.2) $(z_1 - z_2)R_T(z_1)R_T(z_2) = R_T(z_1) - R_T(z_2)$.

Proof. — When $z_1 = z_2$, there is nothing to show. Assume that $z_1 \neq z_2$, and observe that, on $\mathrm{Dom}\,(T)$,

$$(z_1 - z_2)\mathrm{Id}$$
$$= (T - z_2) - (T - z_1)$$
$$= \big[(T - z_1)R_T(z_1)(T - z_2) - (T - z_1)R_T(z_2)(T - z_2)\big]$$
$$= (T - z_1)\big[R_T(z_1) - R_T(z_2)\big](T - z_2).$$

Compose this expression on the left with $R_T(z_1)$ and on the right with $R_T(z_2)$ to get (3.3.1.2). Then, exchanging the roles of z_1 and z_2, we get

$$(z_2 - z_1)R_T(z_2)R_T(z_1) = R_T(z_2) - R_T(z_1)$$
$$= -(z_1 - z_2)R_T(z_1)R_T(z_2).$$

$\square$

3.1.3. About the bounded case. —

***Definition 3.19* (Spectral radius).** — Let $T \in \mathcal{L}(E)$. We let

$$r(T) = \sup_{\lambda \in \mathrm{sp}(T)} |\lambda|.$$

Lemma 3.20. — *Let* $T \in \mathcal{L}(E)$. *The sequence* $(\|T^n\|^{\frac{1}{n}})_{n \in \mathbb{N}^*}$ *is convergent to*

$$\tilde{r}(T) := \inf_{n \in \mathbb{N}^*} \|T^n\|^{\frac{1}{n}}.$$

Proof. — We can assume that $T^n \neq 0$ for all $n \in \mathbb{N}^*$. We let $u_n = \ln \|T^n\|$. We have

$$\forall n, p \in \mathbb{N}^*, \quad u_{n+p} \leqslant u_n + u_p.$$

Let $p \in \mathbb{N}^*$. We write $n = qp + r$ with $r \in [0, p)$. We have

$$u_n \leqslant qu_p + u_r.$$

Thus,

$$\frac{u_n}{n} \leqslant \frac{u_p}{p} + \frac{u_r}{n}.$$

We have, for all $p \in \mathbb{N}^*$,

$$\limsup_{n \to +\infty} \frac{u_n}{n} \leqslant \frac{u_p}{p} \implies \limsup_{n \to +\infty} \|T^n\|^{\frac{1}{n}} \leqslant \|T^p\|^{\frac{1}{p}} .$$

It follows that

$$\limsup_{n \to +\infty} \|T^n\|^{\frac{1}{n}} \leqslant \inf_{n \in \mathbb{N}^*} \|T^n\|^{\frac{1}{n}} \leqslant \liminf_{n \to +\infty} \|T^n\|^{\frac{1}{n}} ,$$

which gives rise to the result. $\qquad\qquad\square$

Proposition 3.21. — — *[Gelfand's Formula, 1941] Let $T \in \mathcal{L}(E)$. Then $r(T) = \tilde{r}(T) \leqslant \|T\|$.*

Proof. — We have $T - z = z(z^{-1}T - \mathrm{Id})$. For all $z \in \mathbb{C} \backslash \{0\}$ with $\|T\|/|z| < 1$, we can define the resolvent $R_T(z)$ as a convergent power series according to
(3.3.1.3)

$$R_T(z) = (T - z)^{-1} = z^{-1}(z^{-1}T - \mathrm{Id})^{-1} = -z^{-1} \sum_{n=0}^{+\infty} T^n z^{-n} .$$

This implies that $\mathrm{sp}(T) \subset B(0, \|T\|)$, and therefore $r(T) \leqslant \|T\|$. Note also that

$$(3.3.1.4) \qquad \forall |z| > \|T\|, \quad \|R_T(z)\| \leqslant (|z| - \|T\|)^{-1} .$$

Now, let $\lambda \in \mathrm{sp}(T)$. Observe that

$$\ker(T - \lambda) \subset \ker(T^n - \lambda^n), \quad \mathrm{ran}(T^n - \lambda^n) \subset \mathrm{ran}(T - \lambda) .$$

Thus, if $T^n - \lambda^n$ is bijective, the same is true for $T - \lambda$. This means that $\lambda^n \in \mathrm{sp}(T^n)$ and thereby $|\lambda|^n \leqslant \|T^n\|$, and then

$$\forall n \in \mathbb{N}, \quad r(T) \leqslant \|T^n\|^{\frac{1}{n}} \implies r(T) \leqslant \tilde{r}(T) .$$

Moreover, R_T is holomorphic on $\{z \in \mathbb{C} : |z| > r(T)\} \subset \rho(T)$. It follows that the function

$$z \mapsto \begin{cases} 0 & \text{if } z = 0 \\ R_T(1/z) = z \sum_{n=0}^{+\infty} T^n z^n & \text{if } |z| < r(T)^{-1} \end{cases}$$

is holomorphic. In view of the Cauchy–Hadamard theorem, its radius of convergence is $\tilde{r}(T)^{-1}$. Therefore, we must have $r(T)^{-1} \leqslant \tilde{r}(T)^{-1}$ or $\tilde{r}(T) \leqslant r(T)$. $\qquad\square$

Proposition 3.22. — *If $T \in \mathcal{L}(E)$, then $\mathrm{sp}(T) \neq \emptyset$.*

Proof. — We use Proposition 3.12 and (3.3.1.4) to see that, if $\rho(T) = \mathbb{C}$, the function $z \mapsto R_T(z)$ is holomorphic and bounded on $\mathbb{C}$. Then, we apply Corollary 3.4 to see that R_T is constant. We again use (3.3.1.4) to notice that R_T goes to 0 at infinity. So $R_T = 0$ and this is a contradiction. $\qquad\square$

3.1.4. Spectrum of the adjoint. —

Proposition 3.23. *— Consider a closed and densely defined operator* $(\mathrm{Dom}\,(T), T)$. *The operator* $T : \mathrm{Dom}\,(T) \to \mathsf{H}$ *is bijective if and only if the adjoint operator* $T^* : \mathrm{Dom}\,(T^*) \to \mathsf{H}$ *is bijective. In this case, the inverse operator* $T^{-1} = \mathsf{H} \to \mathrm{Dom}\,(T)$ *is bounded. Moreover, we have* $(T^*)^{-1} = (T^{-1})^*$.

Proof. — Assume that T is bijective. We can apply Proposition 2.14 to see that the operator $T^{-1} = \mathsf{H} \to \mathrm{Dom}\,(T) \subset \mathsf{H}$ is bounded. For the sake of completeness, we repeat the proof below. With the graph norm $\|\cdot\|_T$ defined as in (2.2.1.1), the application

$$T : \left(\mathrm{Dom}\,(T), \|\cdot\|_T\right) \longrightarrow (\mathsf{H}, \|\cdot\|)$$

is a continuous bijective linear map between Banach spaces. The inverse mapping theorem guarantees that T^{-1} is continuous, and therefore

$$\|T^{-1}(y)\|_T = \|T^{-1}(y)\| + \|T(T^{-1}y)\|$$
$$= \|T^{-1}(y)\| + \|y\| \leqslant C\|y\|.$$

This implies that $T^{-1} = \mathsf{H} \to \mathsf{H}$ is bounded. Its adjoint $(T^{-1})^* : \mathsf{H} \to \mathsf{H}$ is also bounded

$$\forall y \in \mathsf{H}, \quad \|(T^{-1})^* y\| \leqslant C\|y\|.$$

If $x \in \mathrm{Dom}\,(T^*)$ and $v \in \mathsf{H}$, we have

$$\langle (T^{-1})^* T^* x, v \rangle = \langle T^* x, T^{-1} v \rangle = \langle x, TT^{-1} v \rangle = \langle x, v \rangle,$$

so that

$$(T^{-1})^* T^* = \mathrm{Id}_{\mathrm{Dom}\,(T^*)}.$$

Note also that, for all $u \in \mathsf{H}$ and $v \in \mathrm{Dom}\,(T)$,

$$\langle (T^{-1})^* u, Tv \rangle = \langle u, v \rangle,$$

so that $(T^{-1})^* u \in \mathrm{Dom}\,(T^*)$ and $T^*(T^{-1})^* = \mathrm{Id}_{\mathsf{H}}$. Thus, T^* is bijective.

If T^* is bijective, the same reasoning as above shows that T^{**} is bijective. We use Proposition 2.50 to get $T^{**} = \overline{T} = T$. Thus, T is bijective. $\qquad\qquad\qquad\qquad\qquad\qquad\qquad\qquad\qquad\qquad\qquad\square$

Corollary 3.24. — *Let* $(\mathrm{Dom}\,(T), T)$ *be a closed and densely defined operator. Then, we have* $\mathrm{sp}(T^*) = \overline{\mathrm{sp}(T)}$*, where the bar denotes the complex conjugation.*

Proof. — By Proposition 3.23, we have
$$z \in \rho(T^*) \Leftrightarrow T^* - z \text{ is bijective} \Leftrightarrow T - \bar{z} \text{ is bijective} .$$
$$\qquad\qquad\qquad\qquad\qquad\qquad\qquad\qquad\qquad\qquad\qquad\square$$

Exercise 3.25. — In this exercise, we use the notation of Section 2.6.4.2.

   i. What are the eigenvalues of $(\mathsf{H}^1_0(\Omega), d_{\mathbf{A}})$ and of its adjoint?

  ii. Determine the spectrum of the operator $(\mathsf{H}^1_0(\Omega), d_{\mathbf{A}})$, and the spectrum of its adjoint.

3.2. Spectral radius and resolvent bound in the self-adjoint case

Definition 3.26 (**Normal operator**). — Let $T \in \mathcal{L}(\mathsf{H})$. T is normal when $TT^* = T^*T$.

Remark that all Hermitian ($T^* = T$), skew-Hermitian ($T^* = -T$), and unitary ($T^* = T^{-1}$) operators are normal. More generally, any operator T whose adjoint T^* is a polynomial function of T is normal.

Proposition 3.27. — *Let* $T \in \mathcal{L}(\mathsf{H})$ *be a normal operator. Then,*
$$r(T) = \|T\| .$$

Proof. — Let us start to deal with the Hermitian case, that is, when $T = T^*$. For all $S \in \mathcal{L}(\mathsf{H})$, we have
$$\|S\| = \sup_{u \neq 0, v \neq 0} \frac{|\langle Su, v \rangle|}{\|u\| \|v\|} .$$
Replace S by $S = T^2 = T^*T$ to find
$$\|T\|^2 \geqslant \|T^2\| = \sup_{u \neq 0, v \neq 0} \frac{|\langle Tu, Tv \rangle|}{\|u\| \|v\|} \geqslant \sup_{u \neq 0} \frac{\|Tu\|^2}{\|u\|^2} = \|T\|^2 .$$

Thus, we must have $\|T^2\| = \|T\|^2$. Since $T^2 = (T^*)^2 = (T^2)^*$, we can repeat this argument with T^2 to obtain $\|T^4\| = \|T\|^4$ and so on up to $\|T^{2^n}\| = \|T\|^{2^n}$. By Lemma 3.20 and Proposition 3.21, we have

$$r(T) = \lim_{n \to +\infty} \|T^n\|^{\frac{1}{n}} = \lim_{n \to +\infty} \|T^{2^n}\|^{\frac{1}{2^n}} = \|T\|.$$

Let us now assume that T is normal. Observe that T^*T is self-adjoint so that

$$r(T^*T) = \|T^*T\| = \sup_{\|u\|=1, \|v\|=1} \langle T^*Tu, v \rangle = \|T\|^2.$$

Indeed, by using the Cauchy–Schwarz inequality, and a direct comparison of the suprema,

$$\sup_{\|u\|=1, \|v\|=1} \langle Tu, Tv \rangle = \sup_{\|u\|=1} \|Tu\|^2.$$

On the other hand, since T is normal, we have

$$r(T^*T) = \lim_{n \to +\infty} \|(T^*T)^n\|^{\frac{1}{n}} = \lim_{n \to +\infty} \|(T^n)^*(T)^n\|^{\frac{1}{n}}$$

$$= \left(\lim_{n \to +\infty} \|(T)^n\|^{\frac{1}{n}} \right)^2 = r(T)^2,$$

and therefore $\|T\| = r(T)$. $\qquad\qquad\square$

Corollary 3.28. — *Let $T \in \mathcal{L}(\mathsf{H})$ be a normal operator. If $\mathsf{sp}(T) = \{0\}$, then $T = 0$.*

Proposition 3.29. — *Let $T \in \mathcal{L}(\mathsf{H})$ be a normal operator. For all $z \notin \mathsf{sp}(T)$, we have*

$$\|(T - z)^{-1}\| = \frac{1}{\mathsf{dist}(z, \mathsf{sp}(T))}.$$

Proof. — Let $z \notin \mathsf{sp}(T)$ and $\lambda \neq z$. From the identity

$$(T - z)^{-1} - (\lambda - z)^{-1} = (\lambda - z)^{-1}(T - z)^{-1}(\lambda - T),$$

it is easy to deduce that

$$(3.3.2.5) \qquad \mathsf{sp}\left((T - z)^{-1}\right) = \{(\lambda - z)^{-1}, \quad \lambda \in \mathsf{sp}(T)\}.$$

From Proposition 3.27, we know that

$$\|(T - z)^{-1}\| = r\left((T - z)^{-1}\right) = \sup_{\lambda \in \mathsf{sp}(T)} |\lambda - z|^{-1}$$

$$= \mathsf{dist}\left(z, \mathsf{sp}(T)\right)^{-1}.$$

$\square$

Exercise 3.30. — Consider $H = \mathbb{C}^d$ (with $d \geqslant 2$) equipped with the canonical scalar product.

  i. Let $T \in \mathcal{L}(H)$. We assume that $d \geqslant 3$ and that, for all strict subspace F of H such that $T(F) \subset F$, $T_{|F}$ is normal.

      a. Assume that T has at least two distinct eigenvalues. By using the decomposition in characteristic subspaces, show that T is diagonalizable[1]. Prove then that the characteristic subspaces are orthogonal.

      b. Assume that T has only one eigenvalue λ and let $N = T - \lambda \mathrm{Id}$. Prove that $N = 0$.

      c. Conclude that T is normal.

  ii. Let $T \in \mathcal{L}(H)$ be a non-normal operator.

      a. Show that there exists $F \subset H$ of dimension two and invariant by T such that $S := T_{|F}$ is non-normal.

      b. Prove that there exists a (z_n) sequence (in the resolvent set of S) converging to an element λ in the spectrum of S and such that

$$\|(S - z_n)^{-1}\| > \frac{1}{\mathsf{dist}(z_n, \mathsf{sp}(S))}.$$

      c. Deduce that there exists z in the resolvent set of T such that

$$\|(T - z)^{-1}\| > \mathsf{dist}(z, \mathsf{sp}(T))^{-1}.$$

Proposition 3.31. — *Let $(T, \mathrm{Dom}\,(T))$ be a self-adjoint operator. For all $z \notin \mathsf{sp}(T)$, we have*

$$\|(T - z)^{-1}\| = \mathsf{dist}(z, \mathsf{sp}(T))^{-1}.$$

Proof. — Let $z \notin \mathsf{sp}(T)$. We have $(T - z)^{-1} \in \mathcal{L}(H)$ as well as $\left((T - z)^{-1}\right)^* = (T - \bar{z})^{-1}$. Moreover, the operators $(T - \bar{z})^{-1}$ and $(T - z)^{-1}$ commute. Thus, $(T - z)^{-1}$ is normal and

$$\|(T - z)^{-1}\| = r\left((T - z)^{-1}\right) = \mathsf{dist}(z, \mathsf{sp}(T))^{-1}.$$

$\square$

---

[1] *Hint:* use Theorem 6.2, which will be proved later.

3.3. Riesz projections

3.3.1. Properties. —

Proposition 3.32. — *Let us consider an unbounded closed operator* $(T, \mathrm{Dom}(T))$ *and* $\lambda \in \mathbb{C}$ *an isolated element of* $\mathrm{sp}(T)$. *Let* $\Gamma_\lambda \subset \rho(T)$ *be a counterclockwise contour that encircles only* λ *as an element of the spectrum of* T. *Define*

$$(3.3.3.6) \qquad P_\lambda := \frac{1}{2i\pi} \int_{\Gamma_\lambda} (z - T)^{-1} \, \mathrm{d}z .$$

The bounded operator $P_\lambda : \mathsf{H} \to \mathrm{Dom}(T) \subset \mathsf{H}$ *commutes with* T *and does not depend on the choice of* Γ_λ. *The operator* P_λ *is a projection and*

$$(3.3.3.7) \quad P_\lambda - \mathrm{Id} = \frac{1}{2i\pi} \int_{\Gamma_\lambda} (\zeta - \lambda)^{-1} (T - \lambda)(\zeta - T)^{-1} \, \mathrm{d}\zeta .$$

Proof. — Since $\Gamma_\lambda \subset \rho(T)$, we know that $(T - z)^{-1}$ is a bounded operator when $z \in \Gamma_\lambda$. Since the function $z \mapsto (T - z)^{-1}$ is holomorphic on $\rho(T)$, it is continuous on Γ_λ. Thus, the integral defining P_λ can be understood as the limit of a corresponding Riemannian sum. From these Riemannian sums, and using the fact that T is closed, we see that P_λ is valued in $\mathrm{Dom}(T)$.

Since $(T - z)^{-1}$ commutes with T, the same applies to the limit P_λ. Due to the holomorphy of the resolvent $R_T(\cdot)$ on the open connected component of $\rho(T)$ containing Γ_λ, the operator P_λ does not depend on the contour encircling λ. There exists $\tilde{r} > 0$ such that

$$\forall r \in (0, \tilde{r}], \quad P_\lambda = \frac{1}{2i\pi} \int_{C(\lambda, r)} (z - T)^{-1} \, \mathrm{d}z$$

$$= \frac{1}{2i\pi} \int_{C(\lambda, \tilde{r})} (w - T)^{-1} \, \mathrm{d}w .$$

Then, by the resolvent formula, we have

$$P_\lambda^2 = \frac{1}{(2i\pi)^2} \int_{z \in C(\lambda, r)} \int_{w \in C(\lambda, \tilde{r})} R_T(z) R_T(w) \, \mathrm{d}w \, \mathrm{d}z$$

$$= \frac{1}{(2i\pi)^2} \int_{z \in C(\lambda, r)} \int_{w \in C(\lambda, \tilde{r})} \frac{R_T(z) - R_T(w)}{z - w} \, \mathrm{d}z \, \mathrm{d}w .$$

By using the Fubini theorem,

$$
P_\lambda^2 = \frac{1}{(2i\pi)^2} \int_{z\in C(\lambda,r)} R_T(z) \left(\int_{w\in C(\lambda,\tilde{r})} \frac{1}{z-w} \, dw \right) dz
$$
$$
- \frac{1}{(2i\pi)^2} \int_{w\in C(\lambda,\tilde{r})} R_T(w) \left(\int_{z\in C(\lambda,r)} \frac{1}{z-w} \, dz \right) dw \, .
$$

Since the function $z \to (z-w)^{-1}$ is holomorphic in the ball $B(\lambda, r)$, the second line disappears. The first line gives rise to

$$
P_\lambda^2 = \frac{2i\pi}{(2i\pi)^2} \int_{z\in C(\lambda,r)} R_T(z) \, dz = P_\lambda \, .
$$

Remark also that

$$
(\zeta-\lambda)^{-1}(T-\lambda)(\zeta-T)^{-1} = -(\zeta-\lambda)^{-1} + (\zeta-T)^{-1} \, .
$$

After integration along Γ_λ, this leads to (3.3.3.7). $\qquad\square$

Definition 3.33 (Finite algebraic multiplicity)
We say that an isolated element λ of $\mathrm{sp}(T)$ has a finite algebraic multiplicity when the rank of P_λ is finite.

Lemma 3.34. — *Let $(T, \mathrm{Dom}(T))$ be a densely defined unbounded closed operator and λ be an isolated element of $\mathrm{sp}(T)$. Then we have $1 \in \mathrm{sp}(P_\lambda)$ and $1 \in \mathrm{sp}(P_\lambda^*)$. In any case, we have $P_\lambda \neq 0$ and $P_\lambda^* \neq 0$.*

Proof. — Before starting the proof, recall that $\lambda \in \mathrm{sp}(T)$ iff $\overline{\lambda} \in \mathrm{sp}(T^*)$. We have just to consider the following two cases:

i. $T - \lambda$ is injective with a closed range. Since $\lambda \in \mathrm{sp}(T)$, the operator $T - \lambda$ is not surjective, and we cannot have

$$
\overline{\mathrm{ran}\,(T-\lambda)} = \mathrm{ran}\,(T-\lambda) = \mathsf{H} \, .
$$

It follows that

$$
\ker(T^* - \overline{\lambda}) = \mathrm{ran}\,(T-\lambda)^\perp \neq \{0\} \, .
$$

We can select $0 \neq u \in \ker(T^* - \overline{\lambda})$. On the other hand, passing to the adjoint at the level of (3.3.3.7) gives

$$
P_\lambda^* - \mathsf{Id} = -\frac{1}{2i\pi} \int_{\Gamma_\lambda} (\overline{\zeta}-\overline{\lambda})^{-1}(\overline{\zeta}-T^*)^{-1}(T^*-\overline{\lambda}) \, d\zeta
$$

from which we can deduce that $P_\lambda^* u = u$, and thus $1 \in \mathsf{sp}(P_\lambda^*)$.

ii. Or, applying Proposition 2.14, we have

$$\not\exists c > 0, \quad \forall u \in \mathrm{Dom}(T), \quad \|(T - \lambda)u\| \geqslant c\|u\|$$

or equivalently, there exists a Weyl sequence (u_n) associated with λ, that is,

$$\|u_n\| = 1, \quad \lim_{n \to +\infty} (T - \lambda)u_n = 0.$$

In view of Formula (3.3.3.7), we have

$$\|u_n\| = 1, \quad \lim_{n \to +\infty} (P_\lambda - \mathsf{Id})u_n = 0.$$

By Lemma 3.13, we know that $1 \in \mathsf{sp}(P_\lambda)$.

Therefore, we have either $1 \in \mathsf{sp}(P_\lambda)$ or $1 \in \mathsf{sp}(P_\lambda^*)$. But P_λ is a closed and densely defined operator. Thus, by Corollary 3.24, we find that $1 \in \mathsf{sp}(P_\lambda)$ and $1 \in \mathsf{sp}(P_\lambda^*)$. $\qquad\square$

3.3.2. About the finite algebraic multiplicity. —

Proposition 3.35. — *Assume that the Hilbert space* H *is of finite dimension. Fix* $T \in \mathcal{L}(\mathsf{H})$. *Let* $\lambda \in \mathsf{sp}(T)$. *Then,* λ *is an eigenvalue. If* Γ_λ *is a contour encircling only* λ, *then* P_λ *is the projection on the algebraic eigenspace associated with* λ.

Proof. — It is well known that H can be written as a sum of the characteristic subspaces H_j associated with the distinct eigenvalues of T. The characteristic subspaces H_j are stable under T. We can assume that H_1 is associated with λ. There exists a basis of H such that the matrix of T is block diagonal $(T_1, \dots, T_k)$ where the T_j is the (upper triangular) matrix of T_{H_j}. In this adapted basis, the matrix of P_λ is block diagonal $(P_{\lambda,1}, \dots, P_{\lambda,k})$. By holomorphy, we have $P_{\lambda,j} = 0$ when $j \neq 1$. To simplify, assume that $\dim \mathsf{H}_1 = 2$ (the other cases being similar) so that

$$T_1 := \begin{pmatrix} \lambda & 1 \\ 0 & \lambda \end{pmatrix}, \qquad P_{\lambda,1} := \frac{1}{2i\pi} \int_{\Gamma_\lambda} (z - T_1)^{-1} \, dz,$$

where Γ_λ is (for example) the circle of center λ and radius 1. Let $n \in \mathbb{N}$. Recall that

$$\frac{1}{2i\pi} \int_{\Gamma_\lambda} (z - \lambda)^{-n} \, dz = \frac{1}{2\pi} \int_{\theta=0}^{2\pi} e^{i(1-n)\theta} \, d\theta = \begin{cases} 1 & \text{if } n = 1, \\ 0 & \text{if } n \neq 1. \end{cases}$$

It follows that

$$P_{\lambda,1} := \frac{1}{2i\pi} \int_{\Gamma_\lambda} \begin{pmatrix} (z-\lambda)^{-1} & -(z-\lambda)^{-2} \\ 0 & (z-\lambda)^{-1} \end{pmatrix} dz = \begin{pmatrix} 1 & 0 \\ 0 & 1 \end{pmatrix}$$
$$= \text{Id}_{H_1}.$$

The application P_λ is indeed the projection on H_1.   □

Corollary 3.36. — *If $\lambda \in \text{sp}(T)$ is isolated with a finite algebraic multiplicity, then it is necessarily an eigenvalue.*

Proof. — If H is of finite dimension, just apply Proposition 3.35. From now on, we may assume that $\dim H = +\infty$. Note $P = P_\lambda$ the projection defined by (3.3.3.6). Any element $u \in H$ can be uniquely written as $u = Pu + (\text{Id} - P)u$. Thus, $H = \ker P \oplus \text{ran} P$. Moreover, the projection $P = P_\lambda$ commutes with T. It follows that

$$T = T_{|\text{ran} P} \oplus T_{|\ker P}.$$

The spectrum of T is the union of the corresponding spectra, and λ is still isolated in these spectra. By definition, we have

$$\frac{1}{2i\pi} \int_\Gamma (\zeta - T_{|\ker P})^{-1} d\zeta = P_{|\ker P} = 0.$$

In view of Lemma 3.34, this condition is not compatible with the existence of an isolated element inside $\text{sp}(T_{|\ker P})$. Necessarily, λ belongs to the spectrum of the "matrix" $T_{|\text{ran} P}$. It is therefore an eigenvalue of $T_{|\text{ran} P}$, *a fortiori* of T.   □

3.4. Fredholm operators

3.4.1. Definition and first properties. —

Definition 3.37. — Let E and F be two Banach spaces. An application $T \in \mathcal{L}(E, F)$ is said to be Fredholm when $\dim \ker T < +\infty$ and $\text{codim} \, \text{ran} \, T < +\infty$. By definition, we call index of T the following number:

$$\text{ind} \, T = \dim \ker(T) - \text{codim} \, \text{ran} \, (T).$$

The set of the Fredholm operators from E to F is denoted by $\text{Fred}(E, F)$.

Example 3.38. — A bijective operator $T \in \mathcal{L}(E, F)$ is Fredholm of index 0.

Example 3.39. — Consider $\mathsf{H} = \ell^2(\mathbb{N})$ and, for $u \in \mathsf{H}$, define Tu by $(Tu)_n = u_{n+1}$ for all $n \in \mathbb{N}$. T is a Fredholm operator of index 1.

Proposition 3.40. — *Let $T \in \mathrm{Fred}(E, F)$. Then $\mathrm{ran}\, T$ is closed.*

Proof. — Let us write $E = \ker T \oplus \tilde{E}$, with $\tilde{E}$ closed. Then, $T : \tilde{E} \to F$ is injective. Let us also write $F = \mathrm{ran}\, T \oplus \tilde{F}$, with $\tilde{F}$ of finite dimension. Consider a basis $(f_j)_{1 \leqslant j \leqslant n}$ of $\tilde{F}$ and introduce the application

$$S : \tilde{E} \times \mathbb{C}^n \ni (x, v) \mapsto Tx + \sum_{j=1}^{n} v_j f_j \in F.$$

The operator S is continuous and bijective between two Banach spaces. Thus, its inverse is continuous and there exists $C > 0$ such that, for all $f \in F$,

$$\|S^{-1} f\|_{\tilde{E} \times \mathbb{C}^n} \leqslant C \|f\|_F,$$

and, for all $(x, v) \in \tilde{E} \times \mathbb{C}^n$,

$$\|x\|_E + \|v\|_{\mathbb{C}^n} \leqslant C \|S(x, v)\|_F.$$

For $v = 0$, this becomes

$$\|x\|_E \leqslant C \|Tx\|_F.$$

Select a sequence $(y_n)_n \in F^{\mathbb{N}}$ with values in the range of T (that is such that $y_n = Tx_n$ for some $x_n \in E$) converging to some $y \in F$. It gives rise to a Cauchy sequence $(x_n)_n \in E^{\mathbb{N}}$, which tends to some $x \in E$ which is such that $Tx = y$. The set $\mathrm{ran}\, T$ is closed. $\qquad\square$

Definition 3.41. — A closed unbounded operator $T : \mathrm{Dom}\,(T) \subset E \to F$ is said to be Fredholm when T is closed and when $T \in \mathcal{L}((\mathrm{Dom}\,(T), \|\cdot\|_T), F)$ is Fredholm.

Proposition 3.42. — *In the case when E and F have finite dimensions, we have $T \in \mathrm{Fred}(E, F)$ and $\mathrm{ind}\, T = \dim E - \dim F$.*

Proof. — This is an immediate consequence of the dimension formula

$$\dim E = \dim(\ker T) + \dim(\operatorname{ran} T)$$
$$= \dim(\ker T) + \dim F - \operatorname{codim}(\operatorname{ran} T).$$

$\square$

Proposition 3.43. — *Let $T \in \mathcal{L}(E, F)$. Then, T is Fredholm if and only if $\dim \ker T < +\infty$ and $\dim \ker T' < +\infty$, and $\operatorname{ran}(T)$ is closed. In this case, we have*

$$\operatorname{ind} T = \dim \ker(T) - \dim \ker(T').$$

Proof. — When the range of T is closed, by Propositions 2.31 and 2.34, we know that
(3.3.4.8)
$$\ker(T') = \operatorname{ran}(T)^{\perp}, \qquad \dim \operatorname{ran}(T)^{\perp} = \operatorname{codim} \operatorname{ran}(T).$$

$\Longrightarrow$ Let $T \in \operatorname{Fred}(E, F)$. By Proposition 3.40, the range of T is closed. Using (3.3.4.8), we get

$$\dim \ker T' = \dim \operatorname{ran}(T)^{\perp} = \operatorname{codim} \operatorname{ran}(T) < +\infty.$$

$\Longleftarrow$ We can still exploit (3.3.4.8). $\square$

The following consequence can actually be proved directly.

Proposition 3.44. — *Let $(T, \operatorname{Dom}(T))$ be a closed operator on H. T is a Fredholm operator when $\dim \ker(T) < +\infty$, $\dim \ker(T^*) < +\infty$, and $\operatorname{ran}(T)$ is closed. The index of T is*

$$\operatorname{ind} T = \dim \ker(T) - \dim \ker(T^*).$$

A remarkable property is the following.

Proposition 3.45. — *Let $T \in \operatorname{Fred}(E, F)$ with index 0. Then, T is injective if and only if T is surjective.*

3.4.2. Spectrum and Fredholm operators. —

Definition 3.46. — We define

  i. essential spectrum: $\lambda \in \mathsf{sp}_{\mathsf{ess}}(T)$ if $T - \lambda$ viewed as an operator from $\operatorname{Dom}(T)$ into H is not Fredholm with index 0.

ii. discrete spectrum: $\lambda \in \mathsf{sp}_{\mathsf{dis}}(T)$ if λ is isolated in the spectrum of T, with finite algebraic multiplicity and such that ran $(T - \lambda)$ is closed.

Remark 3.47. — In some references, the essential spectrum is defined as the complement (in the spectrum) of the discrete spectrum.

Proposition 3.48. — *We have* $\mathsf{sp}_{\mathsf{ess}}(T) \subset \mathsf{sp}(T)$.

Proof. — The statement is equivalent to $\rho(T) \subset \complement\mathsf{sp}_{\mathsf{ess}}(T)$. Let $\lambda \in \rho(T)$. Then

$$T - \lambda \in \mathcal{L}((\mathrm{Dom}\,(T), \|\cdot\|_T), F)$$

is injective and surjective, and therefore it is a Fredholm operator of index 0. □

Proposition 3.49. — *Let* T *be a self-adjoint operator which is Fredholm. Then, the index of* T *is zero.*

Proof. — This is a direct consequence of Proposition 2.52 because

$$\dim \ker T = \dim \ker T^* = \dim \mathrm{ran}\,(T)^{\perp} = \mathrm{codim}\,\mathrm{ran}\,(T).$$

□

Thus, when $T = T^*$, we have $\lambda \in \mathsf{sp}_{\mathsf{ess}}(T)$ if and only if $T - \lambda$ viewed as an operator from $\mathrm{Dom}\,(T)$ into H is not Fredholm.

Proposition 3.50. — *We have* $\mathsf{sp}_{\mathsf{dis}}(T) \subset \mathsf{sp}_p(T)$.

Proof. — This a consequence of Corollary 3.36. □

Exercise 3.51. — Find an example of an operator $T \in \mathcal{L}(\mathsf{H})$ such that $\mathsf{sp}_{\mathsf{dis}}(T)$ is strictly included in $\mathsf{sp}_p(T)$.

Solution: We can give two typical examples.

i. Come back to Exercise 3.16 for which $\mathsf{sp}_{\mathsf{dis}}(T) = \emptyset$ (since there is no isolated spectral element), whereas $\mathsf{sp}_p(T) = D(0, 1[$.

ii. Take $H = \ell^2(\mathbb{N})$ and $(Tu)_n = (\lambda_n u_n)_n$ for a sequence $(\lambda_n)_n$ satisfying

$$\lambda_0 = 0, \qquad \lim_{n \to +\infty} \lambda_n = 0.$$

Looking at $(\delta_{jn})_j$, we can see that

$$0 \in \{\lambda_n; n \in \mathbb{N}\} \subset \mathsf{sp}_p(T) \subset \mathsf{sp}(T).$$

Nevertheless, $0 \notin \mathsf{sp}_{\mathsf{dis}}(T)$ since the sequence $(\lambda_n)_n \in \mathsf{sp}(T)^{\mathbb{N}}$ goes to zero.

○

3.5. Notes

i. The Riesz projections are described in a concise way in [22, Chapter 6]. However, the Reader should read carefully the proof of [22, Prop. 6.4].

CHAPTER 4

COMPACT OPERATORS

This chapter recalls various elementary facts about *compact operators*. We prove the fundamental fact that $K - z\mathrm{Id}$ is a Fredholm operator when $z \neq 0$ and when $K \in \mathcal{L}(E)$ is compact. This fact has important spectral consequences for compact operators (especially once we will have proved that the index of $K - z\mathrm{Id}$ is actually 0). We also give some criteria to establish that an operator is compact. In practice, these criteria are related to precompact subsets of L^2-spaces, such as balls for the H^1-topology.

4.1. Definition and fundamental properties

Definition 4.1. — Let E and F be two Banach spaces. A linear map T is said to be compact when $T(B_E(0, 1))$ is relatively compact (or, equivalently, precompact) in F.

Proposition 4.2. — *The following assertions are equivalent.*

   i. $T \in \mathcal{K}(E, F)$ *is compact.*
   ii. *For all $B \subset E$ with B bounded, $T(B)$ is relatively compact in F.*
   iii. *For all bounded sequences $(u_n) \in E^{\mathbb{N}}$, (Tu_n) has a convergent subsequence.*

Proposition 4.3. — $\mathcal{K}(E, F)$ *is a closed subspace of $\mathcal{L}(E, F)$.*

Proposition 4.4. — $\mathcal{K}(E, F)$ *is a two-sided ideal of $\mathcal{L}(E, F)$.*

Proposition 4.5. — *If $T \in \mathcal{L}(E, F)$ has finite rank, it is compact.*

© The Author(s), under exclusive license
to Springer Nature Switzerland AG 2021
C. Cheverry and N. Raymond, *A Guide to Spectral Theory*,
Birkhäuser Advanced Texts Basler Lehrbücher,
https://doi.org/10.1007/978-3-030-67462-5_4

The following proposition, which will be proved later, is a consequence of Proposition 4.10.

Proposition 4.6. — *When F is a Hilbert space, $\mathcal{K}(E, F)$ is the closure of the set of finite-rank operators.*

Proposition 4.7. — *If $T \in \mathcal{K}(E, F)$ is compact, it transforms weakly convergent sequences into convergent sequences. The converse is true when E is reflexive.*

Proof. — Let us only give the proof when E is a (separable) Hilbert space.

$\Longrightarrow$ Consider a weakly convergent sequence (u_n) and denote by $u \in E$ its limit. We have

$$\forall \ell \in E', \quad \lim_{n \to +\infty} \ell(u_n) = \ell(u).$$

For all $v \in F'$, we have $T'(v) \in E'$ and then, by definition of the weak convergence,

$$\lim_{n \to +\infty} T'(v)(u_n) = T'(v)(u).$$

Thus, by definition of the adjoint,

$$\lim_{n \to +\infty} v(Tu_n) = v(Tu).$$

This means that (Tu_n) weakly converges to Tu. We want to prove that the convergence is strong.

Let us consider a possible limit point w of (Tu_n). Then, we get $v(w) = v(Tu)$, for all $v \in F'$. The Hahn–Banach theorem gives that $w = Tu$.

Now, we have to show that there is a limit point.

By the Riesz representation theorem, we have, for all $v \in E$,

$$(4.4.1.1) \qquad \lim_{n \to +\infty} \langle u_n, v \rangle = \langle u, v \rangle.$$

By using the Uniform Boundedness Principle (with the continuous linear forms $T_n = \langle \cdot, u_n \rangle$), we deduce that (u_n) is bounded. Since T is compact, (Tu_n) has a convergent subsequence.

Therefore, (Tu_n) strongly converges to Tu (since there is one and only one limit point).

$\Longleftarrow$ Let us assume that $T \in \mathcal{L}(E, F)$ transforms weakly convergent sequences into convergent sequences. Consider a bounded sequence $(u_n) \in E^{\mathbb{N}}$. Since E is a (separable) Hilbert space, by a diagonal argument, $(u_n) \in E^{\mathbb{N}}$ has a weakly convergent subsequence. Thus, up to a subsequence extraction, (Tu_n) converges.

$\square$

Proposition 4.8. — *[Schauder] Let $T \in \mathcal{L}(E, F)$. Then T is a compact operator if and only if $T' \in \mathcal{L}(F', E')$ is a compact operator. When $E \equiv F \equiv \mathsf{H}$, the operator $T \in \mathcal{L}(\mathsf{H})$ is a compact operator if and only if T^* is compact.*

Proof. —
$\Longrightarrow$ Given a sequence $(\ell_n)_n \in (F')^{\mathbb{N}}$ with $\|\ell_n\| \leqslant 1$, it suffices to show that $(T^*\ell_n)$ has a Cauchy subsequence $(T^*\ell_{n_j})_j$. In other words, we must show that, for all $\varepsilon > 0$, we can find $N \in \mathbb{N}^*$ such that

$$N \leqslant j \leqslant k \quad \Longrightarrow \quad \|T^*\ell_{n_j} - T^*\ell_{n_k}\|$$
$$= \sup_{\|x\| \leqslant 1} \|\ell_{n_j}(Tx) - \ell_{n_k}(Tx)\| \leqslant \varepsilon .$$

Let B be the unit ball of E. Introduce the compact set $K := \overline{T(B)}$. Then, the above estimate is a consequence of

$$\sup_{y \in K} \|\ell_{n_j}(y) - \ell_{n_k}(y)\| \leqslant \varepsilon .$$

But the sequence $(\ell_n)_n$, viewed as a family of bounded continuous functions on K, satisfies

$$\sup_n \|\ell_n(y)\| \leqslant \|y\|, \quad \sup_n \|\ell_n(y_1) - \ell_n(y_2)\| \leqslant \|y_1 - y_2\| .$$

It is therefore uniformly bounded pointwise and equicontinuous on K. By the Ascoli theorem, the sequence $(\ell_n)_n$ has a uniformly convergent subsequence, as desired.
$\Longleftarrow$ Given a sequence $(x_n)_n \in E^{\mathbb{N}}$ with $\|x_n\| \leqslant 1$, it suffices to show that (Tx_n) has a Cauchy subsequence $(Tx_{n_j})_j$. In other words, we must show that, for all $\varepsilon > 0$, we can find $N \in \mathbb{N}^*$ such that

$$N \leqslant j \leqslant k \quad \Longrightarrow \quad \|Tx_{n_j} - Tx_{n_k}\|$$
$$= \sup_{\|\ell\|_{F'} \leqslant 1} \|(T'\ell)(x_{n_j}) - (T'\ell)(x_{n_k})\| \leqslant \varepsilon .$$

Let B' be the unit ball of F'. Introduce the compact set $K' := \overline{T'(B')} \subset E'$. Then, the above estimate is a consequence of

$$\sup_{\ell' \in K'} \|\ell'(x_{n_j}) - \ell'(x_{n_k})\| \leqslant \varepsilon.$$

As before, we can consider $(x_n)_n$ as a family of uniformly bounded pointwise and equicontinuous functions on $K'^{(1)}$. By the Ascoli theorem, the sequence $(x_n)_n$ has a converging subsequence, as desired.

The last part of Proposition 4.8 is an immediate consequence of Proposition 2.37.   □

Proposition 4.9. — *Let $K \in \mathcal{K}(E)$ be a compact operator. Then $\mathrm{Id}_E + K$ is Fredholm.*

Proof. — The restriction of K to the subspace $\ker(\mathrm{Id}_E + K)$ coincides with $-\mathrm{Id}$, and it must be compact. By the Riesz theorem, this is possible only if $\dim \ker(\mathrm{Id}_E + K) < +\infty$. By Proposition 4.8, we have $T' \in \mathcal{K}(E')$ and thus $\dim \ker(\mathrm{Id}_{E'} + K') < +\infty$. In view of Proposition 3.43, there remains to show that $\mathrm{ran}\,(\mathrm{Id}_E + K)$ is closed. To this end, let us consider a sequence (u_n) such that $(u_n + Ku_n)$ converges to f. We let

$$d_n = \mathrm{dist}(u_n, \ker(\mathrm{Id}_E + K)).$$

There exists $v_n \in \ker(\mathrm{Id}_E + K)$ such that $d_n = \|u_n - v_n\|$. We have

$$u_n + Ku_n = u_n - v_n + K(u_n - v_n).$$

Assume that (d_n) is not bounded. Up to a subsequence extraction, we can assume that (d_n) tends to $+\infty$. Introduce $w_n := d_n^{-1}(u_n - v_n)$, so that

$$\lim_{n \to +\infty} d_n^{-1}(u_n + Ku_n) = \lim_{n \to +\infty} (w_n + Kw_n)$$

$$= \lim_{n \to +\infty} d_n^{-1} f = 0.$$

---

[1]To do so, consider (T_n) defined by

$$T_n(\ell) = \ell(x_n), \quad \forall \ell \in E',$$

and notice that $\|T_n\| \leqslant \|x_n\| \leqslant 1$.

By compactness of K, we can assume that (Kw_n) converges to some g, and therefore (w_n) converges to $-g$. Since K is continuous, we must have $K(-g) = g$ or $g \in \ker(\mathrm{Id}_E + K)$. But, we know that

$$\mathrm{dist}(w_n, \ker(\mathrm{Id}_E + K)) = 1,$$

and this is a contradiction.

Necessarily, the sequence (d_n) is bounded. Modulo the extraction of a subsequence, we can assume that $K(u_n - v_n)$ converges to some h, and therefore $u_n - v_n$ converges to $f - h$, so that

$$f = f - h + K(f - h) \in \mathrm{ran}\,(\mathrm{Id}_E + K),$$

and the closedness of the range follows. We can conclude with Proposition 3.43. $\square$

In the case when $E = F = \mathsf{H}$, there is a characterization of a compact operator $T \in \mathcal{L}(\mathsf{H})$ as the limit of a sequence $(T_n)_n$ with $T_n \in \mathcal{L}(\mathsf{H})$ of finite rank.

Proposition 4.10. — *Consider a Hilbert basis* $(\psi_n)_{n \in \mathbb{N}}$ *of* H*. Let* $T \in \mathcal{L}(\mathsf{H})$*. For all* $n \in \mathbb{N}$*, define*

$$\rho_n = \sup_{\substack{\psi \in \mathrm{span}\,(\psi_k)_{k \in \{0,\dots,n\}}^{\perp} \\ \|\psi\| = 1}} \|T\psi\|, \qquad T_n = \sum_{k=0}^{n} \langle \cdot, \psi_k \rangle T\psi_k .$$

Then,

   i. $\rho_n = \|T - T_n\|$,
  ii. T *is compact iff* $\lim_{n \to +\infty} \rho_n = 0$.

Proof. — For the first point, we write

$$\|T - T_n\| = \sup_{\psi \in \mathsf{H} \setminus \{0\}} \frac{\|(T - T_n)\psi\|}{\|\psi\|} = \sup_{\psi \in \mathsf{H} \setminus \{0\}} \frac{\|T\Pi_n^{\perp}\psi\|}{\|\psi\|} = \rho_n .$$

Consider the second point. Since (ρ_n) is non-increasing, it converges to some $\rho \geqslant 0$. If $\rho = 0$, by Proposition 4.6, the operator T is compact. Assume that $\rho > 0$. Thus, for all $n \in \mathbb{N}$, there exists $\phi_n \in \mathrm{span}\,(\psi_k)_{k \in \{0,\dots,n\}}^{\perp}$ with $\|\phi_n\| = 1$ and $\|T\phi_n\| \geqslant \rho/2 > 0$.

Then, we notice that $(\phi_n)_n$ weakly converges to 0. Indeed, for all $\psi \in \mathsf{H}$, we have

$$|\langle \phi_n, \psi \rangle| \leqslant \left(\sum_{k=n+1}^{+\infty} |\langle \phi_n, \psi_k \rangle|^2 \right)^{\frac{1}{2}} \left(\sum_{k=n+1}^{+\infty} |\langle \psi, \psi_k \rangle|^2 \right)^{\frac{1}{2}}$$

$$\leqslant \left(\sum_{k=n+1}^{+\infty} |\langle \psi, \psi_k \rangle|^2 \right)^{\frac{1}{2}} \underset{n \to +\infty}{\longrightarrow} 0 \,.$$

The operator T cannot be compact. Otherwise, the sequence $(T\phi_n)$ would converge to 0, which is not the case. $\qquad\square$

Exercise 4.11. — Take $\mathsf{H} = \ell^2(\mathbb{N})$, and consider the operator T : $\mathsf{H} \to \mathsf{H}$ given by

$$Tu = v, \qquad u = (u_n)_n, \qquad v = (v_n)_n \,,$$

with

$$v_n = \begin{cases} 0 & \text{if } n = 0 \,, \\ \dfrac{u_{n-1}}{n} & \text{if } n \in \mathbb{N}^* \,. \end{cases}$$

Prove that T is compact.

Solution: The family $(\delta_{nj})_j$ with $n \in \mathbb{N}$ is a Hilbert basis of $\ell^2(\mathbb{N})$. With T_n defined accordingly, we have by Cauchy–Schwarz inequality

$$\rho_n = \|T - T_n\| \leqslant \left(\sum_{k=n}^{\infty} \frac{1}{(k+1)^2} \right)^{1/2} \underset{n \to +\infty}{\longrightarrow} 0 \,.$$

$\circ$

Lemma 4.12. — *Let $T \in \mathcal{L}(\mathsf{H})$ be a non-negative operator. Then, T is compact iff $T^{\frac{1}{2}}$ is compact.*

Proof. — By Proposition 2.81, we can find $S \geqslant 0$ such that $S^2 = T$. If S is compact, by Proposition 4.4, the operator T is compact. For all $\psi \in \mathsf{H}$, we have

$$\|S\psi\|^2 = \langle T\psi, \psi \rangle \,,$$

and thus

$$\|S\psi\|^2 \leqslant \|\psi\| \|T\psi\|.$$

It follows that

$$\rho_n(S)^2 \leqslant \rho_n(T).$$

By Proposition 4.10, if T is compact, S must be compact. □

Proposition 4.13. — *Consider* $T \in \mathcal{L}(\mathsf{H})$. *Then*, T *is compact iff* $|T|$ *is compact*.

Proof. — If $|T|$ is compact, by using the polar decomposition, T is also compact.

Assume now that T is compact. In particular, the non-negative operator T^*T is compact and so is its square root by Lemma 4.12.

□

4.2. Compactness in L^p spaces

4.2.1. About the Ascoli theorem in L^p spaces. — In order to prove that an operator is compact, the following criterion of relative compactness in $\mathsf{L}^p(\Omega)$ will be useful (see Appendix, Theorem A.11).

Theorem 4.14 (Kolmogorov–Riesz). — *Let* $\Omega \subset \mathbb{R}^N$ *be an open set and* $\mathscr{F}$ *a bounded subset of* $\mathsf{L}^p(\Omega)$, *with* $p \in [1, +\infty)$. *We assume that*

$$(4.4.2.2) \quad \forall \varepsilon > 0, \exists \omega \subset\subset \Omega, \quad \forall f \in \mathscr{F}, \quad \|f\|_{\mathsf{L}^p(\Omega \setminus \omega)} \leqslant \varepsilon,$$

and that

$$(4.4.2.3) \quad \forall \varepsilon > 0, \forall \omega \subset\subset \Omega, \exists \delta > 0, \quad \delta < \mathrm{dist}(\omega, \complement\Omega),$$
$$\forall |h| \leqslant \delta, \quad \forall f \in \mathscr{F}, \|f(\cdot + h) - f(\cdot)\|_{\mathsf{L}^p(\omega)} \leqslant \varepsilon.$$

Then, $\mathscr{F}$ *is relatively compact (or, equivalently, precompact) in* $\mathsf{L}^p(\Omega)$.

Remark 4.15. — To get the control of the translations in practice, we can use Proposition 2.94.

Proof. — Let $\varepsilon > 0$.

1. The equi-integrability condition (4.4.2.2) provides us with $\omega \subset\subset \Omega$ such that

$$\forall f \in \mathscr{F}, \quad \|f\|_{\mathsf{L}^p(\Omega \setminus \omega)} \leqslant \varepsilon.$$

2. Let $\tilde{\Omega}$ be a bounded open set such that $\overline{\omega} \subset \tilde{\Omega} \subset \Omega$. We consider $\mathscr{F}$ the set of the restrictions to $\tilde{\Omega}$, (and extended by 0 outside $\tilde{\Omega}$). Clearly, $\mathscr{F}$ is a bounded subset of $L^p(\mathbb{R}^d)$ and also of $L^1(\mathbb{R}^d)$. The usual convolution argument, combined with (4.4.2.3), gives, for all n such that $B(0, \frac{1}{n}) + \omega \subset \tilde{\Omega}$, and all $g \in \mathscr{F}$,

$$(4.4.2.4) \qquad \|\rho_n \star g - g\|_{L^p(\omega)} \leqslant \varepsilon.$$

3. Let us consider $\mathscr{G} = \rho_n \star \mathscr{F}_{|\overline{\omega}} \subset \mathscr{C}^0(\overline{\omega}, \mathbb{C})$. Note that there exists $C_n > 0$ such that, for all $g \in \mathscr{F}$,

$$\|\rho_n \star g\|_\infty \leqslant \|\rho_n\|_\infty \|g\|_{L^1(\mathbb{R}^d)} \leqslant C_n.$$

Moreover, for all $x_1, x_2 \in \overline{\omega}$, we have

$$|\rho_n \star g(x_1) - \rho_n \star g(x_2)| \leqslant |x_1 - x_2| \|\nabla \rho_n\|_\infty \|g\|_{L^1(\mathbb{R}^d)}$$
$$\leqslant D_n |x_1 - x_2|.$$

Therefore, by the Ascoli theorem, $\mathscr{G}$ is precompact in $\mathscr{C}^0(\overline{\omega}, \mathbb{R})$. It can be covered by finitely many balls of radius $\varepsilon/|\omega|^{\frac{1}{p}}$:

$$\mathscr{G} \subset \bigcup_{j=1}^k B_{L^\infty}(g_j, |\omega|^{-\frac{1}{p}}\varepsilon) \subset \bigcup_{j=1}^k B_{L^p}(g_j, \varepsilon).$$

4. By the triangle inequality and (4.4.2.4), we get

$$\mathscr{F}_{|\omega} = \mathscr{F}_{|\omega} \subset \bigcup_{j=1}^k B_{L^p}(g_j, 2\varepsilon).$$

We extend the g_j by zero outside ω, and we deduce that

$$\mathscr{F} \subset \bigcup_{j=1}^k B_{L^p(\Omega)}(g_j, 3\varepsilon).$$

$\square$

Exercise 4.16. — Consider the operator $\mathscr{L} = -\Delta$ with domain $H^2(\mathbb{R}^d)$ and take $\lambda \in \mathbb{R}_-$.

i. Show that $\lambda \in \rho(\mathscr{L})$.

ii. Consider then a function $V \in \mathscr{C}^\infty(\mathbb{R}^d, \mathbb{C})$ such that ∇V is bounded and $\lim_{|x| \to +\infty} V(x) = 0$. Prove that $V(\mathscr{L} - \lambda)^{-1} :$ $L^2(\mathbb{R}^d) \to L^2(\mathbb{R}^d)$ is compact.

Exercise 4.17. — Consider

$$B^1(\mathbb{R}) = \{\psi \in H^1(\mathbb{R}) : x\psi \in L^2(\mathbb{R})\} \subset L^2(\mathbb{R}).$$

Prove that the injection of $B^1(\mathbb{R})$ in $L^2(\mathbb{R})$ is a compact operator.

4.2.2. Rellich theorems. —

4.2.2.1. *First Rellich theorem.* —

Lemma 4.18. — *Let Ω be an open set in $\mathbb{R}^d$. For all $u \in H_0^1(\Omega)$, consider its extension by zero outside Ω, denoted by $\underline{u}$. Then $\underline{u} \in H_0^1(\mathbb{R}^d)$ and $\|\underline{u}\|_{H^1(\mathbb{R}^d)} = \|u\|_{H^1(\Omega)}$.*

Proof. — Clearly, $\underline{u} \in L^2(\mathbb{R}^d)$ and $\|\underline{u}\|_{L^2(\mathbb{R}^d)} = \|u\|_{L^2(\Omega)}$. We know that, by definition, $\mathscr{C}_0^\infty(\Omega)$ is dense in $H_0^1(\Omega)$. Consider a sequence $(u_n)_{n \in \mathbb{N}} \subset \mathscr{C}_0^\infty(\Omega)$ converging to u in H^1-norm. For all $n \in \mathbb{N}$, we have $\underline{u_n} \in \mathscr{C}_0^\infty(\mathbb{R}^d)$. For all $n, p \in \mathbb{N}$, we have

$$\|u_n - u_p\|_{H^1(\Omega)} = \|\underline{u_n} - \underline{u_p}\|_{H^1(\mathbb{R}^d)}.$$

Thus, $(\underline{u_n})$ is a Cauchy sequence in $H^1(\mathbb{R}^d)$. We deduce that $(\underline{u_n})$ converges in $H^1(\mathbb{R}^d)$ to some $v \in H^1(\mathbb{R}^d)$. We have $v = \underline{u}$ and the equality of the norms. $\square$

Theorem 4.19 (Rellich). — *Let Ω be an open bounded set in $\mathbb{R}^d$. The injection of $H_0^1(\Omega)$ in $L^2(\Omega)$ is compact.*

Proof. — Let us prove that if $(u_n)_{n \in \mathbb{N}}$ weakly converges to u in $H_0^1(\Omega)$, it strongly converges to u in $L^2(\Omega)$. The sequence $(u_n)_{n \in \mathbb{N}}$ is bounded in $H_0^1(\Omega)$. Let $\varepsilon > 0$.

For all $n \in \mathbb{N}$, we let $f_n = \widehat{\underline{u_n}}$ and we define $f = \widehat{\underline{u}}$. By the Parseval formula, it is sufficient to show that f_n converges to f in $L^2(\mathbb{R}^d)$.

We notice that, for all $\xi \in \mathbb{R}^d$,

$$f_n(\xi) = \int_\Omega u_n(x) e^{-ix \cdot \xi} \, dx,$$

so that

$$|f_n(\xi)| \leqslant |\Omega|^{\frac{1}{2}} \|u_n\|_{L^2(\Omega)} \leqslant C.$$

We recall that $(u_n)_{n \in \mathbb{N}}$ weakly converges to u in $\mathsf{H}_0^1(\Omega)$ and, in particular, for all $\varphi \in \mathsf{L}^2(\Omega)$,

$$\int_\Omega u_n \overline{\varphi} \, dx \to \int_\Omega u \overline{\varphi} \, dx \,.$$

We choose $\varphi(x) = e^{ix \cdot \xi}$ and thus, for all $\xi \in \mathbb{R}^d$, $f_n(\xi) \to f(\xi)$.

Moreover, we have

$$\|u_n\|_{\mathsf{H}^1(\Omega)}^2 = \|\underline{u_n}\|_{\mathsf{H}^1(\mathbb{R}^d)}^2 = \int_{\mathbb{R}^d} \langle \xi \rangle^2 |f_n(\xi)|^2 \, d\xi \,.$$

In particular, there exists $R > 0$ such that, for all $n \in \mathbb{N}$,

$$\int_{|\xi|>R} |f_n(\xi)|^2 \, d\xi \leqslant \varepsilon \,.$$

Up to changing R, we also have

$$\int_{|\xi|>R} |f(\xi)|^2 \, d\xi \leqslant \varepsilon \,.$$

Let us now write

$$\|f_n - f\|_{\mathsf{L}^2(\mathbb{R}^d)}^2 = \int_{|x| \leqslant R} |f_n(\xi) - f(\xi)|^2 \, d\xi$$

$$+ \int_{|x|>R} |f_n(\xi) - f(\xi)|^2 \, d\xi \,.$$

We deal with the first integral by using the dominated convergence theorem (the sequence (f_n) is uniformly bounded). □

4.2.2.2. *Second Rellich theorem.* — Assuming that the boundary of Ω is of class $\mathscr{C}^1$, we can also establish a theorem for $\mathsf{H}^1(\Omega)$.

Theorem 4.20. — *Let Ω be a bounded open subset of $\mathbb{R}^d$ with $\mathscr{C}^1$ boundary. The injection of $\mathsf{H}^1(\Omega)$ in $\mathsf{L}^2(\Omega)$ is compact.*

Before starting the proof of Theorem 4.20, one needs to prove a few technical lemmas related to the description of $\partial\Omega$ in local charts. Each lemma has actually an interest of its own.

Lemma 4.21. — *For all $u \in \mathsf{H}^1(\mathbb{R}_+)$, we let*

$$Pu(x', x_d) = u(x', x_d) \,, \quad \text{when } x_d > 0 \,,$$

and

$$Pu(x', x_d) = u(x', -x_d) \,, \quad \text{when } x_d < 0 \,.$$

Then, $Pu \in \mathsf{H}^1(\mathbb{R}^d)$ *and*

$$\|Pu\|_{\mathsf{L}^2(\mathbb{R}^d)}^2 = 2\|u\|_{\mathsf{L}^2(\mathbb{R}_+^d)}^2, \quad \|\nabla Pu\|_{\mathsf{L}^2(\mathbb{R}^d)}^2 = 2\|\nabla u\|_{\mathsf{L}^2(\mathbb{R}_+^d)}^2.$$

Proof. — The first equality easily follows by symmetry. Let us deal with the second one.

Let us show that

$$\partial_j(Pu) = P(\partial_j u), \quad 1 \leqslant j \leqslant d-1, \quad \text{and} \quad \partial_d(Pu) = \tilde{P}(\partial_d u),$$

where

$$\tilde{P}v(x', x_d) = v(x', x_d), \text{ when } x_d > 0,$$

and

$$\tilde{P}v(x', x_d) = -v(x', -x_d), \text{ when } x_d < 0.$$

It will be convenient to use an even cutoff function $0 \leqslant \chi \leqslant 1$ such that

$$\chi(x_d) = 0, \quad \text{for } |x_d| \leqslant 1, \quad \chi(x_d) = 1 \text{ for } |x_d| \geqslant 2.$$

For all $n \in \mathbb{N}$, we let $\chi_n(x_d) = \chi(nx_d)$. Let $\varphi \in \mathscr{C}_0^\infty(\mathbb{R}^d)$.
We write, for all $j \in \{1, \ldots, d\}$,

$$\langle \partial_j(Pu), \varphi \rangle_{\mathscr{D}'(\mathbb{R}^d) \times \mathscr{D}(\mathbb{R}^d)} = -\langle Pu, \partial_j \varphi \rangle_{\mathscr{D}'(\mathbb{R}^d) \times \mathscr{D}(\mathbb{R}^d)}$$

$$= -\int_{\mathbb{R}^d} Pu\, \partial_j \varphi\, \mathrm{d}x.$$

Note that

$$-\int_{\mathbb{R}^d} Pu\, \chi_n \partial_j \varphi\, \mathrm{d}x$$

$$= -\int_{\mathbb{R}_+^d} \chi_n u(\partial_j \varphi(x', x_d) + \partial_j \varphi(x', -x_d))\, \mathrm{d}x.$$

Then, for all $j \in \{1, \ldots, d-1\}$,

$$-\int_{\mathbb{R}^d} Pu\, \chi_n \partial_j \varphi\, \mathrm{d}x$$

$$= -\int_{\mathbb{R}_+^d} \chi_n u \partial_j \left(\varphi(x', x_d) + \varphi(x', -x_d) \right)\, \mathrm{d}x$$

$$= \int_{\mathbb{R}_+^d} \chi_n \partial_j u \left(\varphi(x', x_d) + \varphi(x', -x_d) \right)\, \mathrm{d}x.$$

Then, by dominated convergence, we have

$$\langle \partial_j(Pu), \varphi \rangle_{\mathscr{D}'(\mathbb{R}^d) \times \mathscr{D}(\mathbb{R}^d)}$$

$$= \int_{\mathbb{R}^d_+} \partial_j u \left(\varphi(x', x_d) + \varphi(x', -x_d) \right) \, dx$$

$$= \langle P(\partial_j u), \varphi \rangle_{\mathscr{D}'(\mathbb{R}^d) \times \mathscr{D}(\mathbb{R}^d)} .$$

For $j = d$, we have

$$- \int_{\mathbb{R}^d} Pu \, \chi_n \partial_j \varphi \, dx$$

$$= - \int_{\mathbb{R}^d_+} \chi_n u \partial_j \left(\varphi(x', x_d) - \varphi(x', -x_d) \right) \, dx$$

$$= - \int_{\mathbb{R}^d_+} u \partial_j \left(\chi_n \psi \right) \, dx + \int_{\mathbb{R}^d_+} u \chi'_n \psi \, dx$$

$$= \int_{\mathbb{R}^d_+} \partial_j u \left(\chi_n \psi \right) \, dx + \int_{\mathbb{R}^d_+} u \chi'_n \psi \, dx ,$$

where $\psi(x', x_d) = \varphi(x', x_d) - \varphi(x', -x_d)$. By dominated convergence and the fact that $|\psi(x', x_d)| \leqslant C |x_d|$, we get

$$\langle \partial_j(Pu), \varphi \rangle_{\mathscr{D}'(\mathbb{R}^d) \times \mathscr{D}(\mathbb{R}^d)} = \int_{\mathbb{R}^d_+} \partial_j u \, \psi \, dx$$

$$= \langle \tilde{P}(\partial_j u), \varphi \rangle_{\mathscr{D}'(\mathbb{R}^d) \times \mathscr{D}(\mathbb{R}^d)} .$$

$\square$

Lemma 4.22 (Extension operator, general case)

Let Ω be a bounded open subset of $\mathbb{R}^d$ with $\mathscr{C}^1$ boundary. There exists a bounded operator $P : H^1(\Omega) \to H^1(\mathbb{R}^d)$ such that $Pu_{|\Omega} = u$ and Pu has compact support.

Proof. — Let $C = \{ x \in \mathbb{R}^d : |x'| < 1 , |x_d| < 1 \}$. Since $\partial \Omega$ is of class $\mathscr{C}^1$ and compact, there exists a family of open sets $(U_j)_{1 \leqslant j \leqslant p}$ such that

$$\partial \Omega \subset \bigcup_{j=1}^p U_j ,$$

and $\mathscr{C}^1$-diffeomorphisms $\varphi_j : Q \to U_j$ with $\varphi_j \in \mathscr{C}^1(\overline{Q})$ and $\varphi_j^{-1} \in \mathscr{C}^1(\overline{U_j})$ and $\varphi_j(C_0) = \partial \Omega \cap U_j$. There exists also an open set $U_0 \subset\subset \Omega$ such that

$$\overline{\Omega} \subset U_0 \cup \bigcup_{j=1}^{p} U_j \,.$$

We can apply Lemma 2.97 to get a family of smooth functions with compact supports $(\theta_j)_{0 \leqslant j \leqslant p}$ such that

$$\theta_0 + \sum_{j=1}^{p} \theta_j = 1 \,.$$

Let us consider $u \in H^1(\Omega)$ and write

$$u = u_0 + \sum_{j=1}^{p} u_j \,, \quad u_j = \theta_j u \,.$$

By extending u_0 by zero, we see that $u_0 \in H^1(\mathbb{R}^d)$ and

$$\|u_0\|_{H^1(\mathbb{R}^d)} \leqslant C \|u\|_{H^1(\Omega)} \,.$$

For all $j \in \{1, \dots, p\}$, we let, for all $y \in Q$,

$$v_j(y) = u_j(\varphi_j(y)) \,.$$

Then, we use the extension operator P of Lemma 4.21 and consider Pv_j through the chart φ_j:

$$w_j - (Pv_j) \circ \varphi_j^{-1} = P(u_j \circ \varphi_j) \circ \varphi_j^{-1} \,.$$

Note that $w_j = u_j$ on $U_j \cap \Omega$ and

$$\|w_j\|_{H^1(U_j)} \leqslant C \|u\|_{H^1(U_j \cap \Omega)} \,.$$

We consider the compactly supported function

$$Pu = u_0 + \sum_{j=1}^{p} \theta_j w_j \,,$$

and notice that $Pu \in H^1(\mathbb{R}^d)$ and

$$\forall x \in \Omega \,, \quad Pu(x) = u(x) \,.$$

Moreover, $P : H^1(\Omega) \to H^1(\mathbb{R}^d)$ is bounded. $\qquad\qquad \square$

We can now prove Theorem 4.20.

Proof. — The proof is the same as that of Theorem 4.19. Instead of extending the functions by zero, we use the extension operator P. □

4.3. Operators with compact resolvent

A way to describe the spectrum of unbounded and closed operators is to consider their resolvents (which are bounded) and to prove, in good situations, that they are compact.

Proposition 4.23. — *Let $(T, \operatorname{Dom}(T))$ be a closed operator and $z_0 \in \rho(T)$. If $(T - z_0)^{-1}$ is compact, then, for all $z \in \rho(T)$, the operator $(T - z)^{-1}$ is compact.*

Proof. — The resolvent formula (Proposition 3.18) says that

$$(4.4.3.5) \qquad R_T(z) = R_T(z_0) + (z - z_0) R_T(z) R_T(z_0).$$

The right-hand side is compact because the set of compact operators is an ideal of $\mathcal{L}(\mathsf{H})$. □

Let us provide a useful (topological) criterion for the compactness of a resolvent.

Proposition 4.24. — *A closed operator $(T, \operatorname{Dom}(T))$ has compact resolvent if and only if the injection $\imath : (\operatorname{Dom}(T), \| \cdot \|_T) \hookrightarrow (\mathsf{H}, \| \cdot \|_\mathsf{H})$ is compact.*

Proof. — Assume that the injection is compact. Select $z \in \rho(T)$. Thanks to the closed graph theorem, the application $(T - z)^{-1} : (\mathsf{H}, \| \cdot \|_\mathsf{H}) \to (\operatorname{Dom}(T), \| \cdot \|_T)$ is bounded. Then, the operator $(T - z)^{-1} : (\mathsf{H}, \| \cdot \|_\mathsf{H}) \to (\mathsf{H}, \| \cdot \|_\mathsf{H})$ can be viewed as the composition of the following bounded operators:

$$(\mathsf{H}, \| \cdot \|_\mathsf{H}) \overset{(T-z)^{-1}}{\longrightarrow} (\operatorname{Dom}(T), \| \cdot \|_T) \overset{\imath}{\hookrightarrow} (\mathsf{H}, \| \cdot \|_\mathsf{H}).$$

Again, this is compact because the set of compact operators is an ideal of $\mathcal{L}(\mathsf{H})$.

Conversely, assume that the resolvent is compact. Take $z_0 \in \rho(T)$ and consider

$$\imath\big(B(0, 1]\big) = \{u \in \operatorname{Dom}(T) : \|u\| + \|Tu\| \leqslant 1\}$$
$$\subset \{u \in \operatorname{Dom}(T) : \|u\| + \|(T - z_0)u\| \leqslant 1 + |z_0|\}.$$

Let $u \in \text{Dom}\,(T)$ be such that

$$\|u\| + \|(T - z_0)u\| \leqslant 1 + |z_0|.$$

Then, we have

$$\|u\| \leqslant 1 + |z_0|, \qquad \|v\| \leqslant 1 + |z_0|, \qquad v := (T - z_0)u,$$

meaning that

$$u = (T - z_0)^{-1}v \in (T - z_0)^{-1}(B(0, 1 + |z_0|)).$$

In other words

$$\{u \in \text{Dom}\,(T) : \|u\| + \|(T - z_0)u\| \leqslant 1 + |z_0|\}$$
$$\subset (T - z_0)^{-1}(B(0, 1 + |z_0|)).$$

It suffices to notice that the right-hand side is compact. □

Exercise 4.25. — Let $\Omega \subset \mathbb{R}^d$ be a smooth bounded open set. Prove that the Dirichlet Laplacian on Ω has compact resolvent. *Hint: use Rellich's theorem.*

Corollary 4.26. — *We use the same notation as in Section 2.5. Consider two Hilbert spaces V and H such that $V \subset H$ with continuous injection and with V dense in H. Assume that Q is a continuous, coercive, and Hermitian sesquilinear form on V and let T be the self-adjoint operator associated with Q. Let us denote by $\| \cdot \|_Q$ the norm induced by Q, i.e., $\|u\|_Q = \sqrt{Q(u,u)}$. If $(V, \| \cdot \|_Q) \hookrightarrow (H, \| \cdot \|_H)$ is compact, then T has compact resolvent.*

Proof. — By the Cauchy–Schwarz inequality, we have

$$\|u\|_Q = |\langle Tu, u \rangle|^{1/2} \leqslant \|Tu\|^{1/2}\|u\|^{1/2} \leqslant \frac{1}{\sqrt{2}}\|u\|_T,$$

and therefore $\text{Id} : (\text{Dom}\,T, \| \cdot \|_T) \to (V, \| \cdot \|_Q)$ is bounded, and the conclusion follows from Proposition 4.4. □

Remark 4.27. — The converse is true. See Exercise 6.5.

Exercise 4.28. — Prove that the harmonic oscillator which is defined in Section 2.6.3 has a compact resolvent.

4.4. Notes

i. The proofs of the reminded results in Section 4.1 can be found in [**38**, Chapter 4].

ii. Proposition 4.6 is not true when F is only assumed to be a Banach space (a counter-example has been given by Per Enflo; see [**12**]).

iii. A first version of the Riesz–Kolmogorov theorem is proved by Kolmogorov in [**26**], soon followed by Tamarkin [**43**], and M. Riesz [**36**]. The reader is also invited to discover the article [**19**, Section 4] where some historical references are discussed.

CHAPTER 5

FREDHOLM THEORY

In this chapter, we discuss basic facts about *Fredholm theory*. We show that a Fredholm operator is bijective if and only if some matrix is bijective (and this can only happen for Fredholm operators with index 0). We see that this property implies that the index of a Fredholm operator is locally constant. This fact in mind, we deduce that $K - z\mathrm{Id}$ is a Fredholm operator with index 0 for $K \in \mathcal{L}(E)$ compact and $z \neq 0$. This allows to reduce the spectral analysis of compact operators to finite dimension and basic properties of holomorphic functions. Then, we can get a description of the resolvent of a compact operator near each (isolated) point of its (discrete) spectrum.

5.1. Grushin formalism

In this section, we consider two Banach spaces X_1 and X_2.

Let $T \in \mathcal{L}(X_1, X_2)$ be a Fredholm operator. The finite-dimensional subspace $\ker(T)$ can be complemented by a closed subspace $\tilde{X}_1$, so that $X_1 = \ker(T) \oplus \tilde{X}_1$ with $n_+ = \dim \ker(T)$. We can also find some finite-dimensional subspace $\tilde{X}_2$ with $n_- = \mathrm{codim}\,\mathrm{ran}\,(T) = \dim \tilde{X}_2$ and such that $X_2 = \mathrm{ran}\,(T) \oplus \tilde{X}_2$. We introduce $(k_j)_{1 \leqslant j \leqslant n_+}$ a basis of $\ker(T)$ and $(k'_j)_{1 \leqslant j \leqslant n_-}$ a basis of $\tilde{X}_2$. Let $(k^*_j)_{1 \leqslant j \leqslant n_+}$ be such that

$$k^*_j \in X'_1, \qquad k^*_j(k_i) = \delta_{ij}, \qquad k^*_{j\,|\tilde{X}_1} \equiv 0\,.$$

© The Author(s), under exclusive license
to Springer Nature Switzerland AG 2021
C. Cheverry and N. Raymond, *A Guide to Spectral Theory*,
Birkhäuser Advanced Texts Basler Lehrbücher,
https://doi.org/10.1007/978-3-030-67462-5_5

Define

$$R_- : \mathbb{C}^{n_-} \to \tilde{X}_2, \qquad R_- \alpha = \sum_{j=1}^{n_-} \alpha_j k'_j, \qquad R_- \text{ is bijective,}$$

$$R_+ : X_1 \to \mathbb{C}^{n_+}, \quad R_+(u) = (k^*_j(u))_{1 \leqslant j \leqslant n_+}, \quad \ker R_+ = \tilde{X}_1.$$

The correspondence between the dimensions is as described below, where the symbol $\updownarrow$ means the existence of a diffeomorphism

$$
\begin{array}{ccccccc}
X_1 \times \mathbb{C}^{n_-} = & \tilde{X}_1 & \oplus & \ker(T) & \oplus & \mathbb{C}^{n_-} \\
\updownarrow & \updownarrow & & \updownarrow & & \updownarrow \\
X_2 \times \mathbb{C}^{n_+} = & \mathrm{ran}\,(T) & \oplus & \mathbb{C}^{n_+} & \oplus & \tilde{X}_2
\end{array}
$$

Consider the operator

$$\mathcal{M} : X_1 \times \mathbb{C}^{n_-} \to X_2 \times \mathbb{C}^{n_+}.$$

that is such that for all $(e, c) \in X_1 \times \mathbb{C}^{n_-}$, we have

$$(5.5.1.1) \quad \mathcal{M} \begin{pmatrix} e \\ c \end{pmatrix} = \begin{pmatrix} T & R_- \\ R_+ & 0 \end{pmatrix} \begin{pmatrix} e \\ c \end{pmatrix} = \begin{pmatrix} Te + R_- c \\ R_+ e \end{pmatrix}.$$

The interest of extending the spaces X_1, X_2, and the operator T is to get a bijective operator $\mathcal{M}$.

Lemma 5.1. — *Let $T \in \mathcal{L}(X_1, X_2)$ be a Fredholm operator. The operator $\mathcal{M}$ is bijective with a bounded inverse.*

Proof. — Let $(e, c) \in X_1 \times \mathbb{C}^{n_-}$ be such that $Te + R_- c = 0$ and $R_+ e = 0$. Since $Te \in \mathrm{ran}\,(T)$ and $R_- c \in \tilde{X}_2$, we must have $Te = 0$ and $R_- c = 0$. It follows that T is injective, because

$$c = 0, \qquad e \in \ker R_+ \cap \ker T = \{0\}.$$

Now, let us consider $(f, d) \in X_2 \times \mathbb{C}^{n_+}$. We seek some $(e, c) \in X_1 \times \mathbb{C}^{n_-}$ which is such that

$$\mathcal{M} \begin{pmatrix} e \\ c \end{pmatrix} = \begin{pmatrix} T & R_- \\ R_+ & 0 \end{pmatrix} \begin{pmatrix} e \\ c \end{pmatrix} = \begin{pmatrix} Te + R_- c \\ R_+ e \end{pmatrix} = \begin{pmatrix} f \\ d \end{pmatrix}.$$

We have a unique decomposition of f according to $f = g + f_0$ with $g \in \mathrm{ran}\, T$ and $f_0 \in \tilde{X}_2$. Thus, we must impose $R_- c = f_0$. In view of the definition of R_-, this means to take for c the coordinates of f_0 in the basis $(k'_j)_{1 \leqslant j \leqslant n_-}$. On the other hand, we can write $e = k + e_0$ with $k \in \ker T$ and $e_0 \in \tilde{X}_1$. The constraint

$R_+e = R_+k = d$ implies that the coordinates of k in the basis $(k_j)_{1 \leqslant j \leqslant n_+}$ are d. Then, we are reduced to solve

$$Te_0 = f - R_-c = g \,.$$

Since T induces a bijection from $\tilde{X}_1 \to \operatorname{ran} T$, e_0 is uniquely determined as $e_0 = T^{-1}g$. The operator $\mathcal{M}$ is bijective.

By the open mapping theorem, the inverse $\mathcal{M}^{-1}$ is bounded.

$\square$

Let us now consider the converse.

Lemma 5.2. — *Let $T \in \mathcal{L}(X_1, X_2)$ and consider the operator matrix*

$$(5.5.1.2) \qquad\qquad \mathcal{M} = \begin{pmatrix} T & R_- \\ R_+ & 0 \end{pmatrix},$$

with $R_- : \mathbb{C}^{n_-} \to X_2$ and $R_+ : X_1 \to \mathbb{C}^{n_+}$ bounded. Assume that $\mathcal{M}$ is bijective. We denote by $\mathcal{E}$ its (bounded) inverse:

$(5.5.1.3)$
$$\mathcal{E} = \begin{pmatrix} E & E_+ \\ E_- & E_0 \end{pmatrix}, \qquad \begin{matrix} E \in \mathcal{L}(X_2, X_1), & E_+ \in \mathcal{L}(\mathbb{C}^{n_+}, X_1), \\ E_- \in \mathcal{L}(X_2, \mathbb{C}^{n_-}), & E_0 \in \mathcal{L}(\mathbb{C}^{n_+}, \mathbb{C}^{n_-}). \end{matrix}$$

Then, T is a Fredholm operator and we have $\operatorname{ind} T = \operatorname{ind}(E_0) = n_+ - n_-$. Moreover, the operator T is bijective if and only if E_0 is bijective.

In other words, any operator $T \in \mathcal{L}(X_1, X_2)$ giving rise, through a decomposition like (5.5.1.1), to a bijective operator $\mathcal{M}$ must be Fredholm. Moreover, there is an easy way to test if T is bijective. It suffices to check that $n_+ = n_-$ and to compute the determinant of the matrix E_0.

Proof. — We write that $\mathcal{E}$ is the inverse on the right:

$(5.5.1.4)$ $\qquad\qquad\qquad TE + R_-E_- = \operatorname{Id},$

$(5.5.1.5)$ $\qquad\qquad\qquad\qquad R_+E_+ = \operatorname{Id},$

$(5.5.1.6)$ $\qquad\qquad\qquad TE_+ + R_-E_0 = 0,$

$(5.5.1.7)$ $\qquad\qquad\qquad\qquad R_+E = 0,$

and on the left:

(5.5.1.8) $$ET + E_+R_+ = \mathsf{Id},$$

(5.5.1.9) $$E_-R_- = \mathsf{Id},$$

(5.5.1.10) $$E_-T + E_0R_+ = 0,$$

(5.5.1.11) $$ER_- = 0.$$

From (5.5.1.5) and (5.5.1.9), we get that R_+ and E_- are surjective and that R_- and E_+ are injective. In particular, R_+ has a right inverse R_{+r}^{-1}, and R_- has a left inverse R_{-l}^{-1}.

Assume that T is bijective. From (5.5.1.6), we see that E_0 must be injective; from (5.5.1.10), we deduce that E_0 must be surjective. Thus, E_0 is bijective. Retain that

(5.5.1.12) $$E_0^{-1} = -R_+T^{-1}R_-, \quad E_0 = -R_{-l}^{-1}TR_{+r}^{-1}.$$

Conversely, suppose that E_0 is bijective. Then, using (5.5.1.8) and (5.5.1.10), compute

(5.5.1.13)
$$(E - E_+E_0^{-1}E_-)T = ET - E_+E_0^{-1}(E_-T)$$
$$= \mathsf{Id} - E_+R_+ + E_+E_0^{-1}(E_0R_+) = \mathsf{Id}.$$

The operator $E - E_+E_0^{-1}E_-$ is a left inverse of T. A similar argument based on (5.5.1.4) and (5.5.1.6) shows that it is also a right inverse of T.

From (5.5.1.6), we can check that the injective application E_+ sends $\ker E_0$ into $\ker(T)$. Now, let $v \in \ker(T)$. From (5.5.1.8), we find that $E_+(R_+v) = v$ and from (5.5.1.10), we have $E_0(R_+v) = 0$. This means that v is in the range of the restriction of E_+ to $\ker E_0$. Thus, the application $E_+ : \ker E_0 \to \ker(T)$ is a bijection, and we have

(5.5.1.14) $\dim \ker T = \dim \ker E_0 = n_+ - \dim \operatorname{ran} E_0 < +\infty.$

Let us consider subspaces H and $\tilde{X}_2$ such that

$$\mathbb{C}^{n_-} = \operatorname{ran} E_0 \oplus H, \quad X_2 = \operatorname{ran} T \oplus \tilde{X}_2.$$

We recall that $E_- : X_2 \to \mathbb{C}^{n_-}$ is surjective. On the other hand, from (5.5.1.10), we know that $E_- : \operatorname{ran}(T) \to \operatorname{ran} E_0$. Consider the map

$$E^\sharp : \tilde{X}_2 \longrightarrow H$$
$$x \longmapsto \Pi_H E_-(x).$$

Since E_- is surjective, so is $E^\sharp$. The application $E^\sharp$ is also injective. Indeed, if $\Pi_H E_- v = 0$ with $v \in \tilde{X}_2$, we have $E_- v \in \operatorname{ran} E_0$ so that we can write $E_- v = E_0 w$. From (5.5.1.4) and (5.5.1.6), we get

$$TEv + R_- E_- v = v, \qquad R_- E_0 w = -TE_+ w,$$

which may be combined to deduce that

$$T(Ev - E_+ w) = v \in \operatorname{ran}(T),$$

which, knowing that $v \in \tilde{X}_2$, is possible only if $v = 0$. Thus, $E^\sharp$ is bijective, and we have
(5.5.1.15)
$$\operatorname{codim} \operatorname{ran} T = \dim \tilde{X}_2 = \dim H = n_- - \dim \operatorname{ran} E_0 < +\infty.$$

From (5.5.1.14) and (5.5.1.15), we deduce that T is Fredholm, with $\operatorname{ind} T = \operatorname{ind} E_0 = n_+ - n_-$. $\qquad\square$

5.2. On the index of Fredholm operators

Proposition 5.3. — *Let $T \in \mathcal{L}(X_1, X_2)$ be Fredholm. Then, $T' \in \mathcal{L}(X_2', X_1')$ is Fredholm and $\operatorname{ind} T' = -\operatorname{ind} T$.*

Proof. — Define $\mathcal{M}$ as in (5.5.1.1). From Lemma 5.1, we know that $\mathcal{M}$ is bijective. It follows that $\mathcal{M}'$ is bijective, and we have

$$\mathcal{M}' = \begin{pmatrix} T' & R'_+ \\ R'_- & 0 \end{pmatrix}, \; (\mathcal{M}')^{-1} = (\mathcal{M}^{-1})' = \mathcal{E}' = \begin{pmatrix} E' & E'_- \\ E'_+ & E'_0 \end{pmatrix}.$$

From Lemma 5.2, we deduce that T' is Fredholm and $\operatorname{ind} T' = \operatorname{ind} E'_0 = n_- - n_+$. $\qquad\square$

Proposition 5.4. — *$\operatorname{Fred}(X_1, X_2)$ is an open subset of $\mathcal{L}(X_1, X_2)$, and the index is a continuous function on $\operatorname{Fred}(X_1, X_2)$, locally constant on the connected components of $\operatorname{Fred}(X_1, X_2)$.*

Proof. — Let $T : X_1 \to X_2$ be a Fredholm operator and let $P : X_1 \to X_2$ be a a continuous operator with small norm. The operator $\mathcal{M}$ given by (5.5.1.1) is bijective, and it remains so for small perturbations of the form

$$\begin{pmatrix} T+P & R_- \\ R_+ & 0 \end{pmatrix}, \qquad P \in \mathcal{L}(X_1, X_2), \qquad \|P\| \ll 1.$$

By Lemma 5.2, the operator $T + P$ is Fredholm. $\qquad\qquad\qquad\square$

Proposition 5.5. — *Let $T \in \mathcal{L}(X_1, X_2)$. Then, T is Fredholm if and only if we can find $S \in \mathcal{L}(X_2, X_1)$, $K_1 \in \mathcal{K}(X_1)$, and $K_2 \in \mathcal{K}(X_2)$ such that*

$$(5.5.2.16) \qquad ST = \mathrm{Id}_{X_1} + K_1, \qquad TS = \mathrm{Id}_{X_2} + K_2.$$

Conversely, if we have (5.5.2.16) for some $S \in \mathcal{L}(X_2, X_1)$, then T is Fredholm.

Proof. — If T is Fredholm, we can use Lemmas 5.1 and 5.2. Take $S = E$, $K_1 = -E_+ R_+$, and $K_2 = -R_- E_-$. In view of (5.5.1.4) and (5.5.1.8), we have (5.5.2.16). Moreover, since K_1 and K_2 are of finite rank, they are compact.

Conversely, assume (5.5.2.16). We have $\ker T \subset \ker(ST)$. From Proposition 4.9, we know that

$$\dim \ker T \leqslant \dim \ker(ST) = \dim \ker(\mathrm{Id}_{X_1} + K_1) < +\infty.$$

We have also $\mathrm{ran}\,(TS) \subset \mathrm{ran}\,T$ and, from Proposition 4.9, we deduce that

$$\mathrm{codim}\,\mathrm{ran}\,T \leqslant \mathrm{codim}\,\mathrm{ran}\,(TS) = \mathrm{codim}\,\mathrm{ran}\,(\mathrm{Id}_{X_2} + K_2) < +\infty.$$

Thus, the operator T is Fredholm. $\qquad\qquad\qquad\square$

Remark 5.6. — When $X_1 = X_2 = \mathsf{H}$, Proposition 5.5 is known as the Atkinson theorem. It tells us that a Fredholm operator is invertible in the Calkin algebra $\mathcal{L}(\mathsf{H})/\mathcal{K}(\mathsf{H})$.

Corollary 5.7. — *Let $T \in \mathcal{L}(X_1, X_2)$ and $U \in \mathcal{L}(X_2, X_3)$ be Fredholm operators. Then, UT is a Fredholm operator and*

$$\mathrm{ind}\,(UT) = \mathrm{ind}\,U + \mathrm{ind}\,T.$$

Proof. — From Proposition 5.5, we can write for $(K_1, K_2) \in \mathcal{K}(X_1) \times \mathcal{K}(X_2)$ and $(\tilde{K}_1, \tilde{K}_2) \in \mathcal{K}(X_2) \times \mathcal{K}(X_3)$,

$$ST = \mathrm{Id}_{X_1} + K_1, \quad TS = \mathrm{Id}_{X_2} + K_2,$$
$$\tilde{S}U = \mathrm{Id}_{X_2} + \tilde{K}_1, \quad U\tilde{S} = \mathrm{Id}_{X_3} + \tilde{K}_2.$$

Notice that

$$(S\tilde{S})UT = S(\mathrm{Id}_{X_2} + \tilde{K}_1)T = \mathrm{Id}_{X_1} + K_1 + S\tilde{K}_1 T,$$

$$UT(S\tilde{S}) = U(\mathrm{Id}_{X_2} + K_2)\tilde{S} = \mathrm{Id}_{X_3} + \tilde{K}_2 + UK_2\tilde{S}.$$

and that $K_1 + S\tilde{K}_1$ and $\tilde{K}_2 + UK_2\tilde{S}$ are compact.

From Proposition 5.5, we conclude that UT is Fredholm.

Now, for $t \in \left[0, \frac{\pi}{2}\right]$, consider the operator L_t from $X_2 \times X_1$ to $X_3 \times X_2$,

$$L_t = \begin{pmatrix} U & 0 \\ 0 & \mathrm{Id}_{X_2} \end{pmatrix} \begin{pmatrix} \cos t\,\mathrm{Id}_{X_2} & -\sin t\,\mathrm{Id}_{X_2} \\ \sin t\,\mathrm{Id}_{X_2} & \cos t\,\mathrm{Id}_{X_2} \end{pmatrix} \begin{pmatrix} \mathrm{Id}_{X_2} & 0 \\ 0 & T \end{pmatrix}.$$

This is a product of three Fredholm operators. The map $[0, \frac{\pi}{2}] \ni t \mapsto L_t$ is continuous, and valued in the set of Fredholm operators. We have

$$L_0 = \begin{pmatrix} U & 0 \\ 0 & T \end{pmatrix}, \qquad L_{\frac{\pi}{2}} = \begin{pmatrix} 0 & -UT \\ \mathrm{Id}_{X_2} & 0 \end{pmatrix}.$$

Since the index is locally constant, we must have

$$\mathrm{ind}\, L_0 = \mathrm{ind}\, U + \mathrm{ind}\, T = \mathrm{ind}\, L_{\frac{\pi}{2}} = \mathrm{ind}\,(UT).$$

The conclusion follows. □

Exercise 5.8. — With the notations of Exercise 4.16, prove that $\lambda \notin \mathrm{sp}_{\mathrm{ess}}(\mathscr{L} + V)$.

Corollary 5.9. — *Let $T \in \mathcal{L}(X_1, X_2)$ a Fredholm operator and $K \in \mathcal{K}(X_1, X_2)$. Then, $T + K$ is Fredholm and* $\mathrm{ind}\,(T + K) = \mathrm{ind}\, T$.

Proof. — From Proposition 5.5, we have

$$S(T + K) = ST + SK = \mathrm{Id}_E + K_1 + SK,$$

$$(T + K)S = TS + KS = \mathrm{Id}_F + K_2 + KS.$$

and the operators $K_1 + SK$ and $K_2 + KS$ are compact.

It follows that $T + K$ is Fredholm. The map $[0, 1] \ni s \mapsto T + sK$ is continuous and built with Fredholm operators. Since the index locally constant, we must have $\mathrm{ind}\,(T + K) = \mathrm{ind}\, T$. □

Remark 5.10. — With our definition of the essential spectrum (see Definition 3.46), Corollary 5.9 tells that the essential spectrum is stable under compact perturbations.

5.3. On the spectrum of compact operators

In the next theorem, we recall fundamental facts about compact operators. In particular, we will notice that the non-zero spectrum of a compact operator is discrete.

Theorem 5.11. — (Fredholm alternative). *Let* $T \in \mathcal{L}(E)$ *be a compact operator. Then,*

(i) *If* E *is of infinite dimension, then* $0 \in \mathsf{sp}(T)$.

(ii) *For all* $z \in U = \mathbb{C} \setminus \{0\}$, $T - z$ *is a Fredholm operator of index* 0.

(iii) $\ker(T - \mathsf{Id}) = \{0\}$ *if and only if* $\mathrm{ran}(T - \mathsf{Id}) = E$.

(iv) *The elements of* $\mathsf{sp}(T) \setminus \{0\}$ *are isolated with finite algebraic multiplicity and the only possible accumulation point of the spectrum is* 0.

(v) *The non-zero spectrum of* T *is discrete.*

Proof. —
(i) Assume that $0 \notin \mathsf{sp}(T)$. Then, since the set of compact operators forms a ideal of bounded operators, we find that $\mathrm{Id} = T^{-1} \circ T$ is compact, and therefore $B_E(0, 1]$ is relatively compact. In view of the Riesz theorem, this is not possible if $\dim E = +\infty$.

(ii) For $z \neq 0$, we have $T - z = -z(\mathrm{Id} - T/z)$ with T/z compact. From Proposition 4.9, we know that $T - z$ is a Fredholm operator. From Proposition 5.4, the function $s \mapsto \mathrm{ind}\,(sT - z)$ is continuous on $[0, 1]$, and therefore constant. It follows that $\mathrm{ind}\,(T - z) = \mathrm{ind}\,(-z) = 0$.

(iii) This is obvious since $T - \mathsf{Id}$ is of index 0.

(iv) Recall that $U = \mathbb{C} \setminus \{0\}$. By construction, the set

$$V := \{z \in U : \exists r > 0 : D(z, r) \subset \mathsf{sp}(T)\} = U \cap \overset{\circ}{\mathsf{sp}}(T)$$

is open. Let us prove that it is closed in U. Consider a sequence $(z_n)_n \in V^{\mathbb{N}}$ that converges to some $z_\infty \in U$. The Fredholm operator $T - z_\infty$ can be completed as in the beginning of Section 5.1, see (5.5.1.1), to get a bijective invertible operator

$$\mathcal{M}(z_\infty) = \begin{pmatrix} T - z_\infty & R_-(z_\infty) \\ R_+(z_\infty) & 0 \end{pmatrix}.$$

For $z \in U$, consider

$$\mathcal{M}(z) := \begin{pmatrix} T - z & R_-(z_\infty) \\ R_+(z_\infty) & 0 \end{pmatrix} = \mathcal{M}(z_\infty) - (z - z_\infty)\mathcal{N},$$

with

$$\mathcal{N} = \begin{pmatrix} \mathrm{Id} & 0 \\ 0 & 0 \end{pmatrix}.$$

For $|z - z_\infty| < r$ with $r := \|\mathcal{N}\|^{-1}\|\mathcal{M}(z_\infty)\|$, we find that $\mathcal{M}(\cdot)$ is holomorphic with

$$\mathcal{M}(z)^{-1} = \sum_{j=0}^\infty (z - z_\infty)^j \left(\mathcal{M}(z_\infty)^{-1}\mathcal{N}\right)^j \mathcal{M}(z_\infty)^{-1}.$$

On the other hand, with (5.5.1.3), we have the decomposition

$$\mathcal{M}(z)^{-1} = \mathcal{E}(z) = \begin{pmatrix} E(z) & E_+(z) \\ E_-(z) & E_0(z) \end{pmatrix},$$

where the functions $E(\cdot)$, $E_+(\cdot)$, $E_-(\cdot)$, and $E_0(\cdot)$ are holomorphic. From Lemma 5.2, we know that $T - z$ is not bijective if and only if $\det E_0(z) = 0$. Recall that

(5.5.3.17) $\qquad E_0(z) = -R_{-l}(z_\infty)^{-1}(T - z)R_{+r}(z_\infty)^{-1}$

which clearly indicates that $E_0(\cdot)$ is holomorphic in a neighborhood of z_∞. Its zeros are isolated unless $\det E_0 \equiv 0$. By the definition of z_∞, we deduce that $\det E_0 = 0$ in a neighborhood of the limiting point z_∞. This implies that $z_\infty \in V$. The set V is closed in U.

Since V is open and closed in U, we have $V = U$ or $V = \emptyset$. But

$$V \subset \mathrm{sp}(T) \subset B(0, \|T\|] \subsetneq U.$$

Thus, we have $V = \emptyset$. Now let us consider $\lambda \in \mathrm{sp}(T) \setminus \{0\}$. Then, in a neighborhood of λ, $T - z$ is not bijective if and only if $\det E_0(z) = 0$. Since $V = \emptyset$, $\det E_0$ is not zero near λ and thus (by holomorphy), its zeros are isolated. Finally, we recall (5.5.1.13) and thus we have, near each point of the spectrum in U,

$$(T - z)^{-1} = E(z) - E_+(z)E_0^{-1}(z)E_-(z),$$

and we deduce that the resolvent is meromorphic in U. Since $E(\cdot)$ is holomorphic, we have

$$P_\lambda = \frac{1}{2i\pi} \int_{\Gamma_\lambda} E_+(z) E_0^{-1}(z) E_-(z) \, \mathrm{d}z \,,$$

where Γ_λ is a circle about λ with radius small enough. By using a Laurent expansion of $z \mapsto E_+(z) E_0^{-1}(z) E_-(z)$ near λ (to get its residue at λ), we see that the range of P_λ is finite (since E_0 is a finite matrix, we see that the residue is a finite-rank operator).

(v) follows from (iv) and (ii), and from the definition of the discrete spectrum.   □

Corollary 5.12. — *Let* $(\mathrm{Dom}\,(T), T)$ *be a closed operator. Assume that the resolvent set is not empty and that the resolvent is compact. Then, the spectrum of* T *is discrete.*

Proof. — We can find $z_0 \in \rho(T)$. For $z \neq z_0$, we have

$$T - z = (z_0 - z)\big[(z_0 - z)^{-1} + (T - z_0)^{-1}\big](T - z_0) \,.$$

This shows that $T - z$ is a Fredholm operator with index 0.
   Define $Z = h(z)$ with $h(z) := (z - z_0)^{-1}$. Then

$$T - z = -Z^{-1}\big[R_T(z_0) - Z\big](T - z_0).$$

Since $T - z_0$ is invertible, this indicates that $z \in \mathrm{sp}T$ if and only if $Z \in \mathrm{sp}R_T(z_0)$, that is

(5.5.3.18)
$$\mathrm{sp}(T) = \big\{h^{-1}(Z) = z_0 + Z^{-1} \,;\, Z \in \mathrm{sp}R_T(z_0) \cap \mathbb{C}^*\big\} \,.$$

Since the resolvent $(T - z_0)^{-1}$ is compact, the elements of $\mathrm{sp}R_T(z_0) \cap \mathbb{C}^*$ are isolated. The same applies concerning their images by the diffeomorphism $h^{-1} : \mathbb{C}^* \to \mathbb{C} \setminus \{z_0\}$. Note that the possible accumulation point of the spectrum at 0 is sent to $+\infty$.

   Let $\lambda \in \mathrm{sp}(T)$. Since $\lambda \neq z_0$, we can find some $r > 0$ such that the closed disc $D(\lambda, r]$ of center λ and radius r does not contain z_0, and it contains only λ as element of the spectrum. Let $\Gamma_\lambda := \partial D(\lambda, r]$ be the circle of center λ and radius r. By the resolvent formula (4.4.3.5), we have

$$
\begin{aligned}
P_\lambda &= -\frac{1}{2\pi i} \int_{\Gamma_\lambda} R_T(z)\, \mathrm{d}z \\
&= -\frac{1}{2\pi i} \left(\int_{\Gamma_\lambda} \frac{R_T(z) - R_T(z_0)}{z - z_0}\, \mathrm{d}z \right)(T - z_0).
\end{aligned}
$$

Since $z \mapsto (z-z_0)^{-1}$ is holomorphic in a neighborhood of $D(\lambda, r]$, we get

$$
\begin{aligned}
P_\lambda &= -\frac{1}{2\pi i} \left(\int_{\Gamma_\lambda} \frac{R_T(z)}{z - z_0}\, \mathrm{d}z \right)(T - z_0) \\
&= -\frac{1}{2\pi i} R_T(z_0) \left(\int_{\Gamma_\lambda} -Z^2 \left[R_T(z_0) - Z \right]^{-1} \mathrm{d}z \right)(T - z_0) \\
&= -\frac{1}{2\pi i} R_T(z_0) \left(\int_{h(\Gamma_\lambda)} \left[R_T(z_0) - Z \right]^{-1} \mathrm{d}Z \right)(T - z_0).
\end{aligned}
$$

We recognize on the right-hand side (inside brackets) the Riesz projection $P_{h(\lambda)}$ associated with the compact operator $R_T(z_0)$. Thus, the rank of $P_{h(\lambda)}$ is finite, and so is the rank of P_λ. □

Remark 5.13. — Even if a closed operator has compact resolvent (with a non-empty resolvent set), the discrete spectrum might be finite (and even empty!).

Exercise 5.14. — Consider Exercise 4.11. Determine the elements of $\mathrm{sp}(T) \setminus \{0\}$.

Solution: Since T is compact, any $\lambda \in \mathrm{sp}(T) \setminus \{0\}$ is isolated with finite algebraic multiplicity. By Corollary 3.36, it must be an eigenvalue. Thus, there is $u \neq 0$ such that

$$
0 = \lambda u_0, \qquad u_n = \lambda^{-1} n^{-1} u_{n-1} = \lambda^{-n} (n!)^{-1} u_0.
$$

This is possible only if $u = 0$. Therefore $\mathrm{sp}(T) = \{0\}$.   ○

Exercise 5.15. — [Application to elliptic equations] Let $\Omega \subset \mathbb{R}^n$ with $n \in \mathbb{N}^*$ be a bounded open set with smooth boundary. We work on $L^2(\Omega)$. Define $\mathrm{Dom}\,(T) = H^2(\Omega) \cap H^1_0(\Omega)$ and $T : \mathrm{Dom}\,(T) \to L^2(\Omega)$ given by $T := -\Delta + 1$. We admit that T is a bijection with compact (right) inverse K. For each $\lambda \in \mathbb{R}$, show that either the homogeneous equation $Tu - \lambda u = 0$ has a nontrivial solution, or that the inhomogeneous equation $Tu - \lambda u = f$

possesses a unique solution $u \in \mathrm{Dom}\,(T)$ for each given datum $f \in \mathsf{L}^2(\Omega)$.

Solution: Consider some $\mu \neq 0$. As a consequence of Fredholm alternative, either μ is an eigenvalue of K, or the operator $K - \mu$ is bijective from $\mathsf{L}^2(\Omega)$ to itself.

If μ is an eigenvalue of K, we have $Tu = \mu^{-1}u$ with $u \neq 0$. Then, the homogeneous equation $Tu - \lambda u = 0$ with $\lambda = \mu^{-1}$ has a nontrivial solution.

If $K - \mu$ is bijective, then so is

$$(K - \mu)T = \mathrm{Id} - \mu T = -\mu(T - \mu^{-1})\,.$$

Then, the inhomogeneous equation $Tu - \lambda u = f$ with $\lambda = \mu^{-1}$ has a unique solution $u \in \mathrm{Dom}\,(T)$ for each given $f \in \mathsf{L}^2(\Omega)$.

The case $\lambda = 0$ can be dealt separately. ○

5.4. Study of the complex Airy operator

The *complex Airy operator*

$$\mathscr{L}_{\pm} := D_x^2 \pm ix, \quad x \in \mathbb{R}, \quad D := -i\partial_x$$

appears in many contexts: mathematical physics, fluid dynamics, time dependent Ginzburg–Landau problems, and so on. Consider $\mathsf{H} = \mathsf{L}^2(\mathbb{R})$ equipped with the usual scalar product $\langle \cdot, \cdot \rangle$. We set

$$\mathrm{Dom}(\mathscr{L}_{\pm}) = \left\{ \psi \in \mathsf{L}^2(\mathbb{R})\,;\, (D_x^2 \pm ix)\psi \in \mathsf{L}^2(\mathbb{R}) \right\}.$$

The aim of this paragraph is to examine the properties of the operator $\big(\mathrm{Dom}(\mathscr{L}_{\pm}), \mathscr{L}_{\pm}\big)$ through as a succession of corrected (small) exercises. This is an opportunity to illustrate, review and practice many notions and tools that have been previously introduced.

Exercise 5.16. — Prove that $\mathscr{L}_{\pm}$ is a closed operator.

Solution: Come back to Proposition 2.12, criterion (2.12). Let $(\psi_n) \in \mathrm{Dom}(\mathscr{L}_{\pm})^{\mathbb{N}}$ such that $\psi_n \to \psi$ and $\mathscr{L}_{\pm}\psi_n \to \chi$ in the sense of $\mathsf{L}^2(\mathbb{R})$. We have $\mathscr{L}_{\pm}\psi_n \to \mathscr{L}_{\pm}\psi$ in the sense of distributions. The limit is unique so that $\psi \in \mathrm{Dom}(\mathscr{L}_{\pm})$ with $\mathscr{L}_{\pm}\psi = \chi$. ○

Exercise 5.17. — Prove that, for all $u \in \mathscr{C}_0^{\infty}(\mathbb{R})$, we have
(5.5.4.19)
$$\mathrm{Re}\langle \mathscr{L}_{\pm}u, u \rangle = \|u'\|^2\,, \quad \|\mathscr{L}_{\pm}u\|^2 = \|u''\|^2 + \|xu\|^2 + 2\,\mathrm{Im}\,\langle u', u \rangle\,.$$

Solution: Notice that $D_x^2 = -\partial_x^2$, and therefore (after integration by parts)

$$\text{Re}\langle \mathscr{L}_\pm u, u \rangle = \int (-\partial_x^2 u \pm ixu)\bar{u}\mathrm{d}x = \int |u'|^2\mathrm{d}x.$$

On the other hand
$$\|\mathscr{L}_\pm u\|^2$$

$$= \int (-\partial_x^2 u \pm ixu)(-\partial_x^2 \bar{u} \mp ix\bar{u}) \, \mathrm{d}x$$

$$= \int \left(|u''|^2 + |xu|^2 \pm i(-xu\partial_{xx}^2\bar{u} + x\bar{u}\partial_{xx}^2 u)\right) \mathrm{d}x$$

$$= \int \left(|u''|^2 + |xu|^2 \pm i(u + x\partial_x u)\partial_x\bar{u} \mp i(\bar{u} + x\partial_x\bar{u})\partial_x u\right) \mathrm{d}x$$

$$= \int \left(|u''|^2 + |xu|^2 \pm i(u\partial_x\bar{u} - \bar{u}\partial_x u)\right) \mathrm{d}x.$$

○

Exercise 5.18. — Deduce that there exist two constants $c > 0$ and $C > 0$ such that, for all $u \in \mathscr{C}_0^\infty(\mathbb{R})$,
(5.5.4.20)
$$c\left(\|u\|_{\mathsf{H}^2(\mathbb{R})} + \|xu\|\right) \leqslant \|\mathscr{L}_\pm u\| + \|u\| \leqslant C\left(\|u\|_{\mathsf{H}^2(\mathbb{R})} + \|xu\|\right).$$

Solution: Applying two times the Cauchy–Schwartz inequality and using (5.5.4.19), we first get that

$$|\text{Im}\,\langle u', u \rangle| \leqslant \frac{1}{2}\|u'\|^2 + \frac{1}{2}\|u\|^2 \leqslant \frac{1}{4}\|\mathscr{L}_\pm u\|^2 + \frac{3}{4}\|u\|^2.$$

This also implies that

$$\|u\|_{\mathsf{H}^2(\mathbb{R})}^2 + \|xu\|^2 = \|u''\|^2 + \|u'\|^2 + \|u\|^2 + \|xu\|^2$$
$$\leqslant 2\|\mathscr{L}_\pm u\|^2 + 3\|u\|^2.$$

We can take $c = 1/\sqrt{6}$ for instance. Moreover

$$\frac{1}{2}\|\mathscr{L}_\pm u\|^2 \leqslant \|u''\|^2 + \|xu\|^2 + \frac{3}{2}\|u\|^2,$$

so that

$$\frac{1}{2}\left(\|\mathscr{L}_\pm u\|^2 + \|u\|^2\right) \leqslant 2\left(\|u\|_{\mathsf{H}^2(\mathbb{R})}^2 + \|xu\|^2\right).$$

We can take $C = 2\sqrt{2}$.

○

Exercise 5.19. — Prove the density of $\mathscr{C}_0^\infty(\mathbb{R})$ in $\mathrm{Dom}(\mathscr{L}_\pm)$ for the graph norm

$$\|u\|_{\mathscr{L}_\pm} := \|\mathscr{L}_\pm u\| + \|u\|.$$

Solution: Let $u \in \mathrm{Dom}(\mathscr{L})$. From (the left part of) (5.5.4.20), we know that $u \in \mathrm{H}^2(\mathbb{R})$ and $xu \in \mathrm{L}^2(\mathbb{R})$. Let $\chi \in \mathscr{C}_0^\infty(\mathbb{R})$ be such that $0 \leqslant \chi \leqslant 1$, and $\chi \equiv 1$ on $]-1,1[$, as well as $\chi \equiv 0$ outside the interval $[-2,2]$. For $R > 0$, define $\chi_R(x) := \chi(x/R)$, and introduce $u_R := \chi_R u$. Fix any $\varepsilon > 0$. For R large enough, we obtain that

$$\|xu_R - xu\| \leqslant \varepsilon, \qquad \|u_R - u\|_{\mathrm{H}^2(\mathbb{R})} \leqslant \varepsilon.$$

From (5.5.4.20), this means that $\|u_R - u\|_{\mathscr{L}_\pm} \leqslant \varepsilon$. There remains to approach u_R. Recall that $\mathscr{C}_0^\infty(\mathbb{R})$ is dense in $\mathrm{H}_0^2(]-2R,2R[)$. Thus, we can find a sequence (u_n) in $\mathscr{C}_0^\infty(\mathbb{R})^{\mathbb{N}}$, with supports contained in $]-2R,2R[$, converging to u_R in $\mathrm{H}^2(\mathbb{R})$. Since $|x|$ is bounded by $2R$ on the support of u_n, we find that xu_n goes to xu_R in $\mathrm{L}^2(\mathbb{R})$. The right part of (5.5.4.20) gives the conclusion.   ○

Exercise 5.20. — What is the adjoint of $\mathscr{L}_\pm$?

Solution: First observe that, for all $(\psi,\chi) \in \mathscr{C}_0^\infty(\mathbb{R}) \times \mathscr{C}_0^\infty(\mathbb{R})$,

$$\langle \mathscr{L}_\pm \psi, \chi \rangle = \langle \psi, \mathscr{L}_\pm^* \chi \rangle = \langle \psi, \mathscr{L}_\mp \chi \rangle.$$

Thus, we have $\mathrm{Dom}(\mathscr{L}_\mp) \subset \mathrm{Dom}(\mathscr{L}_\pm^*)$. The rest of the proof follows the same lines as in Section 2.2.4. We find that

$$\mathrm{Dom}(\mathscr{L}_\mp) = \{\psi \in \mathrm{H}^2(\mathbb{R}) : x\psi \in \mathrm{L}^2(\mathbb{R})\} = \mathrm{Dom}(\mathscr{L}_\pm^*)$$

and $\mathscr{L}_\pm^* = \mathscr{L}_\mp$.   ○

Exercise 5.21. — Prove that $\mathscr{L}_\pm + 1$ is bijective.

Solution: We have

$$\langle (\mathscr{L}_\pm + 1)u, (\mathscr{L}_\pm + 1)u \rangle$$
$$= \|\mathscr{L}_\pm u\|^2 + 2\mathrm{Re}\,\langle u, \mathscr{L}_\pm u \rangle + \|u\|^2$$
$$= \|u''\|^2 + \|xu\|^2 + 2\mathrm{Im}\,\langle u', u \rangle + 2\|u'\|^2 + \|u\|^2$$
$$\geqslant \frac{1}{2}\|u\|^2.$$

It is enough to apply Proposition 2.14 to get that $\mathscr{L}_\pm + 1$ is injective with closed range. From Proposition 2.52, we also know that

$$\ker(\mathscr{L}_\pm + 1)^* = \ker(\mathscr{L}_\mp + 1) = \{0\} = \mathrm{ran}\,(\mathscr{L}_\pm + 1)^\perp,$$

and therefore

$$\overline{\mathrm{ran}\,(\mathscr{L}_\pm + 1)} = \mathrm{ran}\,(\mathscr{L}_\pm + 1) = \{0\}^\perp = \mathsf{L}^2(\mathbb{R}).$$

$\circ$

Exercise 5.22. — Prove that the resolvent of $\mathscr{L}_\pm$ is compact.

Solution: We use Proposition 4.24. Let (ψ_n) be a sequence in $\mathrm{Dom}(\mathscr{L}_\pm)^{\mathbb{N}}$ which is bounded for the graph norm. We know that (ψ_n) is bounded in $\mathsf{H}^2(\mathbb{R})$, and that $(x\psi_n)$ is bounded in $\mathsf{L}^2(\mathbb{R})$. By the Rellich theorem and using some exhaustion of $\mathbb{R}$ by compact sets, we can extract a subsequence still denoted by (ψ_n), such that (ψ_n) does converge in L^2 on all compact subsets of $\mathbb{R}$, to some $\psi \in \mathsf{H}^2(\mathbb{R})$ satisfying $x\psi \in \mathsf{L}^2(\mathbb{R})$. On the other hand, we have

$$\int_{B(0,R]^c} |\psi_n - \psi|^2 \, \mathrm{d}x \leqslant R^{-2} \int_{B(0,R]^c} x^2 |\psi_n - \psi|^2 \, \mathrm{d}x$$

$$\leqslant 2R^{-2} \int_{B(0,R]^c} x^2 (|\psi_n|^2 + |\psi|^2) \, \mathrm{d}x.$$

Thus, for all $\varepsilon > 0$, we can find some R large enough such that

$$\lim_{n \to +\infty} \|\psi_n - \psi\|_{\mathsf{L}^2(B(0,R])} = 0, \qquad \|\psi_n - \psi\|_{\mathsf{L}^2(B(0,R]^c)} \leqslant \varepsilon.$$

This means that (ψ_n) goes to ψ in $\mathsf{L}^2(\mathbb{R})$. $\circ$

Exercise 5.23. — Show that, if λ belongs to the spectrum of $\mathscr{L}_\pm$, then, for all $\alpha \in \mathbb{R}$, $\lambda + i\alpha$ also belongs to the spectrum.

Solution: The idea is to conjugate $\mathscr{L}_\pm$ with the (bijective and continuous) translation operator $\tau_a \psi(x) = \psi(x + a)$. Remark that $\tau_a^{-1} \mathscr{L}_\pm \tau_a = \mathscr{L}_\pm \pm ia$. $\circ$

Exercise 5.24. — Determine the spectrum of $\mathscr{L}$.

Solution: It is empty!

Otherwise, the spectrum of $\mathscr{L}$ contains (at least) one point z. In view of Corollary 5.12, it should be discrete. Taking into account Exercise 5.23, it should also contain the whole vertical line passing through z. This is clearly a contradiction. $\circ$

5.5. An application of the Grushin formalism

Lemma 5.25. — *Consider P and Q two projections such that $\|P - Q\| < 1$. Then,*

$$\dim \operatorname{ran} P = \dim \operatorname{ran} Q .$$

Proof. — We let

$$U = QP + (\operatorname{Id} - Q)(\operatorname{Id} - P) \in \mathcal{L}\left(\operatorname{ran}(P), \operatorname{ran}(Q)\right) ,$$
$$V = PQ + (\operatorname{Id} - P)(\operatorname{Id} - Q) \in \mathcal{L}\left(\operatorname{ran}(Q), \operatorname{ran}(P)\right) ,$$

and notice that $UV = VU = \operatorname{Id} - (P - Q)^2$. Thus UV and VU are bijective, and so are U and V. The conclusion follows. $\qquad \square$

Lemma 5.26. — *Let Ω be a non-empty open set of $\mathbb{C}$ and $(\mathcal{B}_z)_{z \in \Omega}$ be a family of bounded operators which is strongly holomorphic in the sense that*

$$\forall \psi \in \mathsf{H}, \quad \Omega \ni z \mapsto \mathcal{B}_z \psi \text{ is holomorphic.}$$

Then, the family is holomorphic, meaning that

$$\Omega \ni z \mapsto \mathcal{B}_z \in \mathcal{L}(\mathsf{H}) \text{ is holomorphic.}$$

Proof. — First, let us show that $\Omega \ni z \mapsto \mathcal{B}_z$ is continuous. Let $z_0 \in \Omega$ and $r > 0$ such that $D(z_0, r) \subset \Omega$. We consider the family of bounded operators

$$\left(\frac{\mathcal{B}_z - \mathcal{B}_{z_0}}{z - z_0} \right)_{z \in D(z_0, r) \setminus \{z_0\}} .$$

It satisfies the assumptions of the Uniform Boundedness Principle and we deduce that $\mathcal{B}$ is continuous at z_0. Then, for all $z \in \Omega$ and a convenient contour Γ, the Cauchy formula gives

$$\mathcal{B}_z \psi = \frac{1}{2i\pi} \int_\Gamma \frac{\mathcal{B}_\zeta \psi}{\zeta - z} \, \mathrm{d}\zeta .$$

The continuity of $\mathcal{B}$ implies that, for all $\psi \in \mathsf{H}$,

$$\int_\Gamma \frac{\mathcal{B}_\zeta \psi}{\zeta - z} \, \mathrm{d}\zeta = \left(\int_\Gamma \frac{\mathcal{B}_\zeta}{\zeta - z} \, \mathrm{d}\zeta \right) \psi .$$

Thus,

$$\mathcal{B}_z = \frac{1}{2i\pi} \int_\Gamma \frac{\mathcal{B}_\zeta}{\zeta - z} \, \mathrm{d}\zeta ,$$

and this formula classically implies the holomorphy. $\qquad \square$

Proposition 5.27. — *Consider a family of closed operators* $(\mathscr{L}_\xi)_{\xi\in\Omega}$ *where* $\Omega \subset \mathbb{C}$ *is a non-empty open set. We assume that this family is analytic in the sense that*

(i) *This common domain is denoted by* D *and assumed to be dense.*

(ii) *For all* $\psi \in D$, $\Omega \ni \xi \mapsto \mathscr{L}_\xi\psi$ *is holomorphic.*

Let $\xi_0 \in \Omega$ *and* $z_0 \in \mathbb{C}$. *We also assume that* $\mathscr{L}_{\xi_0} - z_0$ *is Fredholm, and that*

(iii) $\ker(\mathscr{L}_{\xi_0} - z_0) = \operatorname{span} u_{\xi_0}$ *with* $u_{\xi_0} \neq 0$.

(iv) $\ker(\mathscr{L}_{\xi_0}^* - \overline{z_0}) = \operatorname{span} v_{\xi_0}$ *with* $\langle u_{\xi_0}, v_{\xi_0} \rangle \neq 0$.

Then, there exist open neighborhoods $\mathcal{V}$ *and* $\mathcal{W}$ *of* ξ_0 *and* z_0 *respectively, as well as some holomorphic function* $\mu : \mathcal{V} \to \mathcal{W}$ *such that, for all* $(\xi, z) \in \mathcal{V} \times \mathcal{W}$,

$$z \in \operatorname{sp}(\mathscr{L}_\xi) \Leftrightarrow z = \mu(\xi).$$

Moreover, $\dim \ker(\mathscr{L}_\xi - \mu(\xi)) = 1$, *and we can find an analytic eigenvector* $(u_\xi)_{\xi\in\mathcal{V}}$ *associated with* μ.

Proof. — For all $(\xi, z) \in \Omega \times \mathbb{C}$, we consider

$$\mathscr{M}_{\xi,z} = \begin{pmatrix} \mathscr{L}_\xi - z & \cdot v_{\xi_0} \\ \langle \cdot, u_{\xi_0} \rangle & 0 \end{pmatrix} : D \times \mathbb{C} \to \mathsf{H} \times \mathbb{C}.$$

We can check that $\mathscr{M}_{\xi_0,z_0}$ is bijective and

$$\mathscr{M}_{\xi_0,z_0}^{-1} = \begin{pmatrix} (\mathscr{L}_{\xi_0} - z_0)^{-1}\Pi_{v_{\xi_0}}^\perp & \cdot u_{\xi_0} \\ \langle \cdot, v_{\xi_0} \rangle & 0 \end{pmatrix},$$

where $\Pi_{v_{\xi_0}}^\perp$ is the orthogonal projection on $(\operatorname{span}(v_{\xi_0}))^\perp$. Then, we write

$$\mathscr{M}_{\xi,z} = \mathscr{M}_{\xi_0,z_0} + \begin{pmatrix} \mathscr{L}_\xi - \mathscr{L}_{\xi_0} - z + z_0 & 0 \\ 0 & 0 \end{pmatrix}$$

$$= \left(\operatorname{Id} + \begin{pmatrix} \mathscr{L}_\xi - \mathscr{L}_{\xi_0} - z + z_0 & 0 \\ 0 & 0 \end{pmatrix} \mathscr{M}_{\xi_0,z_0}^{-1} \right) \mathscr{M}_{\xi_0,z_0}$$

$$= (\operatorname{Id} + P) \mathscr{M}_{\xi_0,z_0},$$

with

$$P = \begin{pmatrix} -\Pi_{v_{\xi_0}} + (\mathscr{L}_\xi - z)(\mathscr{L}_{\xi_0} - z_0)^{-1}\Pi_{v_{\xi_0}}^\perp & \cdot(\mathscr{L}_\xi - z)u_{\xi_0} \\ 0 & 0 \end{pmatrix}.$$

Note that $(\xi, z) \mapsto (\mathscr{L}_\xi - z)u_{\xi_0}$ is holomorphic by assumption. Moreover, the operator

$$\mathscr{A}_{\xi,z} = (\mathscr{L}_\xi - z)(\mathscr{L}_{\xi_0} - z_0)^{-1}\Pi^{\perp}_{v_{\xi_0}} : \mathsf{H} \to \mathsf{H}$$

is closed and thus bounded, by the closed graph theorem. We can use Lemma 5.26 to see that $(\xi, z) \mapsto \mathscr{A}_{\xi,z}$ is analytic (as a sum of a function analytic in ξ and a function analytic in z). For (ξ, z) close enough to (ξ_0, z_0), the operator

$$\mathrm{Id} + \begin{pmatrix} -\Pi_{v_{\xi_0}} + (\mathscr{L}_\xi - z)(\mathscr{L}_{\xi_0} - z_0)^{-1}\Pi^{\perp}_{v_{\xi_0}} & \cdot(\mathscr{L}_\xi - z)u_{\xi_0} \\ 0 & 0 \end{pmatrix}$$

is bijective, and so is $\mathscr{M}_{\xi,z}$. Thus, we can write

$$\mathscr{M}^{-1}_{\xi,z} = \begin{pmatrix} E_0(\xi, z) & E_+(\xi, z) \\ E_-(\xi, z) & E_{\pm}(\xi, z) \end{pmatrix},$$

and we have $z \in \mathrm{sp}(\mathscr{L}_\xi)$ if and only if $E_{\pm}(\xi, z) = 0$. The function $E_{\pm}$ is analytic with respect to (ξ, z) near (ξ_0, z_0). By using a Neumann series, we get

$$E_{\pm}(\xi_0, z) = (z - z_0)\langle u_{\xi_0}, v_{\xi_0}\rangle + \mathscr{O}(|z - z_0|^2).$$

In particular,

$$\partial_z E_{\pm}(\xi_0, z_0) = \langle u_{\xi_0}, v_{\xi_0}\rangle \neq 0.$$

With the analytic implicit functions theorem, we deduce the existence of μ.

Now, consider a contour Γ encircling $\mu(\mathcal{V})$ and define, for all $\xi \in \mathcal{V}$,

$$P_\xi = \frac{1}{2i\pi}\int_\Gamma (\zeta - \mathscr{L}_\xi)^{-1}\,\mathrm{d}\zeta.$$

Since the projection-valued application $\xi \mapsto P_\xi$ is continuous (in fact analytic) and P_{ξ_0} is of rank one, so is P_ξ by Lemma 5.25. In particular, $P_\xi u_{\xi_0}$ is a non-zero (analytic) eigenvector associated with $\mu(\xi)$. $\qquad\square$

5.6. Toeplitz operators on the circle

The following presentation is inspired by a course which has been given by G. Lebeau at the École Polytechnique. In this section, we consider $\mathsf{H} = \mathsf{L}^2(\mathbb{S}^1, \mathbb{C})$. If $u \in \mathsf{H}$, we denote by

$(u_n)_{n\in\mathbb{Z}}$ the family of the Fourier coefficients of u:

$$\forall n \in \mathbb{Z}, \quad u_n = \frac{1}{2\pi} \int_0^{2\pi} u(\theta)e^{-in\theta}\,d\theta.$$

We define $P : \mathsf{H} \to \mathsf{H}$ by, for all $u \in \mathsf{H}$, $(Pu)_n = u_n$ if $n \in \mathbb{N}$ and $(Pu)_n = 0$ if $n < 0$. The range of P is called the Hardy space and denoted by $\mathcal{H}^2$.

Definition 5.28. — Let $a \in \mathscr{C}^0(\mathbb{S}^1, \mathbb{C})$, and $M_a : \mathsf{H} \to \mathsf{H}$ be the operator corresponding to the multiplication by a. The operator $T(a) := PM_aP : \mathcal{H}^2 \to \mathcal{H}^2$ is called the *Toeplitz operator* of symbol a.

Lemma 5.29. — *Let $a \in \mathscr{C}^0(\mathbb{S}^1, \mathbb{C})$. We have $T(a) \in \mathcal{L}(\mathcal{H}^2)$ and $\|T\| \leqslant \|a\|_\infty$.*

Define $e_n \in \mathscr{C}^0(\mathbb{S}^1, \mathbb{C})$ by $e_n(\theta) := e^{in\theta}$.

Lemma 5.30. — *Let $n \in \mathbb{Z}$. Then, $[M_{e_n}, P]$ is a finite-rank operator (and thus it is compact).*

Proposition 5.31. — *Let $a \in \mathscr{C}^0(\mathbb{S}^1, \mathbb{C})$. Then, $[M_a, P]$ is a compact operator.*

Proof. — By the Fejér theorem, a can be approximated by trigonometric polynomials in the sup norm. □

Proposition 5.32. — *Let $a, b \in \mathscr{C}^0(\mathbb{S}^1, \mathbb{C})$. Then, there exists $K \in \mathcal{K}(\mathcal{H}^2)$ such that*

$$T(a)T(b) = T(ab) + K.$$

Proof. — We find $K = P[M_a, P]M_bP$. It suffices to use Proposition 5.31. □

Proposition 5.33. — *Let $a \in \mathscr{C}^0(\mathbb{S}^1, \mathbb{C})$. Assume that a does not vanish. Then, $T(a)$ is a Fredholm operator.*

Proof. — It is a consequence of Proposition 5.32 with $b = a^{-1}$. □

Lemma 5.34. — *Let $a \in \mathscr{C}^0(\mathbb{S}^1, \mathbb{C})$. Assume that a vanishes on a non-empty open set. Then, $T(a)$ is not a Fredholm operator.*

Proof. — Let us consider a closed bounded interval $[\gamma_1, \gamma_2] \subset [0, 2\pi]$ with $\gamma_1 < \gamma_2$ and on which a is zero. If $\alpha \in \mathbb{R}$ and if ρ_α is the translation by α defined by $\rho_\alpha u(\theta) = u(\theta - \alpha)$, we have $[\rho_\alpha, P] = 0$. We choose $\alpha = \gamma_2 - \gamma_1$. Then, there exists $n \in \mathbb{N}$ such that $(\rho_\alpha M_a)^n = 0$.

By using commutators (see Proposition 5.31), we see that $(\rho_\alpha T(a))^n$ is compact. If $T(a)$ were Fredholm so would be $(\rho_\alpha T(a))^n$ (see Proposition 5.7) and there would exist $S \in \mathcal{L}(\mathcal{H}^2)$ and $K \in \mathcal{K}(\mathcal{H}^2)$ (see Proposition 5.5) such that

$$S(\rho_\alpha T(a))^n = \mathrm{Id}_{\mathcal{H}^2} + K \,,$$

and thus $\mathrm{Id}_{\mathcal{H}^2}$ would be compact. This would be a contradiction. Therefore $T(a)$ is not Fredholm. $\qquad\square$

Proposition 5.35. — *Let $a \in \mathscr{C}^0(\mathbb{S}^1, \mathbb{C})$. Assume that there exists $\theta_0 \in \mathbb{S}^1$ such that $a(\theta_0) = 0$. Then, $T(a)$ is not a Fredholm operator.*

Proof. — For all $\varepsilon > 0$, there exists $\tilde{a} \in \mathscr{C}^0(\mathbb{S}^1, \mathbb{C})$ such that $\|a - \tilde{a}\|_\infty \leqslant \varepsilon$ and $\tilde{a}$ vanishes in a neighborhood of θ_0. If a were Fredholm, so would be $\tilde{a}$ by Lemma 5.2. With Lemma 5.34, this would be a contradiction. $\qquad\square$

Proposition 5.36. — *Let $a \in \mathscr{C}^1(\mathbb{S}^1, \mathbb{C})$. Assume that a does not vanish. We can write $a(\theta) = r(\theta)e^{i\alpha(\theta)}$, with $r > 0$, α of class $\mathscr{C}^1$. Then*

$$\mathrm{ind}\, T(a) = \mathrm{ind}\, T(e^{i\alpha}) = k := \frac{\alpha(2\pi) - \alpha(0)}{2\pi} = \frac{1}{2i\pi} \int_0^{2\pi} \frac{a'}{a} \, d\theta \,.$$

Proof. — Let us consider the following continuous family $(a_t)_{t \in [0,1]}$:

$$a_t(\theta) = ((1 - t)r(\theta) + t)e^{i\alpha(\theta)} \,.$$

For all $t \in [0, 1]$, the function a_t does not vanish. We see that $(T(a_t))_{t \in [0,1]}$ is a continuous family of Fredholm operators. The index being preserved by perturbation, we get the first equality. For the second one, we consider

$$f_t(\theta) = e^{(1-t)i\alpha(\theta)+ikt\theta} = e^{i\alpha(\theta)+it\int_0^\theta (k - \alpha'(u))\, du} \,.$$

It defines a continuous 2π-periodic function. We get

$$\operatorname{ind} T(e^{i\alpha}) = \operatorname{ind} T(e^{ik\cdot}) = k \,.$$

$\square$

5.7. Notes

i. The Fredholm theory is born with the study of linear integral equations (see [14] and [4, Chapter III]).

ii. This chapter is inspired by [49, Appendix D], see also [41] where the power of the Grushin formalism is explained with various examples. This formalism, inspired by elementary considerations of linear algebra, was originally used to study pseudo-differential operators ([17, 40]).

iii. About Atkinson's theorem, the reader can consult the original article [1].

iv. Section 5.5 can be seen as a very particular case of the perturbation theory described in Kato's book [25, Section VII.2]. The advantage of the Grushin formalism is to provide a more synthetic presentation. Note that Lemma 5.25 can be found in [25, Section I.8].

v. The Reader is invited to compare our presentation to the one in [3, Section VI.3].

CHAPTER 6

SPECTRUM OF SELF-ADJOINT OPERATORS

This chapter is devoted to the special case of self-adjoint operators. We explain that the discrete spectrum and the essential spectrum form a partition of the spectrum, and we give various criteria
(via Weyl sequences) to characterize these spectra. We also prove
the famous min-max theorem, which characterizes the low lying
eigenvalues of a self-adjoint operator bounded from below. This
chapter is illustrated by means of various canonical examples, such
as the Hamiltonian of the hydrogen atom.

6.1. Compact normal operators

Lemma 6.1. — *Let $T \in \mathcal{L}(\mathsf{H})$ be a normal operator.*

i. *If $V \subset \mathsf{H}$ is a subspace such that $T(V) \subset V$, then $T^*(V^\perp) \subset V^\perp$.*

ii. *We have $\ker(T) = \ker(T^*)$.*

Proof. — Assume that V is a subspace such that $T(V) \subset V$. For $u \in V^\perp$ and $v \in V$, we have

$$\langle T^*u, v \rangle = \langle u, Tv \rangle = 0 \,.$$

For the second point, note that, for all $x \in \mathsf{H}$,

$$\|Tx\|^2 = \langle T^*Tx, x \rangle = \langle TT^*x, x \rangle = \|T^*x\|^2 \,.$$

$\square$

© The Author(s), under exclusive license
to Springer Nature Switzerland AG 2021
C. Cheverry and N. Raymond, *A Guide to Spectral Theory*,
Birkhäuser Advanced Texts Basler Lehrbücher,
https://doi.org/10.1007/978-3-030-67462-5_6

Theorem 6.2. — *Assume that* H *is infinite dimensional. Let us consider* $T \in \mathcal{L}(H)$ *to be a compact normal operator. Then, its non-zero spectrum is discrete and* 0 *belongs to the spectrum. Let us consider the sequence of the distinct non-zero eigenvalues* $(\lambda_j)_{1 \leqslant j \leqslant k}$ *(with* $k \in \mathbb{N} \cup \{+\infty\}$*) and let* $\lambda_0 = 0$*. Then, we have the decompositions*

$$(6.6.1.1) \quad H = \bigoplus_{j=0}^{k} \ker(T - \lambda_j), \qquad T = \sum_{j=0}^{k} \lambda_j P_j = \sum_{j=1}^{k} \lambda_j P_j,$$

where P_j *is the orthogonal projection on* $\ker(T - \lambda_j)$.

Proof. — If $\lambda, \mu \in \mathsf{sp}(T) \setminus \{0\}$ with $\lambda \neq \mu$, then the corresponding eigenspaces are orthogonal. Indeed, if $u \in \ker(T - \lambda)$ and $v \in \ker(T - \mu)$, by Lemma 6.1, we have $v \in \ker(T^* - \bar{\mu})$ and

$$0 = \langle (T - \lambda)u, v \rangle = \langle u, (T^* - \overline{\lambda})v \rangle = (\mu - \lambda)\langle u, v \rangle.$$

We consider the Hilbertian sum

$$V = \bigoplus_{j=1}^{k} \ker(T - \lambda_j).$$

Note that $\ker T = \ker T^* \subset V^\perp$. Since T is normal, the subspace V is stable under the action of T^*, so that $V^\perp$ is stable under $(T^*)^* = T$. Thus, we can consider $T_{|V^\perp} \in \mathcal{L}(V^\perp)$. It is a compact normal operator on $V^\perp$. Its non-zero spectrum does not exist. Therefore $T_{|V^\perp} \in \mathcal{L}(V^\perp)$ is a normal operator with zero spectrum. Due to Proposition 3.27, we get $T_{|V^\perp} = 0$. Thus $V^\perp \subset \ker(T)$ and then $V^\perp = \ker T$.

When the number k is finite, the sense of (6.6.1.1) is clear. Assume that $k = +\infty$. We can always adjust the eigenvalues λ_j with $j \geqslant 1$ in such a way that the $|\lambda_j|$ form a decreasing sequence going to zero. Then, introduce

$$T_N := \sum_{j=0}^{N} \lambda_j P_j = \sum_{j=1}^{N} \lambda_j P_j, \qquad R_N := T - T_N.$$

By Proposition 3.27, we know that

$$\|T - T_N\| = r(R_N) = \sup_{j \geqslant N+1} |\lambda_j| \leqslant |\lambda_N| \longrightarrow 0.$$

□

Proposition 6.3. — *Assume that* H *is infinite dimensional. Let us consider a self-adjoint operator* T *with compact resolvent. Then, its spectrum is real, discrete and can be written as a sequence tending to* $+\infty$ *in absolute value.*

Proof. — By Proposition 2.64, the resolvent set contains $+i$ and $-i$. The spectrum is real, and we can use Corollary 5.12 to see that the spectrum of T is discrete. The operator $(T+i)^{-1}$ is compact and normal. By Theorem 6.2, we have (6.6.1.1) for $(T+i)^{-1}$, and thereby

$$Tu_j = \mu_j u_j, \qquad \mu_j = \lambda_j^{-1} - i \in \mathbb{R}, \qquad u_j = P_j u_j, \qquad 1 \leqslant j \leqslant k.$$

Since the λ_j go to zero, the μ_j tend $+\infty$. □

Exercise 6.4. — Let $\Omega \subset \mathbb{R}^d$ be a bounded open set.

  i. Prove that the spectrum of the Dirichlet (resp. Neumann) Laplacian on Ω is real, discrete and can be written as a sequence tending to $+\infty$.

  ii. Consider the case $d = 1$ and $\Omega = (0, 1)$. Show that the Dirichlet Laplacian is bijective. Exhibit a Hilbert basis of $\mathsf{L}^2(\Omega)$ made of functions in $\mathsf{H}^1_0(\Omega)$.

Exercise 6.5. — Prove the statement in Remark 4.27.

Proposition 6.6 (Singular values). — *Let* $T \in \mathcal{L}(\mathsf{H})$ *be a compact operator. Then,*

  (i) T^*T *and* $|T|$ *are compact and self-adjoint. Moreover,* $\ker T^*T = \ker |T| = \ker T$.

  (ii) *If* $(s_n)_{n \geqslant 1}$ *denotes the non-increasing sequence of the positive eigenvalues of* $|T|$, *associated with an orthonormalized family of eigenfunctions* $(\psi_n)_{n \geqslant 1}$, *then the series*

$$\sum_{n \geqslant 1} s_n \langle \cdot, \psi_n \rangle \psi_n$$

converges to $|T|$ *in* $\mathcal{L}(\mathsf{H})$.

  (iii) *The series*

$$\sum_{n \geqslant 1} s_n \langle \cdot, \psi_n \rangle \varphi_n, \qquad \text{with } \varphi_n = s_n^{-1} T \psi_n$$

converges to T in $\mathcal{L}(\mathsf{H})$. The family $(\varphi_n)_{n \geqslant 1}$ is orthonormalized.

(iv) *The series*

$$\sum_{n \geqslant 1} s_n^2 \langle \cdot, \psi_n \rangle \psi_n$$

converges to $T^*T = |T|^2$ in $\mathcal{L}(\mathsf{H})$.

Proof. — Only the third point requires a proof. It is enough to notice that $T^*T\psi_n = s_n^2 \psi_n$. □

6.2. About the harmonic oscillator

6.2.1. Domain considerations. — Consider the operator

$$\mathcal{H}_0 = (\mathscr{C}_0^\infty(\mathbb{R}), -\partial_x^2 + x^2) \, .$$

This operator is essentially self-adjoint as we have seen in Example 2.79. Let us denote by $\mathcal{H}$ its closure. We have

$$\mathrm{Dom}\,(\mathcal{H}) = \mathrm{Dom}\,(\mathcal{H}_0^*) = \{\psi \in \mathsf{L}^2(\mathbb{R}) : (-\partial_x^2 + x^2)\psi \in \mathsf{L}^2(\mathbb{R})\} \, .$$

By using the results of Section 2.6.3, we also see that $\mathcal{H}$ is the operator associated with the sesquilinear form defined by

$$\forall \varphi, \psi \in \mathsf{B}^1(\mathbb{R}) \, , \quad Q(\varphi, \psi) = \int_{\mathbb{R}} \left(\varphi' \overline{\psi'} + x^2 \varphi \overline{\psi} \right) \, \mathrm{d}x \, .$$

We can prove the following separation property.

Proposition 6.7. — *We have*

$$\mathrm{Dom}\,(\mathcal{H}) = \{\psi \in \mathsf{H}^2(\mathbb{R}) : x^2\psi \in \mathsf{L}^2(\mathbb{R})\} \, .$$

Proof. — The proof is another illustration of the difference quotient method. Let $\psi \in \mathrm{Dom}\,(\mathcal{H})$. It is sufficient to prove that $\psi'' \in \mathsf{L}^2(\mathbb{R})$. There exists $f \in \mathsf{L}^2(\mathbb{R})$ such that

$$\forall \varphi \in \mathscr{C}_0^\infty(\mathbb{R}), \qquad \langle \partial_x \psi, \partial_x \varphi \rangle + \langle x\psi, x\varphi \rangle = \langle f, \varphi \rangle \, ,$$

where the bracket is now the L^2-bracket. Since $\psi \in \mathsf{B}^1(\mathbb{R})$ and $\mathscr{C}_0^\infty(\mathbb{R})$ is dense in $\mathsf{B}^1(\mathbb{R})$, we can extend this equality and get

$$\forall \varphi \in \mathsf{B}^1(\mathbb{R}), \qquad \langle \partial_x \psi, \partial_x \varphi \rangle + \langle x\psi, x\varphi \rangle = \langle f, \varphi \rangle \, .$$

Let us define the difference quotient

$$D_h \varphi(x) = \frac{\varphi(x+h) - \varphi(x)}{h}, \quad x \in \mathbb{R}, \quad h \neq 0 \, .$$

If $\varphi \in \mathsf{B}^1(\mathbb{R})$, then $D_h\varphi \in \mathsf{B}^1(\mathbb{R})$. We get

$$\forall \varphi \in \mathsf{B}^1(\mathbb{R}), \qquad \langle \partial_x\psi, \partial_x D_h\varphi \rangle + \langle x\psi, x D_h\varphi \rangle = \langle f, D_h\varphi \rangle \,.$$

It follows that

$$\langle \partial_x\psi, \partial_x D_h\varphi \rangle = -\langle \partial_x D_{-h}\psi, \partial_x\varphi \rangle$$

and

$$\langle x\psi, x D_h\varphi \rangle = -\langle x D_{-h}\psi, x\varphi \rangle - \langle \psi(x-h), x\varphi \rangle - \langle x\psi, \varphi(x+h) \rangle \,.$$

We find, for all $\varphi \in \mathsf{B}^1(\mathbb{R})$ and $h \neq 0$,

$$\langle \partial_x D_{-h}\psi, \partial_x\varphi \rangle + \langle x D_{-h}\psi, x\varphi \rangle$$
$$= -\langle f, D_h\varphi \rangle - \langle \psi(x-h), x\varphi \rangle - \langle x\psi, \varphi(x+h) \rangle \,.$$

Applying this equality to $\varphi = D_{-h}\psi$, we get

$$\langle \partial_x D_{-h}\psi, \partial_x D_{-h}\psi \rangle + \langle x D_{-h}\psi, x D_{-h}\psi \rangle$$
$$= -\langle f, D_h D_{-h}\psi \rangle - \langle \psi(x-h), x D_{-h}\psi \rangle - \langle x\psi, D_{-h}\psi(x+h) \rangle.$$

Then we notice that

$$\begin{aligned}
|\langle f, D_h D_{-h}\psi \rangle| &\leqslant \|f\|_{\mathsf{L}^2(\mathbb{R})} \|D_h D_{-h}\psi\|_{\mathsf{L}^2(\mathbb{R})} \\
&\leqslant \|f\|_{\mathsf{L}^2(\mathbb{R})} \|\partial_x D_{-h}\psi\|_{\mathsf{L}^2(\mathbb{R})} \\
&\leqslant \frac{1}{2}\left(\|f\|_{\mathsf{L}^2(\mathbb{R})}^2 + \|\partial_x D_{-h}\psi\|_{\mathsf{L}^2(\mathbb{R})}^2 \right),
\end{aligned}$$

where we have used Proposition 2.94. We can deal with the other terms in the same way and thus get

$$\begin{aligned}
\|\partial_x D_{-h}\psi\|_{\mathsf{L}^2(\mathbb{R})}^2 &+ \|x D_{-h}\psi\|_{\mathsf{L}^2(\mathbb{R})}^2 \\
&\leqslant \frac{1}{2}\Big(\|f\|_{\mathsf{L}^2(\mathbb{R})}^2 + \|\psi\|_{\mathsf{L}^2(\mathbb{R})}^2 + \|x D_{-h}\psi\|_{\mathsf{L}^2(\mathbb{R})}^2 \\
&\qquad\qquad + \|\psi\|_{\mathsf{B}^1(\mathbb{R})}^2 + |h| \|\psi\|_{\mathsf{H}^1(\mathbb{R})}^2 \Big).
\end{aligned}$$

We deduce that

$$\begin{aligned}
\|D_{-h}\partial_x\psi\|_{\mathsf{L}^2(\mathbb{R})}^2 &+ \|x D_{-h}\psi\|_{\mathsf{L}^2(\mathbb{R})}^2 \\
&\leqslant \|f\|_{\mathsf{L}^2(\mathbb{R})}^2 + \|\psi\|_{\mathsf{L}^2(\mathbb{R})}^2 + \|\psi\|_{\mathsf{B}^1(\mathbb{R})}^2 + |h| \|\psi\|_{\mathsf{H}^1(\mathbb{R})}^2 \,.
\end{aligned}$$

We may again use Proposition 2.94 to conclude that $\partial_x\psi \in \mathsf{H}^1(\mathbb{R})$ and $x\psi \in \mathsf{H}^1(\mathbb{R})$. $\qquad\square$

6.2.2. Spectrum of the harmonic oscillator. — We have seen in Exercise 4.28 that $\mathcal{H}$ has compact resolvent. Actually, one could also directly use Propositions 4.24 and 6.7. Thus, the spectrum is real, discrete, and it is a non-decreasing sequence $(\lambda_n)_{n \geqslant 1}$ tending to $+\infty$ (we repeat the eigenvalue according to its multiplicity). We would like to compute these eigenvalues.

Let us consider the following differential operators (acting on $\mathscr{S}(\mathbb{R})$)

$$a = \frac{1}{\sqrt{2}}(\partial_x + x), \quad c = \frac{1}{\sqrt{2}}(-\partial_x + x).$$

We have

$$2ca = -\partial_x^2 + x^2 - 1, \qquad [a, c] = 1.$$

Lemma 6.8. — *For all $\varphi, \psi \in \mathscr{S}(\mathbb{R})$, we have*

$$\langle a\varphi, \psi \rangle_{L^2(\mathbb{R})} = \langle \varphi, c\psi \rangle_{L^2(\mathbb{R})}.$$

Lemma 6.9. — *For all $n \in \mathbb{N} \setminus \{0\}$,*

$$ac^n = nc^{n-1} + c^n a.$$

Proposition 6.10. — *For all $n \geqslant 1$, we have $\lambda_n = 2n - 1$. In particular, the eigenvalues are simple.*

Proof. — We let $g_0(x) = e^{-x^2/2}$. We check that $ag = 0$. In particular, we have $1 \in \mathsf{sp}(\mathcal{H})$.

For $n \in \mathbb{N}$, we let $g_n = c^n g_0$. By induction, we see that $g_n = H_n g_0$ where H_n is a polynomial of degree n. In particular, the functions g_n are in the domain of the harmonic oscillator.

Let us notice that

$$cac^n = nc^n + c^{n+1}a.$$

We get that

$$\mathcal{H}g_n = (2n + 1)g_n.$$

In particular, $\{2n + 1, n \in \mathbb{N}\} \subset \mathsf{sp}(\mathcal{H})$.

Let us check that $(g_n)_{n \in \mathbb{N}}$ is an orthogonal family. Let $n, m \in \mathbb{N}$ with $n < m$. Let us consider

$$\langle g_n, g_m \rangle_{L^2(\mathbb{R})} = \langle c^n g_0, c^m g_0 \rangle_{L^2(\mathbb{R})} = \langle a^m c^n g_0, g_0 \rangle_{L^2(\mathbb{R})} = 0,$$

where we used Lemmas 6.8 and 6.9, $ag_0 = 0$, and an induction procedure.

Let us check that the family is total. Take $f \in L^2(\mathbb{R})$ such that, for all $n \in \mathbb{N}$, $\langle f, g_n \rangle_{L^2(\mathbb{R})} = 0$. It follows that, for all $n \in \mathbb{N}$,

$$\int_{\mathbb{R}} x^n f(x) e^{-x^2/2} \, \mathrm{d}x = 0 \,.$$

For all $\xi \in \mathbb{R}$, we let

$$F(\xi) = \int_{\mathbb{R}} e^{-ix\xi} f(x) e^{-x^2/2} \, \mathrm{d}x \,.$$

The function F is well defined. Now, we notice that

$$F(\xi) = \int_{\mathbb{R}} \sum_{k=0}^{+\infty} f(x) \frac{(-ix\xi)^k}{k!} e^{-x^2/2} \, \mathrm{d}x \,.$$

By the Fubini–Tonelli theorem, we have

$$\int_{\mathbb{R}} \sum_{k=0}^{+\infty} |f(x)| \frac{(|x\xi|)^k}{k!} e^{-x^2/2} \, \mathrm{d}x$$

$$= \sum_{k=0}^{+\infty} |\xi|^k \int_{\mathbb{R}} |f(x)| \frac{|x|^k}{k!} e^{-x^2/2} \, \mathrm{d}x \,.$$

Then, we have

$$\sum_{k=0}^{+\infty} |\xi|^k \int_{\mathbb{R}} |f(x)| \frac{|x|^k}{k!} e^{-x^2/2} \, \mathrm{d}x$$

$$\leqslant \|f\|_{L^2(\mathbb{R})} \sum_{k=0}^{+\infty} \frac{|\xi|^k}{k!} \left(\int_{\mathbb{R}} |x|^{2k} e^{-x^2} \, \mathrm{d}x \right)^{\frac{1}{2}}$$

$$\leqslant \|f\|_{L^2(\mathbb{R})} \sum_{k=0}^{+\infty} \frac{|\xi|^k}{k!} \max_{x \in \mathbb{R}} \left(|x|^k e^{-x^2/4} \right) \left(\int_{\mathbb{R}} e^{-x^2/2} \, \mathrm{d}x \right)^{\frac{1}{2}}$$

$$\leqslant C \sum_{k=0}^{+\infty} \frac{|\xi|^k}{k!} \max_{x \in \mathbb{R}} \left(|x|^k e^{-x^2/4} \right)$$

$$= C \sum_{k=0}^{+\infty} \frac{1}{k!} \left(\frac{2k}{e} \right)^{\frac{k}{2}} |\xi|^k \,.$$

Since this last power series is convergent (with infinite convergence radius), we can apply the Fubini theorem and we get

$$F(\xi) = \sum_{k=0}^{+\infty} \xi^k \int_{\mathbb{R}} f(x) \frac{(-ix)^k}{k!} e^{-x^2/2} \, dx = 0 \, .$$

Therefore, the Fourier transform of $f e^{-x^2/2}$ is 0 and $f = 0$.

If we denote by $(f_n)_{n \in \mathbb{N}}$ the L^2-normalization of the family $(g_n)_{n \in \mathbb{N}}$, $(f_n)_{n \in \mathbb{N}}$ is a Hilbert basis of $L^2(\mathbb{R})$ such that $\mathcal{H} f_n = (2n+1) f_n$.

Since the spectrum of $\mathcal{H}$ is discrete, we only have to care about the eigenvalues. Let us solve $\mathcal{H}\psi = \lambda\psi$ with $\lambda \in \mathbb{R}$ and $\psi \in$ Dom $(\mathcal{H})$. We write the following decomposition, converging in $L^2(\mathbb{R})$,

$$\psi = \sum_{n \in \mathbb{N}} \langle \psi, f_n \rangle_{L^2(\mathbb{R})} f_n \, .$$

For all $\varphi \in \mathscr{S}(\mathbb{R})$, we have

$$\langle \psi, (\mathcal{H} - \lambda)\varphi \rangle_{L^2(\mathbb{R})} = 0 \, .$$

Thus, by convergence in $L^2(\mathbb{R})$, for all $\varphi \in \mathscr{S}(\mathbb{R})$,

$$\sum_{n \in \mathbb{N}} \langle \psi, f_n \rangle_{L^2(\mathbb{R})} \langle f_n, (\mathcal{H} - \lambda)\varphi \rangle_{L^2(\mathbb{R})} = 0 \, .$$

We choose $\varphi = f_k$ to see that

$$\sum_{n \in \mathbb{N}} \langle \psi, f_n \rangle_{L^2(\mathbb{R})} \langle f_n, ((2k+1) - \lambda) f_k \rangle_{L^2(\mathbb{R})}$$
$$= \langle \psi, f_k \rangle_{L^2(\mathbb{R})} ((2k+1) - \lambda) = 0 \, .$$

If, for all $k \in \mathbb{N}$, $\langle \psi, f_k \rangle_{L^2(\mathbb{R})} = 0$, then $\psi = 0$. Therefore, there exists $k \in \mathbb{N}$ such that $(2k+1) - \lambda = 0$.

We have proved that

$$\mathsf{sp}(\mathcal{H}) = \{2n - 1 , n \in \mathbb{N} \setminus \{0\}\} \, .$$

Let us now prove the statement about the multiplicity. Consider a solution $\psi \in$ Dom $(\mathcal{H})$ of $\mathcal{H}\psi = (2n+1)\psi$. For all $k \in \mathbb{N}$, we get

$$\langle \psi, f_k \rangle_{L^2(\mathbb{R})} ((2k+1) - (2n+1)) = 0 \, .$$

Thus, for $k \neq n$, $\langle \psi, f_k \rangle_{L^2(\mathbb{R})} = 0$. Thus, ψ is proportional to f_n. $\square$

6.3. Characterization of the spectra

6.3.1. Properties. —

Lemma 6.11. *— If T is self-adjoint, we have the equivalence: $\lambda \in$ $\mathsf{sp}(T)$ if and only if there exists a sequence $(u_n) \in \mathrm{Dom}\,(T)$ such that $\|u_n\|_{\mathsf{H}} = 1$, and $(T - \lambda)u_n \underset{n \longrightarrow +\infty}{\longrightarrow} 0$ in H.*

Proof. —

$\Longleftarrow$ By Lemma 3.13, if there exists a sequence $(u_n) \in \mathrm{Dom}\,(T)$ which is such that $\|u_n\|_{\mathsf{H}} = 1$ and $(T - \lambda)u_n \underset{n \rightarrow +\infty}{\rightarrow} 0$, then $\lambda \in \mathsf{sp}(T)$.

$\Longrightarrow$ If $\lambda = \lambda_1 + i\lambda_2 \in \mathbb{C} \setminus \mathbb{R}$, that is $\lambda_2 \neq 0$, by Proposition 2.64 applied to the self-adjoint operator $\lambda_2^{-1}(T - \lambda_1)$, we can infer that $T - \lambda$ is invertible (with bounded inverse), so that $\lambda \notin \mathsf{sp}(T)$, and the criterion cannot be verified due to the contradiction between

$$\|u_n\|_{\mathsf{H}} = 1, \qquad R_T(\lambda)(T - \lambda)u_n = u_n \underset{n \longrightarrow +\infty}{\longrightarrow} 0.$$

We can restrict the discussion to the case $\lambda \in \mathbb{R}$. If there is no sequence $(u_n) \subset \mathrm{Dom}\,(T)$ such that $\|u_n\|_{\mathsf{H}} = 1$ and $(T - \lambda)u_n \underset{n \rightarrow +\infty}{\longrightarrow} 0$, then we can find $c > 0$ such that

$$\|(T - \lambda)u\| \geqslant c\|u\|, \qquad \forall u \in \mathrm{Dom}\,(T).$$

Therefore $T - \lambda$ is injective with closed range. But, since $T - \lambda = (T - \lambda)^*$, we have

$$\mathrm{ran}\,(T - \lambda) = \overline{\mathrm{ran}\,(T - \lambda)} = \left(\ker(T - \lambda)^*\right)^{\perp}$$

$$= \left(\ker(T - \lambda)\right)^{\perp} = \{0\}^{\perp} = \mathsf{H}.$$

This means that $T - \lambda$ is surjective, and therefore $\lambda \in \rho(T)$. $\square$

Lemma 6.12 **(Weyl criterion).** *— If T is self-adjoint, then $\lambda \in$ $\mathsf{sp}_{\mathsf{ess}}(T)$ if and only if there exists a sequence $(u_n) \subset \mathrm{Dom}\,(T)$ such that*

(i) *$\|u_n\|_{\mathsf{H}} = 1$;*

(ii) *(u_n) has no subsequence converging in H;*

(iii) *$(T - \lambda)u_n \underset{n \longrightarrow +\infty}{\rightarrow} 0$ in H.*

This means that the condition (ii) allows to distinguish the essential spectrum.

Proof. — First, remark that if (u_n) with $\|u_n\|_H = 1$ has a subsequence converging in H to u, then we must have $\|u\|_H = 1$. Indeed, we have (after extraction of a subsequence)

$$0 = \lim_{n \to +\infty} \|u_n - u\|^2 = \lim_{n \to +\infty} (1 - \langle u_n, u \rangle - \langle u, u_n \rangle + \|u\|^2)$$
$$= 1 - \|u\|^2 .$$

If $\lambda \notin \mathrm{sp}(T)$, then $T - \lambda$ is bijective and $u_n \to u = 0$, in contradiction with $\|u\|_H = 1$.

If $\lambda \in \mathrm{sp}(T) \setminus \mathrm{sp}_{\mathrm{ess}}(T)$, the operator $T - \lambda$ is Fredholm with index 0. Let $(u_n) \subset \mathrm{Dom}\,(T)$ such that $\|u_n\|_H = 1$ and

$$\lim_{n \to +\infty} (T - \lambda)u_n = 0 .$$

The operator $T - \lambda : \ker(T - \lambda)^\perp \to \mathrm{ran}\,(T - \lambda)$ is injective with closed range. Therefore, there exists $c > 0$ such that, for all $w \in \ker(T - \lambda)^\perp$, $\|(T - \lambda)w\| \geqslant c\|w\|$. We write

$$u_n = v_n + w_n , \qquad v_n \in \ker(T - \lambda) , \qquad w_n \in \ker(T - \lambda)^\perp ,$$

and we have

$$\|(T - \lambda)u_n\|^2 = \|(T - \lambda)v_n\|^2 + \|(T - \lambda)w_n\|^2 .$$

We deduce that $w_n \to 0$. Moreover, (v_n) is bounded in a finite-dimensional space, thus there exists a converging subsequence of (u_n).

Conversely, let us now assume that any sequence $(u_n) \subset \mathrm{Dom}\,(T)$ such that $\|u_n\|_H = 1$ and $\lim_{n \to +\infty}(T - \lambda)u_n = 0$ has a converging subsequence, going in H to some u with $\|u\|_H = 1$. Then $\lambda \in \mathrm{sp}(T)$. Moreover, the kernel $\ker(T - \lambda)$ is finite dimensional. Indeed, if it were of infinite dimension, one could construct an infinite orthonormal family (u_n) in $\ker(T - \lambda)$, and in particular we would get $u_n \rightharpoonup u = 0$ (weak convergence), which is a contradiction.

Let us now check that

$$\exists c > 0 , \quad \forall u \in \ker(T - \lambda)^\perp , \quad \|(T - \lambda)u\| \geqslant c\|u\| .$$

If not, there exists a normalized sequence (u_n) in $\ker(T - \lambda)^\perp$ such that $\|(T - \lambda)u_n\| \to 0$. By assumption, we may assume that

(u_n) converges to some u that necessarily belongs to the space $\ker(T - \lambda)^{\perp}$. But since $T - \lambda$ is closed (because it is self-adjoint), we have $(T - \lambda)u = 0$ so that $u = 0$, and this is a contradiction. Thus, the range of $T - \lambda$ is closed and

(6.6.3.2)

$$\operatorname{codim\,ran}(T - \lambda) = \operatorname{codim\,\overline{ran}}(T - \lambda) = \dim \ker(T^* - \lambda)$$
$$= \dim \ker(T - \lambda)$$

is finite. Thus (see Proposition 3.43) $T - \lambda$ is Fredholm with index 0 and $\lambda \notin \mathsf{sp}_{\mathsf{ess}}(T)$. $\qquad\square$

Lemma 6.13. — *Assume that T is self-adjoint. Then $\lambda \in \mathsf{sp}_{\mathsf{ess}}(T)$ if and only if there exists a sequence $(u_n) \subset \operatorname{Dom}(T)$ such that*

(i) $\|u_n\|_{\mathsf{H}} = 1$;

(ii) *(u_n) converges weakly to 0;*

(iii) $(T - \lambda)u_n \underset{n \longrightarrow +\infty}{\to} 0$ *in* H.

A sequence (u_n) satisfying (i) and converging weakly to 0 has no subsequence converging in H. Otherwise, the limit (of norm 1) would be 0. The criterion (ii) of Lemma 6.13 is stronger than the condition (i) of Lemma 6.12. Thus, Lemma 6.13 is a slight improvement of Lemma 6.12.

Proof. — Let $\lambda \in \mathsf{sp}_{\mathsf{ess}}(T)$. In the case $\dim \ker(T - \lambda \,\mathsf{Id}) = +\infty$, we can select some orthonormal basis (v_n) of $\ker(T - \lambda \,\mathsf{Id})$. The sequence (v_n) is weakly converging to 0 and, as expected, it is such that $(T - \lambda)v_n = 0$.

Now, we consider the case when $\dim \ker(T - \lambda) < +\infty$. By Lemma 6.12, there exists a sequence $(u_n) \subset \operatorname{Dom}(T)$ such that $\|u_n\|_{\mathsf{H}} = 1$ with no converging subsequence such that we have $\lim_{n \to +\infty}(T - \lambda)u_n = 0$ in H. We can write

$$u_n = \tilde{u}_n + k_n, \quad \text{with} \quad \tilde{u}_n \in \ker(T - \lambda)^{\perp}, \quad k_n \in \ker(T - \lambda).$$

We still have

$$(T - \lambda)\tilde{u}_n = (T - \lambda)u_n \underset{n \to +\infty}{\to} 0,$$

and we may assume (up to a subsequence extraction) that (k_n) converges to k. Since (u_n) has no converging subsequence, $(\tilde{u}_n)$

does not converge, and so it does not go to 0. Therefore, up to another extraction, we may assume that

$$\exists \varepsilon_0 > 0, \quad \forall n \in \mathbb{N}, \quad \|\tilde{u}_n\| \geqslant \varepsilon_0.$$

Define $\hat{u}_n = \|\tilde{u}_n\|^{-1}\tilde{u}_n$ so that

$$\|\hat{u}_n\| = 1, \qquad \|(T - \lambda)\hat{u}_n\| \leqslant \varepsilon_0^{-1}\|(T - \lambda)\tilde{u}_n\| \underset{n \to +\infty}{\to} 0.$$

Up to another extraction, we may assume that $(\hat{u}_n)$ converges weakly to some $\hat{u} \in \mathsf{H}$. Then, for all $v \in \mathrm{Dom}T = \mathrm{Dom}T^*$,

$$\langle (T - \lambda)\hat{u}_n, v \rangle = \langle \hat{u}_n, (T - \lambda)v \rangle \underset{n \to +\infty}{\to} 0 = \langle \hat{u}, (T - \lambda)v \rangle.$$

This implies that

$$\hat{u} \in \mathrm{Dom}\,(T^* - \lambda) = \mathrm{Dom}\,(T - \lambda), \qquad \hat{u} \in \ker(T - \lambda).$$

By construction, we have $\hat{u}_n \in \ker(T - \lambda)^\perp$, and thereby

$$\forall n \in \mathbb{N}, \quad \forall v \in \ker(T - \lambda), \quad \langle \hat{u}_n, v \rangle = 0.$$

Passing to the limit, we get

$$\forall v \in \ker(T - \lambda), \quad \langle \hat{u}, v \rangle = 0.$$

Thus, $\hat{u} \in \ker(T-\lambda) \cap \ker(T-\lambda)^\perp = \{0\}$, and thereby $\hat{u} = 0$. We have found a sequence with the required property. For the converse, it is just an application of Lemma 6.12. $\qquad\square$

Definition 6.14. — We call *Fredholm spectrum* $\mathsf{sp}_{\mathrm{fred}}(T)$ of T the complement of the essential spectrum of T in the spectrum of T.

In other words, we have the partition

$$\mathsf{sp}(T) = \mathsf{sp}_{\mathrm{fred}}(T) \cup \mathsf{sp}_{\mathrm{ess}}(T), \qquad \mathsf{sp}_{\mathrm{fred}}(T) \cap \mathsf{sp}_{\mathrm{ess}}(T) = \emptyset.$$

Lemma 6.15. — *Let T be self-adjoint. Any element of the Fredholm spectrum is isolated in $\mathsf{sp}(T)$.*

Proof. — Fix $\lambda \in \mathsf{sp}(T) \setminus \mathsf{sp}_{\mathrm{ess}}(T)$. By Lemma 6.11, there exists a Weyl sequence (u_n) of unit vectors such that $(T - \lambda)u_n \to 0$. By Lemma 6.12, we may assume that (u_n) converges to some u (of norm 1) and we get $(T - \lambda)u = 0$. The eigenvalue λ has finite multiplicity because $T - \lambda$ is Fredholm. Let us prove that it is isolated. If this were not the case, then one could consider a non-constant sequence $(\lambda_n)_n \in \mathrm{sp}(T)^{\mathbb{N}}$ tending to λ. By Lemma 6.11, for all $n \in \mathbb{N}$, we can find u_n satisfying $\|u_n\| = 1$ and

$$\|(T - \lambda_n)u_n\| \leqslant \frac{|\lambda - \lambda_n|}{n},$$

and therefore

$$\|(T - \lambda)u_n\| \leqslant \|(T - \lambda_n)u_n\| + |\lambda_n - \lambda|$$
$$\leqslant \frac{|\lambda - \lambda_n|}{n} + |\lambda - \lambda_n| \longrightarrow 0.$$

By Lemma 6.12, we may assume that (u_n) converges to some $u \in \mathrm{Dom}\,(T)$ with $\|u\| = 1$. It follows that $(T - \lambda)u = 0$, and so

$$\langle (T - \lambda_n)u, u_n \rangle = \langle u, (T - \lambda_n)u_n \rangle = (\lambda - \lambda_n)\langle u, u_n \rangle.$$

By the Cauchy–Schwarz inequality, we have

$$|\langle u, u_n \rangle| \leqslant \frac{\|(T - \lambda_n)u_n\|}{|\lambda - \lambda_n|}\|u\| \leqslant \frac{1}{n}$$

which implies that $\langle u_n, u \rangle \to 0$, and we get $u = 0$, which is a contradiction. $\qquad\square$

Lemma 6.16. — *Let T be self-adjoint. Then, we have the following properties:*

   i. *If $\lambda \in \mathrm{sp}(T)$ is isolated, then it is an eigenvalue.*

   ii. *All isolated eigenvalues of finite algebraic multiplicity belong to the Fredholm spectrum.*

As a consequence, all isolated eigenvalues contained in $\mathrm{sp}_{\mathrm{ess}}(T)$ have infinite multiplicity.

Proof. — Let us prove (i). To this end, consider an isolated point $\lambda \in \mathrm{sp}(T)$. By definition, this means that there exists $\varepsilon_0 > 0$ such that, for all $\mu \neq \lambda$ such that $|\mu - \lambda| \leqslant \varepsilon_0$, we have $\mu \notin \mathrm{sp}(T)$. For all $\varepsilon \in (0, \varepsilon_0)$, we introduce

$$P_\varepsilon = \frac{1}{2i\pi}\int_{\Gamma_\varepsilon} (\zeta - T)^{-1}\,\mathrm{d}\zeta = P,$$

where Γ_ε is the circle of radius ε centered at λ. Since T is closed (and using Riemannian sums), the operator P is valued in $\mathrm{Dom}\,(T)$ and

$$(T - \lambda)P = \frac{1}{2i\pi} \int_{\Gamma_\varepsilon} (T - \lambda)(\zeta - T)^{-1}\,\mathrm{d}\zeta$$

$$= \frac{1}{2i\pi} \int_{\Gamma_\varepsilon} (\zeta - \lambda)(\zeta - T)^{-1}\,\mathrm{d}\zeta\,.$$

Now, we use the resolvent bound to get (as soon as ε_0 is chosen small enough) that

$$\|(T - \zeta)^{-1}\| \leqslant \frac{1}{\mathrm{dist}\,(\zeta, \mathrm{sp}T)} \leqslant \frac{1}{|\lambda - \zeta|}\,.$$

Thus, we infer that $\|(T - \lambda)P\| \leqslant \varepsilon$ for all $\varepsilon \in (0, \varepsilon_0)$. Therefore, $\mathrm{ran}\,P \subset \ker(T - \lambda)$. Note also that if $\psi \in \ker(T - \lambda)$, we have

$$P\psi = \frac{1}{2i\pi} \int_{\Gamma_\varepsilon} (\zeta - T)^{-1}\psi\,\mathrm{d}\zeta = \frac{1}{2i\pi} \int_{\Gamma_\varepsilon} (\zeta - \lambda)^{-1}\psi\,\mathrm{d}\zeta = \psi\,.$$

Thus, $\mathrm{ran}\,P = \ker(T - \lambda)$.

It remains to apply Lemma 3.34 to see that the range of $P = P^*$ is not $\{0\}$.

Let us now consider (ii). Consider an isolated element $\lambda \in \mathrm{sp}(T)$. It is still isolated in $\mathrm{sp}(T_{|\ker(T-\lambda)^\perp})$ while, by using (i), it cannot be an eigenvalue of the restriction $T_{|\ker(T-\lambda)^\perp}$. This implies that $\lambda \in \rho(T_{|\ker(T-\lambda)^\perp})$. Thus, there exists $c > 0$ such that

$$\forall u \in \ker(T - \lambda)^\perp, \qquad \|(T - \lambda)u\| \geqslant c\|u\|\,.$$

We deduce that the range of $T - \lambda$ is closed. Since T is self-adjoint and $\dim\ker(T - \lambda) = \dim\mathrm{ran}\,P < +\infty$, as in (6.6.3.2), we get

$$\dim\,\ker(T - \lambda) < +\infty\,, \qquad \mathrm{codim}\,\mathrm{ran}\,(T - \lambda) < +\infty\,,$$

so that $T - \lambda$ is Fredholm. $\qquad\qquad\square$

Proposition 6.17. — *Let T be self-adjoint. Then, the discrete spectrum coincides with the Fredholm spectrum. We have $\mathrm{sp}_{\mathrm{fred}}(T) = \mathrm{sp}_{\mathrm{dis}}(T)$.*

Proof. — Fix λ in the Fredholm spectrum. Then, by Lemma 6.15, it is isolated. We have seen in the proof of Lemma 6.16 that $\mathrm{ran}\,P \subset \ker(T - \lambda)$. Since the dimension of $\ker(T - \lambda)$ is finite,

λ is of finite algebraic multiplicity. We have also seen in the proof of Lemma 6.12 that the range of $T-\lambda$ is closed. Thus $\lambda \in \mathsf{sp}_{\mathsf{dis}}(T)$. At this stage, we can assert that $\mathsf{sp}_{\mathsf{fred}}(T) \subset \mathsf{sp}_{\mathsf{dis}}(T)$. The converse is guaranteed by ii of Lemma 6.16. $\qquad\square$

Remark 6.18. — In particular, with Lemma 6.16, the discrete spectrum coincides with the set of the isolated eigenvalues of finite multiplicity.

Finally, let us prove another useful property.

Lemma 6.19. — *Let T be self-adjoint. Consider $\lambda \in \mathsf{sp}_{\mathsf{ess}}(T)$. Then, for all $N \in \mathbb{N}^*$ and $\varepsilon > 0$, there exists an orthonormal family $(u_n^\varepsilon)_{1 \leqslant n \leqslant N}$ such that, for all $n \in \{1, \ldots, N\}$,*

$$\|(T - \lambda)u_n^\varepsilon\| \leqslant \varepsilon.$$

Proof. — If λ is isolated, then it is an eigenvalue of infinite multiplicity (see Remark 6.18) and the conclusion follows. Let $\varepsilon \in (0, 1)$. If λ is not isolated, we may consider a sequence of distinct numbers of the spectrum $(\lambda_n)_{n \in \mathbb{N}}$ tending to λ and such that, for all $j, k \in \mathbb{N}$, we have $|\lambda_j - \lambda_k| \leqslant \frac{\varepsilon}{2}$. If $N = 1$, by the Weyl criterion, we get the existence of u_1^ε such that $\|(T - \lambda_1)u_1^\varepsilon\| \leqslant \frac{\varepsilon}{2}$. The conclusion follows for $N = 1$ since $|\lambda - \lambda_1| \leqslant \frac{\varepsilon}{2}$.

Let us now treat the case when $N = 2$. By the Weyl criterion, we can find u_1^ε and $\tilde{u}_2^\varepsilon$ of norm 1 such that

$$\|(T - \lambda_1)u_1^\varepsilon\| \leqslant \frac{\varepsilon}{2}|\lambda_1 - \lambda_2|, \qquad \|(T - \lambda_2)\tilde{u}_2^\varepsilon\| \leqslant \frac{\varepsilon}{2}|\lambda_1 - \lambda_2|.$$

Since T is self-adjoint, we have

$$\langle(\lambda_1 - \lambda_2)u_1^\varepsilon, \tilde{u}_2^\varepsilon\rangle = \langle u_1^\varepsilon, (T - \lambda_2)\tilde{u}_2^\varepsilon\rangle + \langle(\lambda_1 - T)u_1^\varepsilon, \tilde{u}_2^\varepsilon\rangle$$

which implies that $|\langle u_1^\varepsilon, \tilde{u}_2^\varepsilon\rangle| \leqslant \varepsilon$. Setting

$$u_2^\varepsilon = \tilde{u}_2^\varepsilon - \langle\tilde{u}_2^\varepsilon, u_1^\varepsilon\rangle u_1^\varepsilon, \qquad u_1^\varepsilon \perp u_2^\varepsilon, \qquad \sqrt{1 - \varepsilon^2} \leqslant \|u_2^\varepsilon\|,$$

we have

$$\|(T - \lambda_2)u_2^\varepsilon\| \leqslant \frac{\varepsilon}{2}|\lambda_1 - \lambda_2| + \varepsilon\left(|\lambda_1 - \lambda_2| + \frac{\varepsilon}{2}|\lambda_1 - \lambda_2|\right).$$

Up to changing ε, the conclusion follows for $N = 2$.

We leave the case $N \geqslant 3$ to the reader. $\qquad\square$

6.3.2. Determining the essential spectrum: an example. — As in Exercises 4.16 and 5.8, we consider a function $V \in \mathscr{C}^\infty(\mathbb{R}^d, \mathbb{R})$ such that ∇V is bounded and $\lim_{|x| \to +\infty} V(x) = 0$. We are interested in the essential spectrum of the operator $\mathscr{L} + V$ with domain $\mathsf{H}^2(\mathbb{R})$. This operator is self-adjoint. Its spectrum is real. With Exercise 5.8, we have $\mathsf{sp}_{\mathsf{ess}}(\mathscr{L} + V) \subset [0, +\infty)$.

Let us prove that $\mathsf{sp}_{\mathsf{ess}}(\mathscr{L}+V) = [0, +\infty)$. Let us start by showing that $0 \in \mathsf{sp}_{\mathsf{ess}}(\mathscr{L}+V)$. For that purpose, we use Lemma 6.13. Let us consider $\chi \in \mathscr{C}_0^\infty(\mathbb{R}^d)$ such that $\|\chi\|_{\mathsf{L}^2(\mathbb{R}^d)} = 1$. For $n \in \mathbb{N}$, we consider $\chi_n(x) = n^{-\frac{d}{2}} \chi(n^{-1}x - ne_1)$. The sequence (χ_n) is L^2-normalized and converges to 0 weakly. For n large enough, we have

$$\|(\mathscr{L} + V)\chi_n\| \leqslant \|\mathscr{L}\chi_n\| + \|V\chi_n\| = \mathcal{O}(n^{-2}) + o(1) \,.$$

Let us now consider $k \in \mathbb{R}$ and the sequence $\chi_{n,k} = e^{ik\cdot}\chi_n$. We have

$$\|(\mathscr{L} + V - k^2)\chi_{n,k}\| = \|e^{ik\cdot}(\mathscr{L} + V - k^2)\chi_n + [\mathscr{L}, e^{ik\cdot}]\chi_n\| \,.$$

But,

$$e^{-ik\cdot}[\mathscr{L}, e^{ik\cdot}] = k^2 - 2ik\nabla \,,$$

and we deduce that $k^2 \in \mathsf{sp}_{\mathsf{ess}}(\mathscr{L} + V)$, for all $k \in \mathbb{R}$.

6.4. Min-max principle

6.4.1. Statement and proof. — Our aim is to give a standard method to estimate the discrete spectrum and the bottom of the essential spectrum of a self-adjoint operator T on a Hilbert space H. We recall first the definition of the Rayleigh quotients of a self-adjoint operator T.

Definition 6.20. — Let $(\mathrm{Dom}\,(T), T)$ be a self-adjoint operator on H, which is assumed to be semi-bounded from below. The Rayleigh quotients which are associated with T are defined for all positive natural numbers $n \in \mathbb{N}^*$ by

$$\mu_n(T) = \sup_{\psi_1, \ldots, \psi_{n-1}} \quad \inf_{\substack{u \in \mathrm{span}\,(\psi_1, \ldots, \psi_{n-1})^\perp \\ u \in \mathrm{Dom}\,(T), u \neq 0}} \frac{\langle Tu, u \rangle_{\mathsf{H}}}{\langle u, u \rangle_{\mathsf{H}}} \,.$$

Remark 6.21. — Note that T is associated with a quadratic form Q defined by

$$\forall u \in \mathrm{Dom}\,(T)\,, \quad Q(u) = \langle Tu, u \rangle\,.$$

Since Q is bounded from below on $\mathrm{Dom}\,(T)$, we may « close » this quadratic form in the sense that, for some $M > 0$, there exists a vector space, denoted by $\mathrm{Dom}\,(Q)$, containing $\mathrm{Dom}\,(T)$ (as a dense subspace) such that $(\mathrm{Dom}\,(Q), Q + M\| \cdot \|^2)$ is a Hilbert space. From the Lax–Milgram repesentation theorem, the form $Q + M\| \cdot \|^2$ is associated with a self-adjoint operator $\tilde{T}$. We have $T + M\mathrm{Id} \subset \tilde{T}$ and thus, by self-adjointness, $T + M\mathrm{Id} = \tilde{T}$.

With this in mind, we can replace $u \in \mathrm{Dom}\,(T)$ by $u \in \mathrm{Dom}\,(Q)$ and $\langle Tu, u \rangle$ by $Q(u)$ in the definition of $\mu_n(T)$.

Lemma 6.22. — *If T is self-adjoint with non-negative spectrum, then $\mu_1(T) \geqslant 0$.*

Proof. — Let us assume that $\mu_1(T) < 0$. We may define the sesquilinear form

$$Q(u, v) = \langle (T - \mu_1(T))^{-1} u, v \rangle.$$

Then, Q is non-negative. Thus, the Cauchy–Schwarz inequality provides, for $u, v \in \mathsf{H}$,

$$|\langle (T - \mu_1(T))^{-1} u, v \rangle|$$
$$\leqslant \langle (T - \mu_1(T))^{-1} u, u \rangle^{\frac{1}{2}} \langle (T - \mu_1(T))^{-1} v, v \rangle^{\frac{1}{2}}.$$

For $v = (T - \mu_1(T))^{-1} u$, this becomes

$$\|v\|^2 \leqslant \langle v, Tv - \mu_1(T)v \rangle^{\frac{1}{2}} \langle (T - \mu_1(T))^{-1} v, v \rangle^{\frac{1}{2}}\,.$$

By the definition of $\mu_1(T)$, there is a sequence (v_n) which is such that

$$\|v_n\| = 1, \qquad \langle Tv_n, v_n \rangle \to \mu_1(T).$$

We must have

$$1 \leqslant \left(\langle v_n, Tv_n \rangle - \mu_1(T) \right)^{\frac{1}{2}} \|(T - \mu_1(T))^{-1}\|^{\frac{1}{2}},$$

which furnishes a contradiction when $n \to +\infty$. $\qquad\square$

The following statement gives the relation between Rayleigh quotients and eigenvalues.

Theorem 6.23. — *Let T be a self-adjoint operator with domain* $\mathrm{Dom}\,(T)$. *We assume that T is semi-bounded from below. Then the Rayleigh quotients μ_n of T form a non-decreasing sequence and one of the following holds:*

  i. $\mu_n(T)$ *is the n-th eigenvalue counted with mutliplicity of T, and the operator T has only discrete spectrum in* $(-\infty, \mu_n(T)]$.

  ii. $\mu_n(T)$ *is the bottom of the essential spectrum and, for all* $j \geqslant n$, $\mu_j(T) = \mu_n(T)$.

Proof. — By definition, the sequence (μ_n) is non-decreasing. Then, we notice that

$$(6.6.4.3) \qquad a < \mu_n \implies (-\infty, a) \cap \mathsf{sp}_{\mathsf{ess}}(T) = \emptyset.$$

Indeed, if $\lambda \in (-\infty, a)$ were in the essential spectrum, by Lemma 6.19, for all $N \geqslant 1$ and $\varepsilon > 0$, we could find an orthonormal family $(u_j)_{j \in \{1,\ldots,N\}}$ such that $\|(T - \lambda)u_j\| \leqslant \frac{\varepsilon}{\sqrt{N}}$. Then, given $n \geqslant 1$ and taking $N \geqslant n$, for all $(\psi_1, \ldots, \psi_{n-1}) \in \mathsf{H}$, there exists a non-zero u in the intersection $\mathsf{span}\,(u_1, \ldots, u_N) \cap \mathsf{span}\,(\psi_1, \ldots, \psi_{n-1})^\perp$. We write $u = \sum_{j=1}^N \alpha_j u_j$. Then

$$\frac{\langle Tu, u \rangle_{\mathsf{H}}}{\langle u, u \rangle_{\mathsf{H}}} \leqslant \lambda + \frac{\|(T - \lambda)u\|}{\|u\|} \leqslant \lambda + \left(\sum_{j=1}^N \|(T - \lambda)u_j\|^2 \right)^{\frac{1}{2}}$$
$$\leqslant \lambda + \varepsilon.$$

It follows that $\mu_n \leqslant \lambda + \varepsilon$. Moreover, for ε small enough, we get $\mu_n \leqslant a$, which is a contradiction. If γ is the infimum of the essential spectrum (suppose that it is not empty), we have $\mu_n \leqslant \gamma$. Note also that if $\mu_n = +\infty$ for some n, then the essential spectrum is empty. This implies the second assertion.

It remains to prove the first assertion. Assume that $\mu_n < \gamma$. By the same considerations as above, if $a < \mu_n$, the number of eigenvalues (with multiplicity) lying in $(-\infty, a)$ is less than $n-1$. Let us finally show that if $a \in (\mu_n, \gamma)$, then the number of eigenvalues in $(-\infty, a)$ is at least n. If not, the direct sum of eigenspaces associated with eigenvalues below a would be spanned by $\psi_1, \ldots, \psi_{n-1}$

and

$$\mu_n \geqslant \inf_{\substack{u \in \text{span}\,(\psi_1,\ldots,\psi_{n-1})^\perp \\ u \in \text{Dom}\,(T), u \neq 0}} \frac{\langle Tu, u \rangle_{\mathsf{H}}}{\langle u, u \rangle_{\mathsf{H}}} \geqslant a\,,$$

where we have used Lemma 6.22 and the fact that $\text{sp}(T_{|F}) \subset [a, +\infty)$, with

$$F = \text{span}\,(\psi_1, \ldots, \psi_{n-1})^\perp\,.$$

$\square$

An often used consequence of this theorem (or of its proof) is the following proposition.

Proposition 6.24. — *Suppose that there exists $a \in \mathbb{R}$ with $a < \inf \text{sp}_{\text{ess}}(T)$ and an n-dimensional space $\mathsf{V} \subset \text{Dom}\,T$ such that*

$$\langle T\psi, \psi \rangle_{\mathsf{H}} \leqslant a\|\psi\|^2\,, \quad \forall \psi \in \mathsf{V}\,.$$

Then, the n-th eigenvalue exists and satisfies

$$\lambda_n(T) \leqslant a\,.$$

Remark 6.25. — When the Rayleigh quotients are below the essential spectrum, the supremum and the infimum are a maximum and a minimum, respectively.

Proposition 6.26. — *Assume that $\mu_1 < \inf \text{sp}_{\text{ess}}(T)$. Then, μ_1 is a minimum of the first Rayleigh quotient (written with the quadratic form Q). Moreover, any minimizer of the first Rayleigh quotient belongs to the domain of T and is an eigenfunction of T associated with μ_1.*

Proof. — Since any eigenfunction associated with μ_1 is a minimizer, μ_1 is a minimum. Now, consider a minimizer $u_0 \in \text{Dom}(Q)$. We have

$$\mu_1 = \frac{Q(u_0)}{\|u_0\|_{\mathsf{H}}^2}\,.$$

Consider $v \in \text{Dom}(Q)$, and, for t small enough,

$$\varphi : t \mapsto \frac{Q(u_0 + tv)}{\|u_0 + tv\|_{\mathsf{H}}^2}\,.$$

Writing that $\varphi'(0) = 0$, we get, for all $v \in \mathrm{Dom}\,(Q)$,

$$Q(u_0, v) = \mu_1 \langle u_0, v \rangle_{\mathsf{H}}\,.$$

This shows that $u_0 \in \mathrm{Dom}\,(T)$ and then $Tu_0 = \mu_1 u_0$. $\square$

Remark 6.27. — We can extend the result of the last proposition to the other Rayleigh quotients.

Exercise 6.28. — Let $\Omega \subset \mathbb{R}^d$ be an open bounded set. Prove that there exists $c(\Omega) > 0$ such that, for all $\psi \in \mathsf{H}_0^1(\Omega)$,

$$\int_\Omega |\nabla \psi|^2 \, \mathrm{d}x \geqslant c(\Omega) \|\psi\|^2\,.$$

What is the optimal $c(\Omega)$? We will consider the Dirichlet Laplacian on Ω.

Exercise 6.29. — Consider the self-adjoint operator $\mathscr{L}$ associated with the quadratic form

$$\forall \psi \in \mathsf{H}^1(\mathbb{R})\,, \quad Q(\psi) = \int_{\mathbb{R}} |\psi'|^2 + V(x)|\psi|^2 \, \mathrm{d}x\,,$$

where $V \in \mathscr{C}_0^\infty(\mathbb{R}, \mathbb{R})$.

  i. What is the essential spectrum?

  ii. We assume that $\int_{\mathbb{R}} V(x) \, \mathrm{d}x < 0$. Prove that the discrete spectrum is not empty.

6.4.2. Sturm–Liouville's oscillation theorem. — We consider the operator $\mathscr{L} = -\partial_x^2 + V(x)$, with $V \in \mathscr{C}^\infty([0, 1])$, on $[0, 1]$ and domain

$$\mathrm{Dom}\,(\mathscr{L}) = \left\{ \psi \in \mathsf{H}_0^1((0, 1)) : (-\partial_x^2 + V(x))\psi \in \mathsf{L}^2((0, 1)) \right\}\,.$$

$\mathscr{L}$ is a self-adjoint operator with compact resolvent. Therefore, we may consider the non-decreasing sequence of its eigenvalues $(\lambda_n)_{n \geqslant 1}$.

Lemma 6.30. — *The eigenvalues of $\mathscr{L}$ are simple.*

Proof. — It follows from the Cauchy–Lipschitz theorem. $\square$

For all $n \geqslant 1$, let us consider an L^2-normalized eigenfunction u_n associated with λ_n. Notice that $\langle u_n, u_m \rangle = 0$ if $n \neq m$ and that the zeros of u_n are simple and thus isolated.

Theorem 6.31. — *For all $n \geqslant 1$, the function u_n admits exactly $n - 1$ zeros in $(0, 1)$.*

Proof. — Let us denote by Z_n the number of zeros of u_n in $(0, 1)$.

Let us prove that $Z_n \leqslant n - 1$. If the eigenfunction u_n admits at least n zeros in $(0, 1)$, denoted by $z_1, \ldots, z_n$, we let $z_0 = 0$ and $z_{n+1} = 1$. We define $(u_{n,j})_{j=0,\ldots,n}$ by $u_{n,j}(x) = u_n(x)$ for $x \in [z_j, z_{j+1}]$ and $u_{n,j}(x) = 0$ elsewhere. It is clear that these functions belong to the form domain of $\mathscr{L}$ and that they form an orthogonal family. By integrating by parts, we get

$$\forall v \in \operatorname*{span}_{j \in \{0, \ldots, n\}} u_{n,j}, \qquad Q(v, v) \leqslant \lambda_n \|v\|^2_{L^2((0,1))} \, .$$

By the min-max principle, we get $\lambda_{n+1} \leqslant \lambda_n$ and this contradicts the simplicity of the eigenvalues.

Let us now prove that $Z_n \geqslant Z_{n-1} + 1$. It is sufficient to show that if u_{n-1} is zero in z_0 and z_1 (two consecutive zeros, for example, u_{n-1} is positive on (z_0, z_1)), then u_n vanishes in (z_0, z_1). Indeed, this would imply that u_n vanishes at least $Z_{n-1} + 1$ times. For that purpose, we introduce $W(f_1, f_2) = f_1' f_2 - f_1 f_2'$ and compute

$$W(u_{n-1}, u_n)' = (\lambda_n - \lambda_{n-1}) u_{n-1} u_n \, .$$

Assume that u_n does not vanish on (z_0, z_1). For instance $u_n > 0$ on (z_0, z_1). Then, we get $W(u_{n-1}, u_n)' > 0$. We have $W(u_{n-1}, u_n)(z_0) \geqslant 0$ and $W(u_{n-1}, u_n)(z_1) \leqslant 0$, and thus we get a contradiction.

The conclusion follows easily. $\qquad\qquad\qquad\qquad\qquad\qquad\square$

6.4.3. Weyl's law in one dimension. —

6.4.3.1. *Two examples.* —

Definition 6.32. — If $(\mathcal{L}, \operatorname{Dom}(\mathcal{L}))$ is a self-adjoint operator and $E \in \mathbb{R}$, then $\mathsf{N}(\mathcal{L}, E)$ denotes the number of eigenvalues of $\mathcal{L}$ below E.

Let $\mathcal{H}_h^{\mathrm{Dir}} = h^2 D_x^2$ be the Dirichlet Laplacian on $(0, 1)$. Its domain is given by

$$\operatorname{Dom}(\mathcal{H}_h^{\mathrm{Dir}}) = \mathsf{H}^2(0, 1) \cap \mathsf{H}_0^1(0, 1) \, ,$$

and $\mathcal{H}_h^{\mathrm{Dir}}$ has compact resolvent. We can easily compute the eigenvalues:

$$\lambda_n \left(\mathcal{H}_h^{\mathrm{Dir}} \right) = h^2 n^2 \pi^2 \,, \qquad n \in \mathbb{N} \setminus \{0\} \,,$$

so that, for $E > 0$,

$$\mathsf{N} \left(\mathcal{H}_h^{\mathrm{Dir}}, E \right) \underset{h \to 0}{\sim} \frac{\sqrt{E}}{\pi h} = \frac{1}{2\pi h} \int_{\{(x,\xi) \in (0,1) \times \mathbb{R}: \, \xi^2 \leqslant E\}} \mathrm{d}x \, \mathrm{d}\xi \,.$$

In the same way, we can explicitly compute the eigenvalues when $\mathcal{H}_h = h^2 D_x^2 + x^2$. We have

$$\lambda_n \left(\mathcal{H}_h \right) = (2n - 1)h \,, \qquad n \in \mathbb{N} \setminus \{0\} \,,$$

so that, for $E > 0$,

$$\mathsf{N} \left(\mathcal{H}_h, E \right) \underset{h \to 0}{\sim} \frac{E}{2h} = \frac{1}{2\pi h} \int_{\{(x,\xi) \in \mathbb{R}^2: \, \xi^2 + x^2 \leqslant E\}} \mathrm{d}x \, \mathrm{d}\xi \,.$$

From these examples, one could guess the more general formula

$$\mathsf{N} \left(\mathcal{H}_h, E \right) \underset{h \to 0}{\sim} \frac{1}{2\pi h} \int_{\{(x,\xi) \in \mathbb{R}^2: \, \xi^2 + V(x) \leqslant E\}} \mathrm{d}x \, \mathrm{d}\xi$$
$$= \frac{1}{\pi h} \int_{\mathbb{R}} \sqrt{(E - V)_+} \, \mathrm{d}x \,.$$

6.4.3.2. Statement in one dimension. — We propose to prove the following version of the Weyl law in dimension one.

Proposition 6.33. — *Let $V : \mathbb{R} \to \mathbb{R}$ be a piecewise Lipschitzian function with a finite number of discontinuities, and bounded from below. Consider the operator $\mathcal{H}_h = h^2 D_x^2 + V(x)$ (defined via a quadratic form), and $E \in \mathbb{R}$. Assume that*

$$\{x \in \mathbb{R} : V(x) \leqslant E\} = [x_{\min}(E), x_{\max}(E)] \,.$$

Then

$$\mathsf{N}(\mathcal{H}_h, E) \underset{h \to 0}{\sim} \frac{1}{\pi h} \int_{\mathbb{R}} \sqrt{(E - V)_+} \, \mathrm{d}x \,.$$

6.4.4. Proof of the Weyl law. — The following lemma is a consequence of the definition of the Rayleigh quotients.

Lemma 6.34 (Dirichlet–Neumann bracketing)

Let $(s_j)_{j \in \mathbb{Z}}$ be a subdivision of $\mathbb{R}$ and consider the operators (with Dirichlet or Neumann conditions on the points of the subdivision)

$$\mathcal{H}_h^{\mathrm{Dir/Neu}} = \bigoplus_{j \in \mathbb{Z}} \mathcal{H}_{h,j}^{\mathrm{Dir/Neu}},$$

where $\mathcal{H}_{h,j}^{\mathrm{Dir/Neu}}$ is the $\mathrm{Dir/Neu}$ realization of $h^2 D_x^2 + V(x)$ on (s_j, s_{j+1}). We have, in terms of the domains of the quadratic forms,

$$\mathrm{Dom}\left(\mathcal{Q}_h^{\mathrm{Dir}}\right) \subset \mathrm{Dom}\left(\mathcal{Q}_h\right) \subset \mathrm{Dom}\left(\mathcal{Q}_h^{\mathrm{Neu}}\right),$$

and the Rayleigh quotients satisfy, for all $n \geqslant 1$,

$$\mu_n(\mathcal{H}_h^{\mathrm{Neu}}) \leqslant \mu_n(\mathcal{H}_h) \leqslant \mu_n(\mathcal{H}_h^{\mathrm{Dir}}).$$

We can now start the proof of Proposition 6.33.

We consider a subdivision of the real axis $(s_j(h^\alpha))_{j \in \mathbb{Z}}$, which contains the discontinuities of V, for which there exist $c > 0$, $C > 0$ such that, for all $j \in \mathbb{Z}$ and $h > 0$, $ch^\alpha \leqslant s_{j+1}(h^\alpha) - s_j(h^\alpha) \leqslant Ch^\alpha$, where $\alpha > 0$ is to be determined. Denote

$$J_{\min}(h^\alpha) = \min\{j \in \mathbb{Z} : s_j(h^\alpha) \geqslant x_{\min}(E)\},$$

$$J_{\max}(h^\alpha) = \max\{j \in \mathbb{Z} : s_j(h^\alpha) \leqslant x_{\max}(E)\}.$$

For $j \in \mathbb{Z}$, we introduce the Dirichlet (resp. Neumann) realization on $(s_j(h^\alpha), s_{j+1}(h^\alpha))$ of $h^2 D_x^2 + V(x)$ denoted by $\mathcal{H}_{h,j}^{\mathrm{Dir}}$ (resp. $\mathcal{H}_{h,j}^{\mathrm{Neu}}$). The Dirichlet–Neumann bracketing implies that

$$\sum_{j=J_{\min}(h^\alpha)}^{J_{\max}(h^\alpha)} \mathsf{N}(\mathcal{H}_{h,j}^{\mathrm{Dir}}, E) \leqslant \mathsf{N}(\mathcal{H}_h, E) \leqslant \sum_{j=J_{\min}(h^\alpha)-1}^{J_{\max}(h^\alpha)+1} \mathsf{N}(\mathcal{H}_{h,j}^{\mathrm{Neu}}, E).$$

Let us estimate $\mathsf{N}(\mathcal{H}_{h,j}^{\mathrm{Dir}}, E)$. If $Q_{h,j}^{\mathrm{Dir}}$ denotes the quadratic form of $\mathcal{H}_{h,j}^{\mathrm{Dir}}$, we have, for all $\psi \in \mathscr{C}_0^\infty((s_j(h^\alpha), s_{j+1}(h^\alpha)))$,

$$Q_{h,j}^{\mathrm{Dir}}(\psi) \leqslant \int_{s_j(h^\alpha)}^{s_{j+1}(h^\alpha)} h^2 |\psi'(x)|^2 + V_{j,\sup,h} |\psi(x)|^2 \, \mathrm{d}x,$$

where

$$V_{j,\sup,h} = \sup_{x \in (s_j(h^\alpha), s_{j+1}(h^\alpha))} V(x).$$

We infer that

$$N(\mathcal{H}_{h,j}^{\mathrm{Dir}}, E)$$

$$\geqslant \# \left\{ n \geqslant 1 : n \leqslant \frac{1}{\pi h} (s_{j+1}(h^\alpha) - s_j(h^\alpha)) \sqrt{(E - V_{j,\sup,h})_+} \right\}$$

so that

$$N(\mathcal{H}_{h,j}^{\mathrm{Dir}}, E) \geqslant \frac{1}{\pi h} (s_{j+1}(h^\alpha) - s_j(h^\alpha)) \sqrt{(E - V_{j,\sup,h})_+} - 1 \,,$$

and thus

$$\sum_{j=J_{\min}(h^\alpha)}^{J_{\max}(h^\alpha)} N(\mathcal{H}_{h,j}^{\mathrm{Dir}}, E) \geqslant$$

$$\frac{1}{\pi h} \sum_{j=J_{\min}(h^\alpha)}^{J_{\max}(h^\alpha)} (s_{j+1}(h^\alpha) - s_j(h^\alpha)) \sqrt{(E - V_{j,\sup,h})_+}$$

$$- (J_{\max}(h^\alpha) - J_{\min}(h^\alpha) + 1) \,.$$

Let us consider the function

$$f(x) = \sqrt{(E - V(x))_+}$$

and analyze

$$\left| \sum_{j=J_{\min}(h^\alpha)}^{J_{\max}(h^\alpha)} (s_{j+1}(h^\alpha) - s_j(h^\alpha)) \sqrt{(E - V_{j,\sup,h})_+} - \int_{\mathbb{R}} f(x) \, dx \right|$$

$$\leqslant \left| \sum_{j=J_{\min}(h^\alpha)}^{J_{\max}(h^\alpha)} \int_{s_j(h^\alpha)}^{s_{j+1}(h^\alpha)} \sqrt{(E - V_{j,\sup,h})_+} - f(x) \, dx \right|$$

$$+ \int_{s_{J_{\max}}(h^\alpha)}^{x_{\max}(E)} f(x) \, dx + \int_{x_{\min}(E)}^{s_{J_{\min}}(h^\alpha)} f(x) \, dx$$

$$\leqslant \left| \sum_{j=J_{\min}(h^\alpha)}^{J_{\max}(h^\alpha)} \int_{s_j(h^\alpha)}^{s_{j+1}(h^\alpha)} \sqrt{(E - V_{j,\sup,h})_+} - f(x) \, dx \right| + \tilde{C} h^\alpha \,.$$

Using the trivial inequality $|\sqrt{a_+} - \sqrt{b_+}| \leqslant \sqrt{|a - b|}$, we get

$$\left| f(x) - \sqrt{(E - V_{j,\sup,h})_+} \right| \leqslant \sqrt{|V(x) - V_{j,\sup,h}|} \,.$$

Since V is Lipschitzian on $(s_j(h^\alpha), s_{j+1}(h^\alpha))$, we get

$$\left| \sum_{j=J_{\min}(h^\alpha)}^{J_{\max}(h^\alpha)} \int_{s_j(h^\alpha)}^{s_{j+1}(h^\alpha)} \sqrt{(E - V_{j,\sup,h})_+} - f(x)\,\mathrm{d}x \right|$$

$$\leqslant (J_{\max}(h^\alpha) - J_{\min}(h^\alpha) + 1)\tilde{C}h^\alpha h^{\alpha/2}.$$

This leads to the optimal choice $\alpha = \frac{2}{3}$, and we obtain the lower bound

$$\sum_{j=J_{\min}(h^{2/3})}^{J_{\max}(h^{2/3})} \mathsf{N}(\mathcal{H}_{h,j}^{\mathrm{Dir}}, E)$$

$$\geqslant \frac{1}{\pi h}\left(\int_{\mathbb{R}} f(x)\,\mathrm{d}x - \tilde{C}h(J_{\max}(h^{2/3}) - J_{\min}(h^{2/3}) + 1) \right).$$

It follows that

$$\mathsf{N}(\mathcal{H}_h, E)$$

$$\geqslant \frac{1}{\pi h}\left(\int_{\mathbb{R}} f(x)\,\mathrm{d}x - \tilde{C}h^{1/3}(x_{\max}(E) - x_{\min}(E)) - \tilde{C}h \right).$$

We can deal with the Neumann realizations in the same way.

6.4.5. Some exercises. —

Exercise 6.35. — We wish to study the 2D harmonic oscillator $\mathscr{L} = -\Delta + |x|^2$.

  i. Write the operator in radial coordinates.

  ii. Explain how the spectral analysis can be reduced to the study of

$$-\partial_\rho^2 - \rho^{-1}\partial_\rho + \rho^{-2}m^2 + \rho^2,$$

  on $\mathsf{L}^2(\mathbb{R}_+, \rho\mathrm{d}\rho)$ with $m \in \mathbb{Z}$.

  iii. Perform the change of variable $t = \rho^2$.

  iv. For which α is $t \mapsto t^\alpha e^{-t/2}$ an eigenfunction ?

  v. Conjugate the operator by $t^{-m/2}e^{t/2}$. On which space is the new operator $\mathfrak{L}_m$ acting ? Describe the new scalar product.

  vi. Find the eigenvalues (and eigenfunctions) of $\mathfrak{L}_m$ by noticing that $\mathbb{R}_N[X]$ is stable under $\mathfrak{L}_m$. These eigenfunctions are the famous Laguerre polynomials.

vii. Conclude.

Exercise 6.36. — Let $h > 0$. We consider $V \in \mathscr{C}^\infty(\mathbb{R}, \mathbb{R})$. We assume that V has a unique minimum at 0 and that

$$V(0) = 0, \quad V''(0) > 0.$$

We recall that the operator $(-h^2 \partial_x^2 + V(x), \mathscr{C}_0^\infty(\mathbb{R}))$ is essentially self-adjoint, and we denote by $\mathscr{L}_h$ its unique self-adjoint extension.

i. What is the domain of $\mathscr{L}_h$?

ii. Prove that $\mathscr{L}_h$ is unitary equivalent to $\widetilde{\mathscr{L}_h}$, the unique self-adjoint extension of

$$(-h\partial_y^2 + V(h^{\frac{1}{2}}y), \mathscr{C}_0^\infty(\mathbb{R})).$$

iii. Let $n \in \mathbb{N}^*$. We know that there exists a non-zero function $H_n \in \mathscr{S}(\mathbb{R})$ such that

$$-H_n'' + x^2 H_n = (2n - 1)H_n.$$

Find $f_n \in \mathscr{S}(\mathbb{R})$, non-zero, such that

$$\left\| \left(\widetilde{\mathscr{L}_h} - (2n - 1)h\sqrt{\frac{V''(0)}{2}} \right) f_n \right\|_{L^2(\mathbb{R})} \underset{h \to 0}{=} \mathscr{O}(h^{\frac{3}{2}}).$$

iv. Prove that, for all $n \in \mathbb{N}^*$,

$$\text{dist}\left(\text{sp}(\mathscr{L}_h), (2n - 1)h\sqrt{\frac{V''(0)}{2}} \right) = \mathscr{O}(h^{\frac{3}{2}}).$$

v. Thanks to Weyl sequence, show that $[V_\infty, +\infty) \subset \text{sp}_{\text{ess}}(\mathscr{L}_h)$.

vi. Let $\lambda < V_\infty$.

    a. Explain why there exists a function $\chi \in \mathscr{C}_0^\infty(\mathbb{R})$ and $c > 0$ such that $V - \lambda + \chi \geqslant c$. Notice that $\{V \leqslant \lambda\}$ is compact.

    b. We consider $\mathscr{M}_{h,\lambda} = h^2 D_x^2 + V - \lambda + \chi$ (with the same domain as $\mathscr{L}_h$). Prove that $\mathscr{M}_{h,\lambda}$ is bijective and give an upper bound for the norm of its inverse.

    c. Let $(u_k)_{k \in \mathbb{N}}$ be a bounded sequence in $L^2(\mathbb{R})$. Prove that $v_k = \chi \mathscr{M}_{h,\lambda}^{-1} u_k$ is bounded in $H^1(\mathbb{R})$ and that it is equi-L^2-integrable. What can we say about the operator $\chi \mathscr{M}_{h,\lambda}^{-1}$?

    d. Establish that $\mathscr{L}_h - \lambda$ is a Fredholm operator with index 0.

vii. What is the essential spectrum?

viii. Show that, if h is small enough, the operator $\mathscr{L}_h$ has discrete spectrum and give an upper bound of the smallest eigenvalue.

6.5. On the ground-state energy of the hydrogen atom

Let us consider the following quadratic form:

$$\forall \psi \in H^1(\mathbb{R}^3), \quad Q(\psi) = \int_{\mathbb{R}^3} \left(|\nabla \psi|^2 - \frac{1}{|x|} |\psi|^2 \right) \, \mathrm{d}x.$$

It is not immediately clear that the non-positive term is well-defined.

Proposition 6.37. — *There exists $C > 0$ such that, for all $\psi \in H^1(\mathbb{R}^3)$, we have*

(6.6.5.4)
$$\int_{\mathbb{R}^3} \frac{1}{|x|} |\psi|^2 \, \mathrm{d}x \leqslant C \|\psi\|^2_{H^1(\mathbb{R}^3)}.$$

Moreover, there exists $C > 0$ such that, for all $\psi \in H^1(\mathbb{R}^3)$,

(6.6.5.5)
$$Q(\psi) \geqslant \frac{1}{2} \|\nabla \psi\|^2_{L^2(\mathbb{R}^3)} - C \|\psi\|^2_{L^2(\mathbb{R}^3)}.$$

In particular, up to shifting Q, it is a coercive quadratic form on $H^1(\mathbb{R}^3)$.

Proof. — We have

$$\int_{\mathbb{R}^3} \frac{1}{|x|} |\psi|^2 \, \mathrm{d}x \leqslant \int_{B(0,1)} \frac{1}{|x|} |\psi|^2 \, \mathrm{d}x + \|\psi\|^2_{L^2(\mathbb{R}^3)}$$

$$\leqslant C \|\psi\|^2_{L^4(\mathbb{R}^3)} + \|\psi\|^2_{L^2(\mathbb{R}^3)},$$

where we used the Cauchy–Schwarz inequality and the fact that $\frac{1}{|x|} \in L^2_{loc}(\mathbb{R}^d)$. Let us now use a classical Sobolev embedding theorem $H^1(\mathbb{R}^3) \subset L^p(\mathbb{R}^3)$ for $p \in [2, 6]$. In particular, there exists $C > 0$ such that, for all $\psi \in H^1(\mathbb{R}^3)$,

$$\|\psi\|^2_{L^4(\mathbb{R}^3)} \leqslant C \|\psi\|^2_{H^1(\mathbb{R}^3)} = C(\|\psi\|^2_{L^2(\mathbb{R}^3)} + \|\nabla \psi\|^2_{L^2(\mathbb{R}^3)}).$$

Consider $\varphi \in H^1(\mathbb{R}^3)$ and $\alpha > 0$. Inserting $\psi(\cdot) = \varphi(\alpha \cdot)$, we get

$$\|\varphi\|^2_{L^4(\mathbb{R}^3)} \leqslant C\left(\alpha^{-\frac{3}{2}}\|\varphi\|^2_{L^2(\mathbb{R}^3)} + \alpha^{\frac{1}{2}}\|\nabla\varphi\|^2_{L^2(\mathbb{R}^3)}\right).$$

When $\|\nabla\varphi\|_{L^2(\mathbb{R}^3)} \neq 0$, we choose

$$\alpha = \frac{\|\varphi\|_{L^2(\mathbb{R}^3)}}{\|\nabla\varphi\|_{L^2(\mathbb{R}^3)}},$$

and get

$$\|\varphi\|^2_{L^4(\mathbb{R}^3)} \leqslant C\|\varphi\|^{\frac{1}{2}}_{L^2(\mathbb{R}^3)}\|\nabla\varphi\|^{\frac{3}{2}}_{L^2(\mathbb{R}^3)}.$$

This last estimate is actually true for all $\varphi \in H^1(\mathbb{R}^3)$. It follows that

$$\int_{\mathbb{R}^3} \frac{1}{|x|}|\psi|^2 \, dx \leqslant C\|\psi\|^{\frac{1}{2}}_{L^2(\mathbb{R}^3)}\|\nabla\psi\|^{\frac{3}{2}}_{L^2(\mathbb{R}^3)} + \|\psi\|^2_{L^2(\mathbb{R}^3)}.$$

We recall the Young inequality

$$ab \leqslant \varepsilon^{-p}\frac{a^p}{p} + \varepsilon^q\frac{b^q}{q},$$

with

$$a, b \geqslant 0, \quad \varepsilon > 0, \quad p \in (1, +\infty), \quad \frac{1}{p} + \frac{1}{q} = 1.$$

Choosing $p = 4$ and $q = \frac{4}{3}$, we get

$$\int_{\mathbb{R}^3} \frac{1}{|x|}|\psi|^2 \, dx \leqslant C\left(\varepsilon^{-4}\frac{\|\psi\|^2_{L^2(\mathbb{R}^3)}}{4} + \frac{3}{4}\varepsilon^{\frac{4}{3}}\|\nabla\psi\|^2_{L^2(\mathbb{R}^3)}\right)$$
$$+ \|\psi\|^2_{L^2(\mathbb{R}^3)}.$$

Choosing ε such that $\frac{3C}{4}\varepsilon^{\frac{4}{3}} = \frac{1}{2}$, the conclusion follows. $\square$

We may consider $\mathscr{L}$ the operator associated with Q and given by the Lax–Milgram theorem. In Quantum Mechanics, the (Schrödinger) operator $\mathscr{L}$ describes the hydrogen atom. The infimum of its spectrum, denoted by E, is sometimes called « ground-energy ». Let us compute its value. From the min-max theorem, we have

$$E = \inf_{\psi \in H^1(\mathbb{R}^3)\setminus\{0\}} \frac{\int_{\mathbb{R}^3}\left(|\nabla\psi|^2 - \frac{1}{|x|}|\psi|^2\right)\,dx}{\|\psi\|^2_{L^2(\mathbb{R}^3)}}.$$

Proposition 6.38. — *We have*

$$E \leqslant -\frac{1}{4}.$$

Proof. — Let us consider the test function

$$\psi(x) = e^{-\alpha|x|}, \quad \alpha > 0.$$

We use the spherical coordinates

$$x = r \sin\theta \cos\phi, \quad y = r \sin\theta \sin\phi, \quad z = r \cos\theta,$$

with $(r, \theta, \phi) \in (0, +\infty) \times [0, \pi) \times [0, 2\pi)$. We get

$$\int_{\mathbb{R}^3} e^{-2\alpha|x|}\, \mathrm{d}x = 4\pi \int_0^{+\infty} r^2 e^{-2\alpha r}\, \mathrm{d}r.$$

In the same way,

$$\int_{\mathbb{R}^3} \left(|\nabla\psi|^2 - \frac{1}{|x|}|\psi|^2 \right) \mathrm{d}x = 4\pi \int_0^{+\infty} (\alpha^2 r^2 - r)e^{-2\alpha r}\, \mathrm{d}r.$$

Integrating by parts, one easily gets

$$\int_0^{+\infty} r^2 e^{-2\alpha r}\, \mathrm{d}r = \frac{1}{4\alpha^3}, \quad \int_0^{+\infty} r e^{-2\alpha r}\, \mathrm{d}r = \frac{1}{4\alpha^2}.$$

We deduce that

$$\frac{\int_{\mathbb{R}^3} \left(|\nabla\psi|^2 - \frac{1}{|x|}|\psi|^2 \right) \mathrm{d}x}{\|\psi\|_{\mathsf{L}^2(\mathbb{R}^3)}^2} = \alpha(\alpha - 1).$$

With $\alpha = \frac{1}{2}$, we deduce the result. $\square$

Lemma 6.39. — *The subspace $\mathscr{C}_0^\infty(\mathbb{R}^3 \setminus \{0\})$ is dense in $\mathsf{H}^1(\mathbb{R}^3)$.*

Proof. — We know that $\mathscr{C}_0^\infty(\mathbb{R}^3)$ is dense in $\mathsf{H}^1(\mathbb{R}^3)$. Let $\psi \in \mathsf{H}^1(\mathbb{R}^3)$ and $\varepsilon > 0$. Consider $\varphi \in \mathscr{C}_0^\infty(\mathbb{R}^3)$ such that

$$\|\psi - \varphi\|_{\mathsf{H}^1(\mathbb{R}^3)} \leqslant \varepsilon.$$

Let χ be a non-negative smooth function such that $0 \notin \operatorname{supp}(\chi)$ and $\chi(x) = 1$ for all x such that $|x| \geqslant 1$. Let us consider

$$\varphi_n(x) = \chi(nx)\varphi(x).$$

We have

$$\|\varphi_n - \varphi\|^2_{H^1(\mathbb{R}^3)}$$
$$= \|\varphi_n - \varphi\|^2_{L^2(\mathbb{R}^3)} + \|\nabla\varphi_n - \nabla\varphi\|^2_{L^2(\mathbb{R}^3)}$$
$$= \|\varphi_n - \varphi\|^2_{L^2(\mathbb{R}^3)} + \|\chi(n\cdot)\nabla\varphi - \nabla\varphi + n\varphi\nabla\chi(n\cdot)\|^2_{L^2(\mathbb{R}^3)}.$$

By the dominated convergence theorem,

$$\lim_{n\to+\infty}\|\varphi_n-\varphi\|^2_{L^2(\mathbb{R}^3)} = 0, \quad \lim_{n\to+\infty}\|\chi(n\cdot)\nabla\varphi-\nabla\varphi\|^2_{L^2(\mathbb{R}^3)} = 0.$$

We have

$$\|n\varphi\nabla\chi(n\cdot)\|^2_{L^2(\mathbb{R}^3)} = n^{-1}\int_{\mathbb{R}^3}|\nabla\chi(y)|^2|\varphi(n^{-1}y)|^2\,\mathrm{d}y$$
$$\leqslant n^{-1}\|\nabla\chi\|^2_{L^2(\mathbb{R}^3)}\|\varphi\|^2_\infty.$$

Therefore, for n large enough, we get

$$\|\varphi_n - \varphi\|_{H^1(\mathbb{R}^3)} \leqslant \varepsilon.$$

The conclusion easily follows. $\qquad\square$

Lemma 6.40. — *We have*

$$E = \inf_{\psi\in\mathscr{C}_0^\infty(\mathbb{R}^3\setminus\{0\})\setminus\{0\}} \frac{\int_{\mathbb{R}^3}\left(|\nabla\psi|^2 - \frac{1}{|x|}|\psi|^2\right)\,\mathrm{d}x}{\|\psi\|^2_{L^2(\mathbb{R}^3)}}.$$

Proof. — This follows from (6.6.5.4) and Lemma 6.39. $\qquad\square$

Proposition 6.41. — *We have*

$$E = -\frac{1}{4}.$$

Proof. — Let $\varepsilon > 0$. There exists $\psi \in \mathscr{C}_0^\infty(\mathbb{R}^3 \setminus \{0\})$ such that

$$E \geqslant \frac{\int_{\mathbb{R}^3}\left(|\nabla\psi|^2 - \frac{1}{|x|}|\psi|^2\right)\,\mathrm{d}x}{\|\psi\|^2_{L^2(\mathbb{R}^3)}} - \varepsilon.$$

We use again the spherical coordinates, and we let

$$\tilde{\psi}(r,\theta,\phi) = \psi(r\sin\theta\cos\phi, r\sin\theta\sin\phi, r\cos\theta).$$

Letting $U = (0, +\infty) \times [0, \pi) \times [0, 2\pi)$, we have

$$Q(\psi)$$

$$= \int_U \left(|\partial_r \tilde{\psi}|^2 + \frac{|\partial_\theta \tilde{\psi}|^2}{r^2} + \frac{|\partial_\phi \tilde{\psi}|^2}{r^2 \sin^2 \theta} - \frac{|\tilde{\psi}|^2}{r} \right) r^2 \sin \theta \, dr \, d\theta \, d\phi$$

$$\geqslant \int_0^\pi \int_0^{2\pi} \left(\int_0^{+\infty} \left[|\partial_r \tilde{\psi}|^2 - \frac{|\tilde{\psi}|^2}{r} \right] r^2 \, dr \right) \sin \theta \, d\theta \, d\phi \, .$$

Let us consider the quadratic form, in the ambient Hilbert space $L^2(\mathbb{R}_+, r^2 \, dr)$, defined by, for all $f \in \mathscr{C}_0^\infty(\mathbb{R}_+)$,

$$q(f) = \int_0^{+\infty} \left(|f'|^2 - r^{-1}|f|^2 \right) r^2 \, dr \, .$$

Let us show that q is bounded from below by $-\frac{1}{4}$. In fact, we can write

$$q(f) = \int_0^{+\infty} \left| f' + \frac{1}{2} f \right|^2 r^2 \, dr - \frac{1}{4} \int_0^{+\infty} |\psi|^2 r^2 \, dr \, ,$$

since

$$\frac{1}{2} \mathrm{Re} \int_0^{+\infty} (|f|^2)' r^2 \, dr = - \int_0^{+\infty} \frac{|f|^2}{r} r^2 \, dr \, .$$

We deduce that

$$Q(\psi) \geqslant -\frac{1}{4} \|\psi\|_{L^2(\mathbb{R}^3)}^2 \, .$$

Thus, for all $\varepsilon > 0$,

$$E \geqslant -\frac{1}{4} - \varepsilon \, .$$

$\square$

Actually, the spirit of the proof of Proposition 6.41 can also be used as follows.

Proposition 6.42 (Hardy–Leray inequality)
For all $\psi \in H^1(\mathbb{R}^3)$,

(6.6.5.6)
$$\frac{1}{4} \int_{\mathbb{R}^3} \frac{|\psi|^2}{|x|^2} \, dx \leqslant \|\nabla \psi\|_{L^2(\mathbb{R}^3)}^2 \, .$$

Proof. — Let us consider first $\psi \in \mathscr{C}_0^\infty(\mathbb{R}^3 \setminus \{0\})$. By using the spherical coordinates, we have

$$
\|\nabla \psi\|^2_{L^2(\mathbb{R}^3)}
$$

$$
= \int_U \left(|\partial_r \tilde{\psi}|^2 + \frac{|\partial_\theta \tilde{\psi}|^2}{r^2} + \frac{|\partial_\phi \tilde{\psi}|^2}{r^2 \sin^2 \theta} \right) r^2 \sin \theta \, dr \, d\theta \, d\phi
$$

$$
\geqslant \int_0^\pi \int_0^{2\pi} \left(\int_0^{+\infty} |r \partial_r \tilde{\psi}|^2 \, dr \right) \sin \theta \, d\theta \, d\phi .
$$

Expanding a square and using an integration by parts, we get

$$
\int_0^{+\infty} \left| r \partial_r \tilde{\psi} + \frac{\tilde{\psi}}{2} \right|^2 \, dr
$$

$$
= \int_0^{+\infty} |r \partial_r \tilde{\psi}|^2 \, dr + \frac{1}{4} \int_0^{+\infty} |\tilde{\psi}|^2 \, dr + \frac{1}{2} \int_0^{+\infty} r \partial_r (|\tilde{\psi}|^2) \, dr
$$

$$
= \int_0^{+\infty} |r \partial_r \tilde{\psi}|^2 \, dr - \frac{1}{4} \int_0^{+\infty} |\tilde{\psi}|^2 \, dr .
$$

We infer that

$$
\int_0^{+\infty} |r \partial_r \tilde{\psi}|^2 \, dr \geqslant \frac{1}{4} \int_0^{+\infty} r^{-2} |\tilde{\psi}|^2 r^2 \, dr .
$$

We deduce that (6.6.5.6) holds for all $\psi \in \mathscr{C}_0^\infty(\mathbb{R}^3 \setminus \{0\})$.

Now, recall Lemma 6.39, and take $\psi \in H^1(\mathbb{R}^d)$. There exists a sequence $(\psi_n) \subset \mathscr{C}_0^\infty(\mathbb{R}^3 \setminus \{0\})$ converging to ψ in $H^1(\mathbb{R}^3)$-norm. We have

$$
\frac{1}{4} \int_{\mathbb{R}^3} \frac{|\psi_n|^2}{|x|^2} \, dx \leqslant \|\nabla \psi_n\|^2_{L^2(\mathbb{R}^3)} .
$$

The right-hand side converges to $\|\nabla \psi\|^2_{L^2(\mathbb{R}^3)}$. Since (ψ_n) converges to ψ in $L^2(\mathbb{R}^3)$, it has a subsequence $(\psi_{\varphi(n)})$ converging to ψ almost everywhere. By the Fatou lemma, the conclusion follows. □

Proposition 6.43. — *The ground-energy E belongs to the discrete spectrum of $\mathscr{L}$.*

Proof. — Let us actually prove that all the negative eigenvalues belong to the discrete spectrum. Let us use the Weyl criterion.

Consider $\lambda < 0$ in the spectrum of $\mathscr{L}$. Let $(u_n) \subset \text{Dom}(\mathscr{L})$ be such that $\|u_n\| = 1$ and

$$\lim_{n \to +\infty} (\mathscr{L} - \lambda)u_n = 0.$$

Taking the scalar product with u_n, and using the definition of $\mathscr{L}$, we have

$$\lim_{n \to +\infty} Q(u_n) = \lambda.$$

There exists $R_0 > 0$ such that, for all $|x| \geqslant R_0$,

$$\frac{1}{|x|} - \lambda \geqslant \frac{|\lambda|}{2}.$$

Let $\varepsilon > 0$. There exists $R > 0$ such that, for all $n \geqslant 1$,

$$\int_{|x| \geqslant R} |u_n|^2 \, dx \leqslant \varepsilon.$$

From (6.6.5.5), we deduce that (u_n) is bounded in $H^1(\mathbb{R}^3)$. Thanks to the Kolmogorov–Riesz theorem, we infer that $\{u_n, n \geqslant 1\}$ is relatively compact in $L^2(\mathbb{R}^3)$. In particular, (u_n) has a converging subsequence. We deduce that λ belongs to the discrete spectrum. $\qquad\square$

It requires a little more work to prove that E is a simple eigenvalue. Let us first describe the domain of $\mathscr{L}$.

Proposition 6.44. — *We have*

$$\text{Dom}\,(\mathscr{L}) = H^2(\mathbb{R}^d).$$

Proof. — We recall that

$$\text{Dom}\,(\mathscr{L}) = \{\psi \in H^1(\mathbb{R}^3) : H^1(\mathbb{R}^3) \ni \varphi \mapsto Q(\varphi, \psi)$$

$$\text{is continuous for the } L^2(\mathbb{R}^3)\text{-topology}\}.$$

Then, due to Proposition 6.42, the fact that $\psi \in \text{Dom}\,(\mathscr{L})$ is equivalent to the fact that $\psi \in H^1(\mathbb{R}^3)$ and the continuity of

$$H^1(\mathbb{R}^3) \ni \varphi \mapsto \int_{\mathbb{R}^3} \nabla\varphi \, \overline{\nabla\psi} \, dx.$$

Taking $\varphi \in \mathscr{C}_0^\infty(\mathbb{R}^3)$ and using the Riesz representation theorem, we see that

$$\text{Dom}\,(\mathscr{L}) = \{\psi \in H^1(\mathbb{R}^3) : -\Delta\psi \in L^2(\mathbb{R}^3)\} = H^2(\mathbb{R}^3).$$

□

Consider our test function of Proposition 6.38, $\psi_0(x) = (8\pi)^{-\frac{1}{2}}e^{-\frac{|x|}{2}}$. It belongs to $H^2(\mathbb{R}^3)$, and a computation in spherical coordinates gives

$$\mathscr{L}\psi_0 = -\frac{1}{4}\psi_0.$$

Let us now turn to the proof of the simplicity of the smallest eigenvalue.

Proposition 6.45. — *For all $\psi \in H^1(\mathbb{R}^3)$, we have $|\psi| \in H^1(\mathbb{R}^d)$ and*

$$\|\nabla\psi\|_{L^2(\mathbb{R}^3)} \leqslant \||\nabla|\psi|\|_{L^2(\mathbb{R}^3)}.$$

Proof. — Consider $\varphi \in \mathscr{C}_0^\infty(\mathbb{R}^3)$ and $\psi \in H^1(\mathbb{R}^d)$. We may consider a sequence $(\psi_n) \subset \mathscr{C}_0^\infty(\mathbb{R}^3)$ such that $\lim_{n\to+\infty}\psi_n = \psi$ in $H^1(\mathbb{R}^3)$. For all $\varepsilon > 0$ and $z \in \mathbb{C}$, we let

$$|z|_\varepsilon = \sqrt{|z|^2 + \varepsilon^2} - \varepsilon (\leqslant |z|).$$

By dominated convergence, we have

$$\int_{\mathbb{R}^3} \nabla\varphi\,|\psi|\,\mathrm{d}x = \lim_{\varepsilon\to 0}\int_{\mathbb{R}^3} \nabla\varphi\,|\psi|_\varepsilon\,\mathrm{d}x$$

$$= \lim_{\varepsilon\to 0}\lim_{n\to+\infty}\int_{\mathbb{R}^3} \nabla\varphi\,|\psi_n|_\varepsilon\,\mathrm{d}x$$

$$= -\lim_{\varepsilon\to 0}\lim_{n\to+\infty}\int_{\mathbb{R}^3} \varphi\,\nabla|\psi_n|_\varepsilon\,\mathrm{d}x$$

$$= -\lim_{\varepsilon\to 0}\lim_{n\to+\infty}\int_{\mathbb{R}^3} \varphi\,\frac{\mathrm{Re}\,(\overline{\psi_n}\nabla\psi_n)}{\sqrt{|\psi_n|^2 + \varepsilon^2}}\,\mathrm{d}x.$$

By using that $\lim_{n\to+\infty}\nabla\psi_n = \nabla\psi$ in $L^2(\mathbb{R}^d)$, we get

$$\int_{\mathbb{R}^3} \nabla\varphi\,|\psi|\,\mathrm{d}x = -\lim_{\varepsilon\to 0}\lim_{n\to+\infty}\int_{\mathbb{R}^3} \varphi\,\frac{\mathrm{Re}\,(\overline{\psi_n}\nabla\psi)}{\sqrt{|\psi_n|^2 + \varepsilon^2}}\,\mathrm{d}x.$$

Up to a subsequence, we may assume that $\lim_{n\to+\infty}\psi_n = \psi$ almost everywhere. By dominated convergence, we find

$$\int_{\mathbb{R}^3} \nabla\varphi\,|\psi|\,\mathrm{d}x = -\lim_{\varepsilon\to 0}\int_{\mathbb{R}^3} \varphi\,\frac{\mathrm{Re}\,(\overline{\psi}\nabla\psi)}{\sqrt{|\psi|^2 + \varepsilon^2}}\,\mathrm{d}x,$$

and then

$$\int_{\mathbb{R}^3} \nabla \varphi \, |\psi| \, dx = - \int_{\mathbb{R}^3} \varphi \operatorname{Re} \left(\frac{\overline{\psi}}{|\psi|} \nabla \psi \right) \, dx \, .$$

This shows that

$$\nabla |\psi| \in \mathsf{L}^2(\mathbb{R}^d) \, , \quad \text{and} \quad \nabla |\psi| = \operatorname{Re} \left(\frac{\overline{\psi}}{|\psi|} \nabla \psi \right) \, ,$$

and the inequality follows. □

Lemma 6.46. — *Let $\psi \in \mathsf{H}^2(\mathbb{R}^d)$ be an eigenfunction of $\mathscr{L}$ that is associated with $-\frac{1}{4}$. Then, $|\psi|$ is an eigenfunction of $\mathscr{L}$ associated with $-\frac{1}{4}$.*

Proof. — From Proposition 6.45, and the min-max theorem, we have

$$-\frac{1}{4} = \frac{Q(\psi)}{\|\psi\|^2_{\mathsf{L}^2(\mathbb{R}^3)}} \geqslant \frac{Q(|\psi|)}{\||\psi|\|^2_{\mathsf{L}^2(\mathbb{R}^3)}}$$

$$\geqslant \min_{\psi \in \mathsf{H}^1(\mathbb{R}^3) \setminus \{0\}} \frac{Q(u)}{\|u\|^2_{\mathsf{L}^2(\mathbb{R}^3)}} = -\frac{1}{4} \, .$$

From Proposition 6.26, we deduce that $|\psi| \in \operatorname{Dom}(\mathscr{L}) = \mathsf{H}^2(\mathbb{R}^d)$ and

$$\mathscr{L}|\psi| = -\frac{1}{4}|\psi| \, .$$

□

Proposition 6.47. — *Let $\psi \in \mathsf{H}^2(\mathbb{R}^3)$ be an eigenfunction of $\mathscr{L}$ associated with $-\frac{1}{4}$. Then, ψ is continuous and does not vanish.*

Proof. — The continuity of ψ follows from $\mathsf{H}^2(\mathbb{R}^3) \subset \mathscr{C}^0(\mathbb{R}^3)$. From Lemma 6.46, we have $u := |\psi| \in \mathsf{H}^2(\mathbb{R}^3)$ and

$$-\Delta u - \frac{1}{|x|} u = -\frac{1}{4} u \, ,$$

or

$$-\Delta u + \frac{1}{4} u = \frac{1}{|x|} u \, .$$

From Proposition 6.42, we have $v := \frac{1}{|x|} u \in L^2(\mathbb{R}^3)$. Using the Fourier transform, we get

$$\mathscr{F}u = \hat{u} = \frac{1}{|\xi|^2 + \frac{1}{4}} \hat{v} \in L^1(\mathbb{R}^3) .$$

We recall that, for all $\omega > 0$,

$$\mathscr{F}^{-1} \left(\frac{1}{|\xi|^2 + \omega^2} \right) = \frac{e^{-\omega|x|}}{4\pi|x|} .$$

By the inverse Fourier transform, we get

$$u(x) = \int_{\mathbb{R}^3} \frac{1}{4\pi|x-y|} e^{-|x-y|/2} v(y) \, dy .$$

Since $v \geqslant 0$ and $v \neq 0$, we get that, for all $x \in \mathbb{R}^3$, $u(x) > 0$. □

Corollary 6.48. — *The spectral subspace* $\ker \left(\mathscr{L} + \frac{1}{4} \right)$ *is of dimension* 1. *Moreover,*

$$\ker \left(\mathscr{L} + \frac{1}{4} \right) = \mathrm{span} \, (8\pi)^{-\frac{1}{2}} e^{-\frac{|\cdot|}{2}} .$$

Proof. — Consider ψ_1 and ψ_2 two independent eigenfunctions. We can find a linear combination of them vanishing at 0. This is impossible by Proposition 6.47. □

6.6. Notes

  i. The interested Reader can read [**4**, Chapter VI] where the min-max theorem is proved and illustrated in the context of differential equations.

  ii. The Weyl law (see the original reference [**47**]) in higher dimensions is proved in [**4**, Chapter VI, §4]; see also the detailed proof in [**35**, Chapter XIII, Section 15].

  iii. The Sobolev embedding used in the proof of Proposition 6.37 is proved in [**3**, Theorem IX.9 & Corollary IX.10].

  iv. In the proof of Proposition 6.42, we used [**37**, Theorem 3.12].

v. Proposition 6.42 was proved for the first time by Leray in [**30**, Chapitre III], and it actually implies Proposition 6.37.

vi. It is possible to give a complete description of the discrete (and essential) spectrum of the hydrogen atom by means of the famous spherical harmonics and the Laguerre polynomials; see [**45**, Section 10.2].

CHAPTER 7

HILLE-YOSIDA AND STONE'S THEOREMS

This chapter is about the relation between $\mathscr{C}^0$-groups and their generators. In particular, we explain why to each unitary $\mathscr{C}^0$-group we may associate a unique self-adjoint operator. More importantly, we prove that any self-adjoint operator generates a unitary $\mathscr{C}^0$-group which solves an evolution equation, (*e.g.*, the Schrödinger equation).

7.1. Semigroups

Definition 7.1. — Let E be a Banach space. A $\mathscr{C}^0$-semigroup is a family $(T_t)_{t \geqslant 0}$ of bounded operators on E such that

  i. for all $s, t \geqslant 0$, $T_t T_s = T_{t+s}$,

  ii. $T_0 = \mathrm{Id}$,

  iii. for all $x \in E$, the application $\mathbb{R}_+ \ni t \mapsto T_t x$ is continuous.

Exercise 7.2. — Consider the vector space

$$\mathscr{C}_{ub} := \big\{ f : [0, +\infty[\longrightarrow \mathbb{R} \, ; \, f \text{ is bounded and unif. continuous} \big\}$$

equipped with the sup norm. This is a Banach space. Then, for all $t \in \mathbb{R}_+$, define the translation operator

$$
\begin{aligned}
T_t : \mathscr{C}_{ub} &\longrightarrow \mathscr{C}_{ub} \\
f(\cdot) &\longmapsto T_t f(\cdot) := f(t + \cdot).
\end{aligned}
$$

Show that the family $(T_t)_{t \geqslant 0}$ is a $\mathscr{C}^0$-semigroup.

© The Author(s), under exclusive license
to Springer Nature Switzerland AG 2021
C. Cheverry and N. Raymond, *A Guide to Spectral Theory*,
Birkhäuser Advanced Texts Basler Lehrbücher,
https://doi.org/10.1007/978-3-030-67462-5_7

Solution: We have

$$\|T(t)f\|_{\mathsf{L}^\infty(\mathbb{R}_+)} = \|f\|_{\mathsf{L}^\infty([t,+\infty[)} \leqslant \|f\|_{\mathsf{L}^\infty(\mathbb{R}_+)},$$

with equality when the support of f is contained in $[t, +\infty[$. It follows that $T(t)$ is a bounded operator satisfying $\|T(t)\| = 1$. The items i. and ii. follow directly from the definition, whereas iii. is a consequence of the uniform continuity of $f \in \mathscr{C}_{ub}$.   ∘

Lemma 7.3. — *Let $(T_t)_{t \geqslant 0}$ be a $\mathscr{C}^0$-semigroup. Then, there exist $M \geqslant 0$ and $\omega \geqslant 0$ such that*

(7.7.1.1) $\forall t \geqslant 0, \quad \|T_t\| \leqslant Me^{\omega t}.$

Proof. — For all $t \geqslant 0$, we have

$$\|T_t\| \leqslant \|T_1\|^{\lfloor t \rfloor} \sup_{s \in [0,1]} \|T_s\|.$$

Now, for all $x \in E$, the family $(\|T_s x\|)_{s \in [0,1]}$ is bounded (by continuity of the semigroup on the compact $[0, 1]$). Since E is a Banach space, we can use the Uniform Boundedness Principle to deduce that $(T_s)_{s \in [0,1]}$ is bounded. The conclusion follows with

$$\omega = \ln \|T_1\|, \qquad M = \sup_{s \in [0,1]} \|T_s\|.$$

□

Definition 7.4. — Let $(T_t)_{t \geqslant 0}$ be a $\mathscr{C}^0$-semigroup. The *infinitesimal generator* generated by this semigroup is the unbounded operator $\big(\mathrm{Dom}\,(A), A\big)$ defined by

$$\mathrm{Dom}\,(A) := \left\{ x \in E;\, \lim_{t \to 0^+} t^{-1}(T_t - \mathrm{Id})x \text{ exists} \right\},$$

as well as

$$\forall x \in \mathrm{Dom}\,(A), \qquad Ax := \lim_{t \to 0^+} t^{-1}(T_t - \mathrm{Id})x.$$

Observe that $\mathrm{Dom}\,(A)$ is indeed a linear subspace, and that A is by construction a linear map. In particular, we have $0 \in \mathrm{Dom}\,(A)$ and, of course, $A0 = 0$.

Exercise 7.5. — Consider Exercise 7.2. Show that

$$\mathrm{Dom}\,(A) = \{ f \in \mathscr{C}_{ub};\, f' \in \mathscr{C}_{ub} \}, \qquad Af = f'.$$

Solution: For $f \in \mathrm{Dom}\,(A)$, we must have

$$Af = \lim_{t \to 0^+} t^{-1}\big[f(t + \cdot) - f(\cdot)\big] = f'(\cdot) \in \mathscr{C}_{ub},$$

which gives rise to

$$\mathrm{Dom}\,(A) \subset \{f \in \mathscr{C}_{ub}\,;\, f' \in \mathscr{C}_{ub}\}, \qquad Af = f'.$$

Conversely, assume that $f \in \mathscr{C}_{ub}$ is such that $f' \in \mathscr{C}_{ub}$. Then

$$\big\|t^{-1}\big[f(t + \cdot) - f(\cdot)\big] - f'(\cdot)\big\|_{\mathsf{L}^\infty(\mathbb{R}_+)}$$

$$= \sup_{x \in \mathbb{R}_+} \frac{1}{t}\left|\int_x^{x+t} \big[f'(\tau) - f'(x)\big]\,\mathrm{d}\tau\right|.$$

$$\leqslant \sup_{|\tau - x| \leqslant t} |f'(\tau) - f'(x)| = o(1).$$

And therefore $f \in \mathrm{Dom}\,(A)$ with $Af = f'$.   ○

Let us now discuss some properties of A. In the following, the integrals can be understood in the Riemannian sense.

Proposition 7.6. — *Let $(T_t)_{t \geqslant 0}$ be a $\mathscr{C}^0$-semigroup and A its generator. Then,*

(i) *for all $x \in E$ and $t \geqslant 0$, we have*

$$\lim_{\varepsilon \to 0} \frac{1}{\varepsilon} \int_t^{t+\varepsilon} T_s x\,\mathrm{d}s = T_t x.$$

(ii) *for all $x \in E$ and $t \geqslant 0$, we have*

$$\int_0^t T_s x\,\mathrm{d}s \in \mathrm{Dom}\,(A), \qquad A \int_0^t T_s x\,\mathrm{d}s = (T_t - \mathrm{Id})x.$$

(iii) *for all $x \in \mathrm{Dom}\,(A)$ and $t \geqslant 0$, we have $T_t x \in \mathrm{Dom}\,(A)$. The application $t \mapsto T_t x$ is of class $\mathscr{C}^1$ with*

$$\frac{\mathrm{d}(T_t x)}{\mathrm{d}t} = A T_t x = T_t A x.$$

(iv) *for all $x \in \mathrm{Dom}\,(A)$, for all $s, t \geqslant 0$, we have*

$$(T_t - T_s)x = \int_s^t A T_\tau x\,\mathrm{d}\tau.$$

Proof. — The point (i) follows from the continuity. For the point (ii), we write, for all $\varepsilon > 0$,

$$\varepsilon^{-1}(T_\varepsilon - \mathrm{Id}) \int_0^t T_s x \, \mathrm{d}s = \frac{1}{\varepsilon} \int_0^t T_{s+\varepsilon} x \, \mathrm{d}s - \frac{1}{\varepsilon} \int_0^t T_s x \, \mathrm{d}s$$

$$= \frac{1}{\varepsilon} \int_t^{t+\varepsilon} T_u x \, \mathrm{d}u - \frac{1}{\varepsilon} \int_0^\varepsilon T_u x \, \mathrm{d}u \,.$$

Thus, we can take the limit $\varepsilon \to 0$, and the equality follows.

Let us consider (iii). Let $\varepsilon > 0$, $x \in \mathrm{Dom}\,(A)$, and $t \geqslant 0$. We have

$$\frac{1}{\varepsilon}(T_\varepsilon - \mathrm{Id})T_t x = T_t \left(\frac{1}{\varepsilon}(T_\varepsilon - \mathrm{Id})x \right).$$

The right-hand side has a limit when $\varepsilon \to 0^+$, which is $T_t(Ax)$. By definition of $\mathrm{Dom}\,(A)$, we have $T_t x \in \mathrm{Dom}\,(A)$. Moreover, due to the continuity of T_t, this furnishes

$$\lim_{\varepsilon \to 0+} \frac{T_{\varepsilon+t} x - T_t x}{\varepsilon} = \frac{\mathrm{d}(T_t x)}{\mathrm{d}t} = A T_t x = T_t A x \,.$$

Thus, by definition of $\mathrm{Dom}\,(A)$, we get $T_t x \in \mathrm{Dom}\,(A)$ and $A T_t x = T_t A x$. We have to check the derivability on the left at $t > 0$. We write

$$\varepsilon^{-1}(T_t x - T_{t-\varepsilon} x) = T_{t-\varepsilon}(\varepsilon^{-1}(T_\varepsilon x - x))$$

$$= T_{t-\varepsilon}(Ax) + T_{t-\varepsilon}(\varepsilon^{-1}(T_\varepsilon x - x) - Ax) \,.$$

Since $t \mapsto \|T_t\|$ is locally bounded (by Lemma 7.3), the conclusion follows. The point (iv) follows from the point (iii). $\qquad \square$

Proposition 7.7. — *Let $(T_t)_{t\geqslant 0}$ be a $\mathscr{C}^0$-semigroup and A its generator. Then, $\mathrm{Dom}\,(A)$ is dense and A is closed.*

Proof. — For $\varepsilon > 0$, we let $R_\varepsilon = \varepsilon^{-1} \int_0^\varepsilon T_s x \, \mathrm{d}s$. Let $x \in E$. We have $R_\varepsilon x \in \mathrm{Dom}\,(A)$ and $\lim_{\varepsilon \to 0} R_\varepsilon x = x$. Thus, $\mathrm{Dom}\,(A)$ is dense. Then, we consider $(x_n) \in \mathrm{Dom}\,(A)^{\mathbb{N}}$ such that $x_n \to x$ and $A x_n \to y$. For all $t \geqslant 0$, we have

$$(T_t - \mathrm{Id})x_n = \int_0^t T_s A x_n \, \mathrm{d}s \,,$$

and thus, since $s \mapsto \|T_s\|$ is locally bounded,

$$(T_t - \mathrm{Id})x = \int_0^t T_s y \, \mathrm{d}s \,.$$

Dividing by t and taking the limit $t \to 0^+$ we find that $x \in$ Dom (A) and $y = Ax$. $\qquad\square$

With ω as in (7.7.1.1), introduce

$$\Lambda_\omega := \{\lambda \in \mathbb{C}; \operatorname{Re}\lambda > \omega\}.$$

Observe that

$$\| e^{-\lambda t}T_t x \| \leqslant Me^{-(\operatorname{Re}\lambda - \omega)t} \| x \|.$$

Given $\lambda \in \Lambda_\omega$, we define the *Laplace transform*

$$E \ni x \longmapsto R_\lambda x := \int_0^{+\infty} e^{-\lambda t}T_t x \, dt.$$

It is a bounded operator satisfying $\|R_\lambda\| \leqslant M(\operatorname{Re}\lambda - \omega)^{-1}$.

Exercise 7.8. — Show that the map $R : \Lambda_\omega \longrightarrow \mathcal{L}(E)$ satisfies the resolvent formula (3.3.1.2).

Solution: Let us just give hints (there are some intermediate computations to be done). Remark that

$$\frac{R_\lambda - R_\mu}{\mu - \lambda}$$

$$= \int_0^{+\infty} e^{-(\mu-\lambda)\tau} R_\lambda d\tau - \int_0^{+\infty} e^{-(\mu-\lambda)\tau} R_\mu \, d\tau$$

$$= \int_0^{+\infty} \int_0^{+\infty} e^{-(\mu-\lambda)\tau} e^{-\lambda r}T_r \, dr \, d\tau$$

$$- \int_0^{+\infty} \int_0^{+\infty} e^{-(\mu-\lambda)(\tau+r)} e^{-\lambda r}T_r \, dr \, d\tau$$

$$= \int_0^{+\infty} \int_0^r e^{-(\mu-\lambda)s} ds \, e^{-\lambda r}T_r \, dr$$

$$= \int_0^{+\infty} \int_0^{+\infty} e^{-\lambda t} e^{-\mu s}T_{t+s} \, dt \, ds = R_\lambda R_\mu.$$

$\circ$

Lemma 7.9. — *Let $(T_t)_{t \geqslant 0}$ be a $\mathscr{C}^0$-semigroup and A its generator. Then, $\Lambda_\omega \subset \rho(A)$. More precisely, we have*

$$\forall \lambda \in \Lambda_\omega, \quad \forall x \in E, \quad (\lambda - A)^{-1}x = R_\lambda x.$$

Proof. — For all $\varepsilon > 0$, we write

$$\varepsilon^{-1}(T_\varepsilon - \mathrm{Id})R_\lambda x = \varepsilon^{-1} \int_0^{+\infty} e^{-\lambda t}(T_{t+\varepsilon}x - T_t x)\,\mathrm{d}t\,.$$

Thus,

$$\varepsilon^{-1}(T_\varepsilon - \mathrm{Id})R_\lambda x$$
$$= \varepsilon^{-1}e^{\varepsilon\lambda} \int_\varepsilon^{+\infty} e^{-\lambda t}T_t x\,\mathrm{d}t - \varepsilon^{-1} \int_0^{+\infty} e^{-\lambda t}T_t x\,\mathrm{d}t\,,$$

so that

$$\varepsilon^{-1}(T_\varepsilon - \mathrm{Id})R_\lambda x$$
$$= \varepsilon^{-1}(e^{\varepsilon\lambda} - 1) \int_0^{+\infty} e^{-\lambda t}T_t x\,\mathrm{d}t - \varepsilon^{-1} \int_0^\varepsilon e^{-\lambda t}T_t x\,\mathrm{d}t\,.$$

This proves that $R_\lambda x \in \mathrm{Dom}\,(A)$, that $AR_\lambda x = \lambda R_\lambda x - x$, that is $(\lambda - A)R_\lambda = \mathrm{Id}$. On the other hand, for all $x \in \mathrm{Dom}\,(A)$, we have

$$R_\lambda A x = \int_0^{+\infty} e^{-\lambda t}T_t A x\,dt = \int_0^{+\infty} e^{-\lambda t}\frac{d}{dt}T_t x\,dt$$
$$= \left[e^{-\lambda t}T_t x\right]_0^{+\infty} + \lambda \int_0^{+\infty} e^{-\lambda t}T_t x\,dt = -x + \lambda R_\lambda x.$$

In other words, we also have $R_\lambda(\lambda - A) = \mathrm{Id}_{\mathrm{Dom}\,(A)}$. □

7.2. Hille–Yosida's theorem

Definition 7.10. — A contraction on E is a linear map such that $\|T\| \leqslant 1$.

Theorem 7.11 (**Hille–Yosida's theorem**). — *An operator A is the infinitesimal generator of a contraction semigroup $(T_t)_{t \geqslant 0}$ if and only if*

   i. *A is closed and $\mathrm{Dom}\,(A)$ is dense,*

   ii. *$(0, +\infty) \subset \rho(A)$ and, for all $\lambda > 0$, $\|(A - \lambda)^{-1}\| \leqslant \lambda^{-1}$.*

7.2.1. Necessary condition. — If A is the infinitesimal generator of a contraction semigroup $(T_t)_{t \geqslant 0}$, we have already seen that A is closed, that $\mathrm{Dom}\,(A)$ is dense, that $M = 1$ and that $\omega = 0$. In view of Lemma 7.9, we have ii.

7.2.2. Sufficient condition. — Let us now assume that A is closed and $\mathrm{Dom}\,(A)$ is dense and that $(0, +\infty) \subset \rho(A)$ and, for all $\lambda > 0$, $\|(A - \lambda)^{-1}\| \leqslant \lambda^{-1}$. The idea is to approximate A by a bounded operator and use the exponential. For $\lambda > 0$, we let $S_\lambda = \lambda(\lambda - A)^{-1}$ and $A_\lambda = AS_\lambda$. For $x \in \mathrm{Dom}\,(A)$, we have

$$\lambda(\lambda - A)^{-1}x - (\lambda - A)(\lambda - A)^{-1}x = S_\lambda x - x = (\lambda - A)^{-1}Ax,$$

so that

$$\lim_{\lambda \to \pm\infty} S_\lambda x = x.$$

On the other hand, we have

$$\|x - S_\lambda x\| \leqslant \|x - \tilde{x}\| + \|S_\lambda \tilde{x} - S_\lambda x\| + \|\tilde{x} - S_\lambda \tilde{x}\|.$$

Since $\mathrm{Dom}\,(A)$ is dense, for all $\varepsilon \in \mathbb{R}_+^*$, we can find $\tilde{x} \in \mathrm{Dom}\,(A)$ such that $\|x - \tilde{x}\| \leqslant \varepsilon$. Knowing that $\|S_\lambda\| \leqslant 1$, the preceding inequality gives rise to

$$\|x - S_\lambda x\| \leqslant \varepsilon + \varepsilon + \lim_{\lambda \to -\infty} \|\tilde{x} - S_\lambda \tilde{x}\| = 2\varepsilon.$$

Thus, for all $x \in E$, we have

$$\lim_{\lambda \to \pm\infty} S_\lambda x = x.$$

Since $S_\lambda A = AS_\lambda$ on $\mathrm{Dom}\,(A)$, we deduce that

$$(7.7.2.2) \qquad \forall x \in \mathrm{Dom}\,(A), \quad \lim_{\lambda \to +\infty} A_\lambda x = Ax.$$

Observe that

$$\begin{aligned}
\lambda^2(\lambda - A)^{-1} &= \lambda\big[(A - \lambda)(\lambda - A)^{-1} + Id + \lambda(\lambda - A)^{-1}\big] \\
&= \lambda\big[(A - \lambda + \lambda)(\lambda - A)^{-1} + Id\big] \\
&= \lambda A(\lambda - A)^{-1} + \lambda = A_\lambda + \lambda.
\end{aligned}$$

It follows that A_λ is a bounded operator. Moreover, for all $t \geqslant 0$ and $\lambda > 0$, we have

$$e^{tA_\lambda} = e^{-t\lambda + t\lambda^2(\lambda - A)^{-1}},$$

as well as

$$(7.7.2.3) \quad \|e^{tA_\lambda}\| = e^{-t\lambda}\|e^{t\lambda^2(\lambda - A)^{-1}}\| \leqslant e^{-t\lambda}e^{t\lambda\|S_\lambda\|} \leqslant 1.$$

Then, we write

$$e^{tA_\lambda}x - e^{tA_\mu}x = e^{tA_\mu}\left(e^{t(A_\lambda - A_\mu)}x - x\right)$$

$$= \int_0^1 \frac{\mathrm{d}}{\mathrm{d}s}\left[e^{tsA_\lambda}e^{t(1-s)A_\mu}x\right]\,\mathrm{d}s$$

$$= \int_0^1 e^{tsA_\lambda}e^{t(1-s)A_\mu}\,t(A_\lambda - A_\mu)x\,\mathrm{d}s\,.$$

In view of (7.7.2.3), this gives rise to

$$\|e^{tA_\lambda}x - e^{tA_\mu}x\| \leqslant t\|A_\lambda x - A_\mu x\|\,.$$

Applying (7.7.2.2), for all $t \in \mathbb{R}_+$ and all $x \in \mathrm{Dom}\,(A)$, the family $(e^{tA_\lambda}x)_\lambda$ is of Cauchy type when λ goes to $+\infty$, and therefore it has a limit. By density of $\mathrm{Dom}\,(A)$ and since $\|e^{tA_\lambda}\| \leqslant 1$, this limit exists for all $x \in E$. Thus, we can define

$$T_t x = \lim_{\lambda \to +\infty} e^{tA_\lambda}x\,, \qquad \|T_t x\| \leqslant \limsup_{\lambda \to +\infty} \|e^{tA_\lambda}x\|\,.$$

From (7.7.2.3), we can deduce that $(T_t)_{t\geqslant 0}$ is a contraction $\mathscr{C}^0$-semigroup. Let us consider B its generator. Let $x \in \mathrm{Dom}\,(A)$ and $\varepsilon > 0$. We have

$$\varepsilon^{-1}(T_\varepsilon - \mathrm{Id})x = \lim_{\lambda \to +\infty} \varepsilon^{-1}(e^{\varepsilon A_\lambda} - \mathrm{Id})x$$

$$= \lim_{\lambda \to +\infty} \varepsilon^{-1}\int_0^\varepsilon e^{sA_\lambda}A_\lambda x\,\mathrm{d}s$$

$$= \varepsilon^{-1}\int_0^\varepsilon T_s Ax\,\mathrm{d}s\,.$$

We deduce that $x \in \mathrm{Dom}\,(B)$ and $Bx = Ax$. Thus $A \subset B$.

It remains to show that $\mathrm{Dom}\,(A) = \mathrm{Dom}\,(B)$. We do it by contradiction. Assume that we can find $x \in \mathrm{Dom}\,(B) \setminus \mathrm{Dom}\,(A)$. Since $1 \in \rho(A)$, we have $(1 - A)\mathrm{Dom}\,(A) = E$. But we have also $1 \in \rho(B)$ so that $(1 - B)\mathrm{Dom}\,(B) = E$. Consider $(1 - B)x$. We can find $\tilde{x} \in \mathrm{Dom}\,(A)$ such that $(1-B)x = (1-A)\tilde{x} = (1-B)\tilde{x}$. By construction $\tilde{x} \neq x$, which contradicts the injectivity of $1 - B$.

7.3. Stone's theorem

In this section, we work on a Hilbert space H.

***Theorem 7.12* (Stone's theorem).** — *Let $\mathscr{L}$ be a self-adjoint operator. There exists a unique $\mathscr{C}^0$-unitary (on* H*) group $(U_t)_{t \in \mathbb{R}}$ such that*

(i) $U_t : \mathrm{Dom}\,(\mathscr{L}) \to \mathrm{Dom}\,(\mathscr{L})$,

(ii) *for all* $u \in \mathrm{Dom}\,(\mathscr{L})$, $U_t u \in \mathscr{C}^1(\mathbb{R}, \mathsf{H}) \cap \mathscr{C}^0(\mathbb{R}, \mathrm{Dom}\,(\mathscr{L}))$,

(iii) *for all* $u \in \mathrm{Dom}\,(\mathscr{L})$, $\frac{\mathrm{d}}{\mathrm{d}t} U_t u = i\mathscr{L} U_t u = i U_t \mathscr{L} u$,

(iv) $U_0 = \mathrm{Id}$.

We let $U_t = e^{it\mathscr{L}}$ *for all* $t \in \mathbb{R}$.

Conversely, if $(U_t)_{t \in \mathbb{R}}$ is a $\mathscr{C}^0$-unitary group, then, there exists a unique self-adjoint operator $\mathscr{L}$ such that, for all $t \in \mathbb{R}$, $U_t = e^{it\mathscr{L}}$. The domain is

$$(7.7.3.4)\quad \mathrm{Dom}\,(\mathscr{L}) = \left\{ u \in \mathsf{H} : \sup_{0 < t \leqslant 1} t^{-1} \|U_t u - u\| < +\infty \right\}.$$

7.3.1. Necessary condition. — Let $\mathscr{L}$ be a self-adjoint operator. The operator $\mathscr{L}$ is closed with dense domain. For all $\lambda > 0$, we have already seen that $\pm i\mathscr{L} - \lambda$ is bijective and that we have $\|(\pm i\mathscr{L} - \lambda)^{-1}\| \leqslant \lambda^{-1}$. Therefore, the operators $\pm i\mathscr{L}$ are the generators of $\mathscr{C}^0$-semigroups $(U_t^{\pm})_{t \geqslant 0}$. We have

$$\frac{\mathrm{d}}{\mathrm{d}t} U_t^- U_t^+ u = -i\mathscr{L} U_t^- U_t^+ u + U_t^- i\mathscr{L} U_t^+ u = 0.$$

We get that, for all $t \geqslant 0$, $U_t^- U_t^+ u = u$. We let $U_t = U_t^+$ for $t \geqslant 0$ and $U_t = U_{-t}^-$ for $t < 0$. $(U_t)_{t \in \mathbb{R}}$ is a $\mathscr{C}^0$-group. We have, for all $t \in \mathbb{R}$, $U_t' = i\mathscr{L} U_t$. For all $u \in \mathrm{Dom}\,(\mathscr{L})$, we have

$$\frac{\mathrm{d}}{\mathrm{d}t} \|U_t u\|^2 = \langle i\mathscr{L} U_t u, U_t u \rangle + \langle U_t u, i\mathscr{L} U_t u \rangle = 0.$$

Thus, $(U_t)_{t \in \mathbb{R}}$ is unitary.

7.3.2. Sufficient condition. — Let $(U_t)_{t \in \mathbb{R}}$ be a $\mathscr{C}^0$-unitary group. Let us write the generator of the $\mathscr{C}^0$-unitary semigroup $(U_t)_{t \geqslant 0}$ as $i\mathscr{L}$. Applying Hille–Yosida's theorem, the operator $\mathscr{L}$ is closed, and it has a dense domain. Differentiating $U_t U_{-t} = \mathrm{Id}$, we get that

$$0 = i\mathscr{L} U_t U_{-t} + U_t \frac{dU_{-t}}{dt} = U_t \left[i\mathscr{L} U_{-t} + \frac{dU_{-t}}{dt} \right],$$

and therefore $-i\mathscr{L}$ is the generator of $(U_{-t})_{t\geqslant 0}$. Applying again Hille–Yosida's theorem, we know that $1 \in \rho(-i\mathscr{L})$ or that $-i\mathscr{L} - 1 = -i(\mathscr{L} - i)$ is invertible. In particular, this implies that $\text{ran}\,(\mathscr{L} - i) = H$ and that $\ker(\mathscr{L}^* + i) = \{0\}$. Then, differentiating $\|U_t u\|^2 = \|u\|^2$, we get easily that $\mathscr{L}$ is symmetric. From Proposition 2.64, we deduce that $\mathscr{L}$ is self-adjoint with

$$\text{Dom}\,(\mathscr{L}) = \left\{u \in H : \lim_{t\to 0+} t^{-1}(U_t u - u) \text{ exists}\right\}$$

$$\subset \left\{u \in H : \sup_{0<t\leqslant 1} t^{-1}\|U_t u - u\| < +\infty\right\}.$$

Then, take $u \in H$ such that

$$\sup_{0<t\leqslant 1} t^{-1}\|U_t u - u\| < +\infty,$$

and consider $v \in \text{Dom}\,(\mathscr{L})$. We have

$$|\langle u, \mathscr{L}v\rangle| = \lim_{t\to 0^+} \frac{1}{t}|\langle u, U_t v - v\rangle|$$

$$= \lim_{t\to 0^+} \frac{1}{t}|\langle U_{-t}u - u, v\rangle| \leqslant C\|v\|.$$

This shows that $u \in \text{Dom}\,(\mathscr{L}^*) = \text{Dom}\,(\mathscr{L})$.

Exercise 7.13. — Consider a self-adjoint operator $(\text{Dom}\,(\mathscr{L}), H)$. Let $U : H \to H$ be a unitary transform, and let us consider the operator $(\text{Dom}\,(\widetilde{\mathscr{L}}), \widetilde{\mathscr{L}})$ defined by

$$\text{Dom}\,(\widetilde{\mathscr{L}}) = U\text{Dom}\,(\mathscr{L}), \quad \text{and} \quad \widetilde{\mathscr{L}} = U\mathscr{L}U^{-1}.$$

   i. Show that $\widetilde{\mathscr{L}}$ is self-adjoint.

  ii. Prove that, for all $t \in \mathbb{R}$, $e^{it\widetilde{\mathscr{L}}} = Ue^{it\mathscr{L}}U^{-1}$.

7.4. Notes

   i. This chapter has been inspired by [**48**, Chapter IX].

  ii. A direct proof of the Stone theorem can be found in [**35**, Section VIII.4].

 iii. We used integrals of functions valued in a Banach space. In this chapter, these integrals may be understood in the Riemannian sense, since we only deal with continuous functions. Nevertheless, if one wants to use, for instance, the dominated convergence theorem and the Fubini theorem (as we will in the next chapter), it is more convenient to use the Bochner integral (see the original reference [2]).

CHAPTER 8

ABOUT THE SPECTRAL MEASURE

The purpose of this chapter is to introduce the Reader to the notion of spectral measure associated with a self-adjoint operator. Let $\mathscr{L}$ be a self-adjoint operator on H. Given $f : \mathbb{R} \to \mathbb{C}$, we would like to define functions $f(\mathscr{L})$ of $\mathscr{L}$ with the following properties:

(i) $f(\mathscr{L}) : \mathrm{Dom}\,\big(f(\mathscr{L})\big) \to \mathsf{H}$,

(ii) $[f(\mathscr{L}), \mathscr{L}] = 0$,

(iii) $f(\mathscr{L}) + g(\mathscr{L}) = (f + g)(\mathscr{L}) = g(\mathscr{L}) + f(\mathscr{L})$ on $\mathrm{Dom}\,(f(\mathscr{L})) \cap \mathrm{Dom}\,(g(\mathscr{L}))$,

(iv) $f(\mathscr{L})g(\mathscr{L}) = (fg)(\mathscr{L}) = g(\mathscr{L})f(\mathscr{L})$ on
$$\{u \in \mathrm{Dom}\,(g(\mathscr{L})); g(\mathscr{L})u \in \mathrm{Dom}\,(f(\mathscr{L}))\},$$

(v) $f(\mathscr{L})^* = \overline{f}(\mathscr{L})$.

We make the construction progressively by dealing with less and less regular functions $f(\cdot)$. The framework is the Schwartz class $\mathscr{S}(\mathbb{R})$ in Section 8.1, the set $\mathsf{L}^\infty(\mathbb{R})$ of bounded Borelian functions in Section 8.2, and just Borelian functions in Section 8.3. A key step of the construction is to give a definition of the *spectral measure* associated with $\mathscr{L}$. This measure may be decomposed thanks to the Lebesgue theorem, and so the Hilbert space H can be. This allows to define the corresponding classical spectral subspaces (absolutely continuous, singular continuous, pure point) and the corresponding spectra. We also provide the Reader with some criteria to characterize the absolute continuity of the spectrum.

© The Author(s), under exclusive license
to Springer Nature Switzerland AG 2021
C. Cheverry and N. Raymond, *A Guide to Spectral Theory*,
Birkhäuser Advanced Texts Basler Lehrbücher,
https://doi.org/10.1007/978-3-030-67462-5_8

8.1. A functional calculus based on the Fourier transform

We denote by $\mathscr{F}$ the Fourier transform and by $\mathscr{F}^{-1}$ its inverse, which are defined on $\mathscr{S}(\mathbb{R})$ by

$$\mathscr{F}\psi(\xi) = \int_{\mathbb{R}} \psi(x)e^{-ix\xi}\,\mathrm{d}x\,,$$

and

$$\psi(x) = \mathscr{F}^{-1}\mathscr{F}\psi(x) = \frac{1}{2\pi}\int_{\mathbb{R}} \mathscr{F}\psi(\xi)e^{ix\xi}\,\mathrm{d}\xi\,.$$

We can construct a functional calculus by using the inverse Fourier transform.

Definition 8.1. — Let $\mathscr{L}$ be a self-adjoint operator. For all $f \in \mathscr{S}(\mathbb{R})$ and $u \in \mathsf{H}$, we let

$$(8.8.1.1) \qquad f(\mathscr{L})u = \frac{1}{2\pi}\int_{\mathbb{R}} \mathscr{F}f(t)e^{it\mathscr{L}}u\,\mathrm{d}t,$$

where the $\mathscr{C}^0$-unitary group $(e^{it\mathscr{L}})_{t\in\mathbb{R}}$ is given by Stone's Theorem 7.12.

Note that the integral inside (8.8.1.1) is absolutely convergent. We find $f(\mathscr{L}) \in \mathcal{L}(\mathsf{H})$ with

$$\|f(\mathscr{L})\| \leqslant \frac{1}{2\pi}\int_{\mathbb{R}} |\mathscr{F}f(t)|\,\mathrm{d}t < +\infty.$$

Exercise 8.2. — Consider Exercise 7.13, and prove that, for all $f \in \mathscr{S}(\mathbb{R})$,

$$f(U\mathscr{L}U^{-1}) = U f(\mathscr{L})U^{-1}\,.$$

Proposition 8.3. — *For all $f, g \in \mathscr{S}(\mathbb{R})$, we have* (i)–(v).

Proof. — Let us only prove (iii). We recall that

$$\mathscr{F}(fg) = \mathscr{F}f \star \mathscr{F}g\,.$$

Then, we write

$$
\begin{aligned}
f(\mathscr{L})(g(\mathscr{L})u) &= \frac{1}{2\pi} \int_{\mathbb{R}} e^{it\mathscr{L}} \mathscr{F}f(t) g(\mathscr{L})u \, \mathrm{d}t \\
&= \frac{1}{2\pi} \int_{\mathbb{R}} e^{it\mathscr{L}} \mathscr{F}f(t) \int_{\mathbb{R}} e^{i\tau\mathscr{L}} \mathscr{F}g(\tau)u \, \mathrm{d}\tau \, \mathrm{d}t \\
&= \frac{1}{2\pi} \int_{\mathbb{R}} \int_{\mathbb{R}} e^{i(t+\tau)\mathscr{L}} \mathscr{F}f(t)\mathscr{F}g(\tau)u \, \mathrm{d}\tau \, \mathrm{d}t \\
&= \frac{1}{2\pi} \int_{\mathbb{R}} \int_{\mathbb{R}} e^{it\mathscr{L}} \mathscr{F}f(t-\tau)\mathscr{F}g(\tau)u \, \mathrm{d}\tau \, \mathrm{d}t \\
&= \frac{1}{2\pi} \int_{\mathbb{R}} e^{it\mathscr{L}} \mathscr{F}f \star \mathscr{F}g(t)u \, \mathrm{d}t \\
&= \frac{1}{2\pi} \int_{\mathbb{R}} e^{it\mathscr{L}} \mathscr{F}(fg)(t)u \, \mathrm{d}t \\
&= (fg)(\mathscr{L})u \, .
\end{aligned}
$$

$\square$

We introduce $\mathcal{A} := \mathscr{S}(\mathbb{R}) \oplus \mathbb{C}$. Let $f \in \mathcal{A}$ with $f = f_0 + \lambda_0$. We extend the functional calculus by adding the constants. Given f as above, we define

$$
f(\mathscr{L}) = f_0(\mathscr{L}) + \lambda_0 \operatorname{Id} \in \mathcal{L}(\mathsf{H}) \, .
$$

Proposition 8.4. — *For all $f, g \subset \mathcal{A}$, we have (i)–(v).*

Lemma 8.5. — *Let $f \in \mathcal{A}$ with $f \geqslant 0$. Then, we have, for all $u \in \mathsf{H}$,*

$$
\langle f(\mathscr{L})u, u \rangle \geqslant 0 \, .
$$

Proof. — Let $\varepsilon > 0$. The function $(\varepsilon + f)^{\frac{1}{2}}$ belongs to $\mathcal{A}$ (the regularity is guaranteed by the shift in ε). We have

$$
(\varepsilon + f)^{\frac{1}{2}}(\mathscr{L})(\varepsilon + f)^{\frac{1}{2}}(\mathscr{L}) = (\varepsilon + f)(\mathscr{L}) \, .
$$

Thus, since $(\varepsilon + f)^{\frac{1}{2}}(\mathscr{L})$ is symmetric, for all $u \in \mathsf{H}$,

$$
\langle u, (\varepsilon + f)(\mathscr{L})u \rangle = \|(\varepsilon + f)^{\frac{1}{2}}(\mathscr{L})u\|^2 \geqslant 0 \, .
$$

Then, we take the limit $\varepsilon \to 0$.

$\square$

Lemma 8.6. — *For all $f \in \mathcal{A}$, we have $\|f(\mathscr{L})\| \leqslant \|f\|_\infty$.*

Proof. — Let us consider $g = \|f\|_\infty^2 - |f|^2 \in \mathcal{A}$. We get, for all $u \in \mathsf{H}$,

$$\langle g(\mathscr{L})u, u \rangle \geqslant 0 \,,$$

so that

$$0 \leqslant \langle |f|^2(\mathscr{L})u, u \rangle \leqslant \|f\|_\infty^2 \|u\|^2 \,.$$

But, we have

$$\langle |f|^2(\mathscr{L})u, u \rangle = \langle (\overline{f}f)(\mathscr{L})u, u \rangle = \langle \overline{f}(\mathscr{L})f(\mathscr{L})u, u \rangle$$
$$= \langle f(\mathscr{L})^* f(\mathscr{L})u, u \rangle = \|f(\mathscr{L})u\|^2 \,.$$

$\square$

Lemma 8.7. — *Consider* $\chi \in \mathscr{C}_0^\infty(\mathbb{R}, \mathbb{R})$ *such that* $0 \leqslant \chi \leqslant 1$ *equal to* 1 *in a neighborhood of* 0. *For* $R > 0$, *we let* $\chi_R(\cdot) = \chi(R^{-1}\cdot)$. *Then, for all* $u \in \mathsf{H}$,

$$\lim_{R \to +\infty} \chi_R(\mathscr{L})u = u \,.$$

Proof. — By definition, we have

$$2\pi \chi_R(\mathscr{L})u = \int_{\mathbb{R}} \mathscr{F}\chi_R(t)e^{it\mathscr{L}}u \, dt = \int_{\mathbb{R}} R(\mathscr{F}\chi)(Rt)e^{it\mathscr{L}}u \, dt$$
$$= \int_{\mathbb{R}} (\mathscr{F}\chi)(t)e^{it\mathscr{L}/R}u \, dt \,.$$

We have, by continuity of the group, for all $t \in \mathbb{R}$,

$$\lim_{R \to +\infty} e^{it\mathscr{L}/R}u = u \,.$$

Moreover,

$$\|(\mathscr{F}\chi)(t)e^{it\mathscr{L}/R}u\| \leqslant |(\mathscr{F}\chi)(t)|\|u\|, \qquad \mathscr{F}\chi(\cdot) \in \mathsf{L}^1(\mathbb{R}) \,.$$

Therefore, we can use the dominated convergence theorem (or notice directly that the convergence is uniform on the compact subsets) to get

$$\lim_{R \to +\infty} \chi_R(\mathscr{L})u = \frac{1}{2\pi} \int_{\mathbb{R}} \mathscr{F}\chi(t)u \, dt = \chi(0)u = u \,.$$

$\square$

8.2. Where the spectral measure comes into play

Given $f \in \mathcal{S}(\mathbb{R})$, we have defined $f(\mathcal{L}) \in \mathcal{L}(\mathsf{H})$. We would like to extend this definition to the case of bounded functions. To this end, the idea in Sect. 8.2.1 is to test $f(\mathcal{L})$ against vectors in order to recover linear forms which, in view of Lemma 8.6, are continuous on $\mathscr{C}_0^0(\mathbb{R})$. In Sect. 8.2.2, the spectral measure is defined.

8.2.1. Extending a map. —

Definition 8.8. — For all $f \in \mathcal{S}(\mathbb{R})$ and $u, v \in \mathsf{H}$, we let

$$\omega_{u,v}(f) = (f(\mathcal{L})u, v).$$

We would like to extend this formula to the set $\mathscr{C}_{\to 0}^0(\mathbb{R})$ of continuous functions tending to zero at infinity.

Lemma 8.9. — *The following holds.*

  i. *For all $f \in \mathcal{S}(\mathbb{R})$, $\omega_{\cdot,\cdot}(f)$ is a continuous sesquilinear form on H and*

$$\|\omega_{\cdot,\cdot}(f)\| \leq \|f\|_\infty.$$

  ii. *For all $u \in \mathsf{H}$, the linear form*

$$\omega_{u,u} : \mathcal{S}(\mathbb{R}) \ni f \mapsto \omega_{u,u}(f) \in \mathbb{C}$$

is non-negative and continuous for the topology of $\|\cdot\|_\infty$.

  iii. *If $\mathcal{S}(\mathsf{H} \times \mathsf{H}, \mathbb{C})$ denotes the set of the continuous sesquilinear form on H, the map*

$$(\mathcal{S}(\mathbb{R}), \|\cdot\|_\infty) \ni f \mapsto \omega_{\cdot,\cdot}(f) \in (\mathcal{S}(\mathsf{H} \times \mathsf{H}, \mathbb{C}), \|\cdot\|)$$

is linear and continuous. It can be uniquely extended as a continuous linear map on $(\mathscr{C}_{\to 0}^0(\mathbb{R}), \|\cdot\|_\infty)$. Keeping the same notation $\omega_{\cdot,\cdot}(f)$ for the extended map, we have

$$\forall f \in \mathscr{C}_{\to 0}^0(\mathbb{R}), \quad \|\omega_{\cdot,\cdot}(f)\| \leq \|f\|_\infty,$$

and, for all $f \in \mathscr{C}_{\to 0}^0(\mathbb{R})$, with $f \geq 0$, we have $\omega_{\cdot,\cdot}(f) \geq 0$.

Proposition 8.10. — *Let $f \in \mathscr{C}_{\to 0}^0(\mathbb{R})$. There exists a unique bounded operator, denoted by $f(\mathcal{L})$, such that, for all $u, v \in \mathsf{H}$,*

$$\langle f(\mathcal{L})u, v \rangle = \omega_{u,v}(f).$$

We have (i)–(v). *Moreover, we have*

$$\|f(\mathscr{L})\| \leqslant \|f\|_\infty .$$

Exercise 8.11. — Let us recall Exercise 8.2, and prove that, for all $f \in \mathscr{C}^0_{\to 0}(\mathbb{R})$, we have

$$f(U \mathscr{L} U^{-1}) = U f(\mathscr{L}) U^{-1} .$$

8.2.2. Riesz theorem and spectral measure. — Let us now recall a classical representation theorem.

Theorem 8.12 (F. Riesz). — *Let X be a separated and locally compact topological space. Let ω be a non-negative form on $\mathscr{C}^0_0(X)$. Then, there exists a σ-algebra $\mathcal{M}$ containing the Borelian sets of X and a unique non-negative measure μ on $\mathcal{M}$ such that*

$$\forall f \in \mathscr{C}^0_0(\mathbb{R}), \quad \omega(f) = \int_X f \, d\mu .$$

This measure μ is regular in the sense that, for all $\Omega \in \mathcal{M}$,

$$\mu(\Omega) = \inf\{\mu(V) : V \text{ open set s.t. } \Omega \subset V\} ,$$
$$\mu(\Omega) = \sup\{\mu(K) : K \text{ compact set s.t. } K \subset \Omega\} .$$

In view of (iii) of Lemma 8.9, we can apply this theorem to $X = \mathbb{R}$ and $\omega_{u,u}$. By this way, we get a non-negative measure $\mu_{u,u}$ and a σ-algebra $\mathcal{M}_{u,u}$.

Definition 8.13. — The measure $\mu_{u,u}$ is called the *spectral measure* associated with $\mathscr{L}$ and u.

At this stage, we have

$$\forall f \in \mathscr{C}^0_0(\mathbb{R}), \quad \langle f(\mathscr{L})u, u \rangle - \int_{\mathbb{R}} f \, d\mu_{u,u} .$$

Now, we let

$$\mathcal{M} = \bigcap_{u \in \mathsf{H}} \mathcal{M}_{u,u} .$$

It is still a σ-algebra containing the Borelian sets.

Lemma 8.14. — *For all $u \in \mathsf{H}$, the measure $\mu_{u,u}$ is finite, and $\mu_{u,u}(\mathbb{R}) = \|u\|^2$.*

Proof. — We recall Lemma 8.7. Let $u \in H$. We use the function χ_R. We have, for all $R > 0$,

$$\omega_{u,u}(\chi_R) \leqslant \|u\|^2 \,,$$

and

$$\lim_{R \to +\infty} \omega_{u,u}(\chi_R) = \|u\|^2 \,.$$

Moreover, we have

$$\omega_{u,u}(\chi_R) = \int_{\mathbb{R}} \chi_R(\lambda) \, \mathrm{d}\mu_{u,u}(\lambda) \,.$$

With the Fatou Lemma, we get

$$\mu_{u,u}(\mathbb{R}) \leqslant \liminf_{R \to +\infty} \int_{\mathbb{R}} \chi_R(\lambda) \, \mathrm{d}\mu_{u,u}(\lambda) \leqslant \|u\|^2 < +\infty \,.$$

Thus, the measure $\mu_{u,u}$ is finite. It remains to use the dominated convergence theorem to see that

$$\|u\|^2 = \lim_{R \to +\infty} \omega_{u,u}(\chi_R) = \mu_{u,u}(\mathbb{R}) \,.$$

$\square$

Definition 8.15. — Let Ω be a Borelian set. We consider the application $q : H \to \mathbb{R}_+$ defined by

$$H \ni u \mapsto \int_{\mathbb{R}} \mathbb{1}_\Omega \, \mathrm{d}\mu_{u,u} = \mu_{u,u}(\Omega) \,.$$

Lemma 8.16. — q_Ω *is a continuous quadratic form.*

Proof. — Note that $0 \leqslant \mu_{u,u}(\Omega) \leqslant \|u\|^2$. In particular, once we will have proved that q_Ω is a quadratic form, it will be a continuous quadratic form (by using the polarization formula).

Since, for all $u \in H$, $\mu_{u,u}$ is a measure, we only have to prove the result when Ω is an open set or even when Ω is an interval in the form $[a, b]$. In this case, we introduce the sequence of continuous and piecewise affine functions (f_n) such that $f_n(x) = 1$ on $[a, b]$, $f_n(x) = 0$ for $x \leqslant a - \frac{1}{n}$ and $x \geqslant b + \frac{1}{n}$. By dominated convergence, we have

$$\lim_{n \to +\infty} \langle f_n(\mathscr{L})u, u \rangle = \lim_{n \to +\infty} \int_{\mathbb{R}} f_n \, \mathrm{d}\mu_{u,u} = \mu_{u,u}(\Omega) \,,$$

and the conclusion follows from the polarization formula. $\square$

Proposition 8.17. — *Let* $f : \mathbb{R} \to \mathbb{C}$ *be a bounded Borelian function. Then there exists a unique continuous sesquilinear form* $\tilde{\omega}_{\cdot,\cdot}(f)$ *on* H *such that*

$$\forall u \in \mathsf{H}, \quad \tilde{\omega}_{u,u}(f) = \int_{\mathbb{R}} f \, d\mu_{u,u} .$$

Proof. — With Lemma 8.16, this result is known for $f = \mathbb{1}_{\Omega}$, for all Borelian set Ω. From the measure theory, one knows that all bounded Borelian function is a uniform limit of step functions. This implies that $u \mapsto \int_{\mathbb{R}} f \, d\mu_{u,u}$ is a quadratic form. It is continuous since $\left| \int_{\mathbb{R}} f \, d\mu_{u,u} \right| \leqslant \|f\|_{\infty} \|u\|^2$. $\qquad\square$

From this proposition, we can define $f(\mathscr{L})$ via the Riesz representation theorem.

Proposition 8.18. — *Let* $f : \mathbb{R} \to \mathbb{C}$ *be a bounded Borelian function. There exists a unique bounded operator, denoted by* $f(\mathscr{L})$, *such that, for all* $u \in \mathsf{H}$,

$$\langle f(\mathscr{L})u, u \rangle = \int_{\mathbb{R}} f \, d\mu_{u,u} .$$

When $f \in \mathscr{C}^0_{\to 0}$ or $f \in \mathcal{A}$, we recover the same $f(\mathscr{L})$ as before.

Exercise 8.19. — Extend the result of Exercise 8.11 to f bounded and Borelian.

Proposition 8.20. — *Let* f *be a non-negative bounded Borelian function. We have*

$$\|f(\mathscr{L})\| \leqslant \|f\|_{\infty} .$$

Proof. — For all $u \subset \mathsf{H}$, we have

$$0 \leqslant \langle f(\mathscr{L})u, u \rangle \leqslant \|f\|_{\infty} \|u\|^2 .$$

$\qquad\square$

Proposition 8.21. — *Let* $t \in \mathbb{R}$ *and consider* $f(\cdot) = e^{it\cdot}$. *We have* $f(\mathscr{L}) = e^{it\mathscr{L}}$. *In particular,*

$$(8.8.2.2) \quad \forall u \in \mathsf{H}, \forall t \in \mathbb{R}, \quad \langle e^{it\mathscr{L}} u, u \rangle = \int_{\mathbb{R}} e^{it\lambda} \, d\mu_{u,u}(\lambda) .$$

Proof. — Let us consider $\rho \in \mathscr{C}_0^\infty(\mathbb{R})$ such that $0 \leqslant \rho \leqslant 1$, $\mathrm{supp}(\rho) \subset [-1, 1]$ and $\int_\mathbb{R} \rho(x)\,\mathrm{d}x = 2\pi$. We introduce $\chi \in \mathscr{S}(\mathbb{R})$ such that $\mathscr{F}\chi = \rho$. For all $n \in \mathbb{N}^*$, we let

$$\rho_n(\cdot) = n\rho(n\cdot) = \mathscr{F}(\chi(n^{-1}\cdot))\,.$$

Note that

$$\chi(n^{-1}x) = (2\pi)^{-1} \int_\mathbb{R} \rho_n(\xi)e^{ix\xi}\,\mathrm{d}\xi = (2\pi)^{-1} \int_\mathbb{R} \rho(\xi)e^{i\xi\frac{x}{n}}\,\mathrm{d}\xi\,.$$

Thus, $\lim_{n\to+\infty} \chi(n^{-1}x) = 1$ and $\|\chi(n^{-1}\cdot)\|_\infty \leqslant 1$.

Let us consider $f_n(\cdot) = \chi(n^{-1}\cdot)e^{it\cdot} \in \mathscr{S}(\mathbb{R})$. For all $u \in \mathrm{H}$, we have

$$\langle f_n(\mathscr{L})u, u \rangle = \int_\mathbb{R} f_n\,\mathrm{d}\mu_{u,u}\,.$$

By the dominated convergence theorem, we have

$$\lim_{n\to+\infty} \int_\mathbb{R} f_n\,\mathrm{d}\mu_{u,u} = \int_\mathbb{R} e^{it\lambda}\,\mathrm{d}\mu_{u,u}(\lambda)\,.$$

But, we also have

$$f_n(\mathscr{L})u = (2\pi)^{-1} \int_\mathbb{R} \mathscr{F}f_n(\lambda)e^{i\lambda\mathscr{L}}u\,\mathrm{d}\lambda$$

$$= (2\pi)^{-1} \int_\mathbb{R} \rho_n(\lambda - t)e^{i\lambda\mathscr{L}}u\,\mathrm{d}\lambda\,,$$

and then

$$f_n(\mathscr{L})u = (2\pi)^{-1}e^{it\mathscr{L}} \int_\mathbb{R} \rho(\lambda)e^{in^{-1}\lambda\mathscr{L}}u\,\mathrm{d}\lambda\,,$$

so that

$$\lim_{n\to+\infty} f_n(\mathscr{L})u = e^{it\mathscr{L}}u\,.$$

Therefore, we have, for all $u \in \mathrm{H}$,

$$\langle e^{it\mathscr{L}}u, u \rangle = \int_\mathbb{R} e^{it\lambda}\,\mathrm{d}\mu_{u,u}(\lambda)\,.$$

$\square$

8.3. Spectral projections

8.3.1. Properties. —

Definition 8.22. — Let Ω be a Borelian set. We let $E_\Omega = \mathbb{1}_\Omega(\mathscr{L}) \in \mathcal{L}(\mathsf{H})$.

Proposition 8.23. — *There holds:*

(i) $E_\emptyset = 0$ *and* $E_\mathbb{R} = \mathrm{Id}$.

(ii) *For all open set* Ω, E_Ω *is an orthogonal projection.*

(iii) *For all open sets* Ω_1 *and* Ω_2, $E_{\Omega_1} E_{\Omega_2} = E_{\Omega_1 \cap \Omega_2}$.

(iv) *Let* $\Omega = \bigcup_{j \in \mathbb{N}} \Omega_j$ *be a partition with open sets. Then, for all* $u \in \mathsf{H}$,

$$\lim_{N \to +\infty} \sum_{j=0}^{N} E_{\Omega_j} u = E_\Omega u.$$

Proof. — For the first point, we use Lemma 8.14. Let $V \subset \mathbb{R}$ be an open set. By using an exhaustion by compact sets of V and Urysohn's lemma, we can construct a non-decreasing sequence $(f_n) \subset \mathscr{C}_0^0(\mathbb{R})$ such that $f_n f_m = f_n$ for all $m \geqslant n$ and $\lim_{n \to +\infty} f_n = \mathbb{1}_V$. For all $u \in \mathsf{H}$, we have

$$\langle f_n(\mathscr{L})u, u \rangle = \int_\mathbb{R} f_n \, \mathrm{d}\mu_{u,u},$$

and thus, by Beppo Levi's theorem,

$$\lim_{n \to +\infty} \langle f_n(\mathscr{L})u, u \rangle = \langle \mathbb{1}_V(\mathscr{L})u, u \rangle.$$

This implies that, for all $u, v \in \mathsf{H}$,

$$\lim_{n \to +\infty} \langle f_n(\mathscr{L})u, v \rangle = \langle \mathbb{1}_V(\mathscr{L})u, v \rangle.$$

We have, for all $m \geqslant n$,

$$\langle f_m(\mathscr{L})u, f_n(\mathscr{L})^* u \rangle = \langle (f_n f_m)(\mathscr{L})u, u \rangle = \langle f_n(\mathscr{L})u, u \rangle.$$

Taking the limit $m \to +\infty$, we get

$$\langle f_n(\mathscr{L}) \mathbb{1}_V(\mathscr{L})u, u \rangle = \langle f_n(\mathscr{L})u, u \rangle,$$

so that, for all $u \in \mathsf{H}$,

$$\langle \mathbb{1}_V(\mathscr{L})^2 u, u \rangle = \langle \mathbb{1}_V(\mathscr{L})u, u \rangle.$$

Thus, $\mathbb{1}_V(\mathscr{L})^2 = \mathbb{1}_V(\mathscr{L})$, and it is clear that the operator $\mathbb{1}_V(\mathscr{L})$ is self-adjoint (since $\mathbb{1}_V$ is real-valued $\overline{f_n} = f_n$). If V_1 and V_2 are two open sets, we easily get, by considering associated sequences of functions,

$$\mathbb{1}_{V_1}(\mathscr{L})\mathbb{1}_{V_2}(\mathscr{L}) = \mathbb{1}_{V_1 \cap V_2}(\mathscr{L}).$$

Let us prove (iv). Take $u \in \mathsf{H}$. For all $N \geqslant 1$, we have

$$\left\| \sum_{j=1}^{N} E_{\Omega_j} u - E_\Omega u \right\|^2 = \mu_{u,u}(\Omega) - \sum_{j=1}^{N} \mu_{uu}(\Omega_j),$$

and we get the desired convergence.   □

Proposition 8.24. — *For all* $f, g \in \mathscr{B}_b(\mathbb{R}, \mathbb{C})$, *we have* $f(\mathscr{L})g(\mathscr{L}) = (fg)(\mathscr{L})$.

Proof. — Let us denote by $\mathscr{O}$ the class of open sets of $\mathbb{R}$. Let $V \in \mathscr{O}$. Consider the set

$$\mathscr{A} = \{W \subset \mathbb{R} : \mathbb{1}_V(\mathscr{L})\mathbb{1}_W(\mathscr{L}) = (\mathbb{1}_V \mathbb{1}_W)(\mathscr{L})\}.$$

We have $\mathscr{O} \subset \mathscr{A}$. It is clear that $\mathscr{O}$ is a π-system[1]. Moreover, we can show that $\mathscr{A}$ is a λ-system[2] by using similar arguments as in the proof of Proposition 8.23. The monotone class theorem shows that the smallest λ-system containing $\mathscr{O}$ is the σ-algebra generated by $\mathscr{O}$, *i.e.*, the Borelian σ-algebra $\mathscr{B}(\mathbb{R})$. In particular, we deduce that

$$\mathscr{B}(\mathbb{R}) \subset \mathscr{A}.$$

Playing the same game with $V \in \mathscr{B}(\mathbb{R})$, we get that

$$\forall V, W \in \mathscr{B}(\mathbb{R}), \quad \mathbb{1}_V(\mathscr{L})\mathbb{1}_W(\mathscr{L}) = (\mathbb{1}_V \mathbb{1}_W)(\mathscr{L}).$$

We can extend this formula by linearity to all step functions f and g, we have

$$f(\mathscr{L})g(\mathscr{L}) = (fg)(\mathscr{L}).$$

Since all bounded Borelian functions can be uniformly approximated by sequences of step functions, we deduce the result.   □

Corollary 8.25. — *There holds:*

(i) $E_\emptyset = 0$ *and* $E_\mathbb{R} = \mathrm{Id}$.

---

[1]it is stable under taking finite intersections

[2]it is stable under taking non-decreasing unions and by proper differences

(ii) *For all Borelian set Ω, E_Ω is an orthogonal projection.*

(iii) *For all Borelian sets Ω_1 and Ω_2, $E_{\Omega_1} E_{\Omega_2} = E_{\Omega_1 \cap \Omega_2}$.*

(iv) *Let $\Omega = \bigcup_{j \in \mathbb{N}} \Omega_j$ be a Borelian partition. For all $u \in \mathsf{H}$,*

$$\lim_{N \to +\infty} \sum_{j=0}^{N} E_{\Omega_j} u = E_\Omega u \, .$$

Proposition 8.26. — *For all bounded Borelian functions, we have the properties* (i)–(v) *of the introduction, and*

$$\|f(\mathscr{L})\| \leqslant \|f\|_\infty \, .$$

Proof. — Let us check (ii). Let $u \in \mathrm{Dom}\,(\mathscr{L})$ and $\varepsilon > 0$. Then, we have, with the multiplication property (iv) and Proposition 8.21,

$$\frac{e^{i\varepsilon \mathscr{L}} - \mathrm{Id}}{\varepsilon} f(\mathscr{L}) u = f(\mathscr{L}) \frac{e^{i\varepsilon \mathscr{L}} - \mathrm{Id}}{\varepsilon} u \, .$$

The conclusion follows by taking the limit $\varepsilon \to 0$.

The last inequality comes from the fact that, for all $u \in \mathsf{H}$,

$$\|f(\mathscr{L})u\|^2 = \langle f(\mathscr{L})^* f(\mathscr{L})u, u \rangle = \langle \overline{f}(\mathscr{L}) f(\mathscr{L})u, u \rangle$$

$$= \langle (\overline{f}f)(\mathscr{L})u, u \rangle = \int_{\mathbb{R}} |f|^2 \, \mathrm{d}\mu_{u,u} \, .$$

$\square$

Proposition 8.27. — *Let Ω be a bounded Borelian set. Then, for all $u \in \mathsf{H}$, we have $\mathbb{1}_\Omega(\mathscr{L})u \in \mathrm{Dom}\,(\mathscr{L})$.*

Proof. — For all $\varepsilon > 0$ and $u \in \mathsf{H}$, we have, by Propositions 8.21 and 8.26,

$$\left\| \frac{e^{i\varepsilon \mathscr{L}} - \mathrm{Id}}{\varepsilon} \mathbb{1}_\Omega(\mathscr{L})u \right\|^2 = \int_\Omega \left| \frac{e^{i\varepsilon \lambda} - 1}{\varepsilon} \right|^2 \mathrm{d}\mu_{u,u}$$

$$\leqslant \int_\Omega |\lambda|^2 \, \mathrm{d}\mu_{u,u} < +\infty \, .$$

$\square$

8.3.2. Extension to unbounded functions. —

Definition 8.28. — Let $f : \mathbb{R} \to \mathbb{C}$ be a Borelian function. We let

$$\mathrm{Dom}\,(f(\mathscr{L})) = \{u \in \mathsf{H} : \int_{\mathbb{R}} |f|^2 \, d\mu_{u,u} < +\infty\}.$$

For all $u \in \mathrm{Dom}\,(\mathscr{L})$, we let

$$f(\mathscr{L})u = \lim_{n \to +\infty} f_n(\mathscr{L})u,$$

with $f_n(\lambda) = f(\lambda)\mathbb{1}_{|f| \leqslant n}(\lambda)$.

Note that this definition is consistent since, for all $u \in \mathrm{Dom}\,(\mathscr{L})$, and all $m \geqslant n$,

$$\|(f_n(\mathscr{L}) - f_m(\mathscr{L}))u\|^2 = \int_{\mathbb{R}} |f_n - f_m|^2 \, d\mu_{u,u}$$

$$= \int_{\{|f|>n\}} |f|^2 \, d\mu_{u,u}.$$

Lemma 8.29. — Let $f : \mathbb{R} \to \mathbb{C}$ be a Borelian function. Then $\mathrm{Dom}\,(f(\mathscr{L}))$ is dense.

Proof. — For all $\varphi \in \mathsf{H}$, we let $\varphi_n = \mathbb{1}_{|f| \leqslant n}(\mathscr{L})\varphi$. The sequence $(\varphi_n)_{n \in \mathbb{N}}$ converges to φ.

For all $k \in \mathbb{N}$, we have

$$\|f_k(\mathscr{L})\varphi_n\|^2 = \int_{\mathbb{R}} |f_k|^2 \, d\mu_{\varphi_n,\varphi_n} = \int_{\mathbb{R}} |f_k|^2 \mathbb{1}_{|f| \leqslant n} \, d\mu_{\varphi,\varphi}$$

$$= \int_{\mathbb{R}} |f|^2 \mathbb{1}_{|f| \leqslant k} \mathbb{1}_{|f| \leqslant n} \, d\mu_{\varphi,\varphi}.$$

Thus, for $k \geqslant n$, we have

$$\int_{\mathbb{R}} |f_k|^2 \, d\mu_{\varphi_n,\varphi_n} \leqslant n^2 \|\varphi\|^2.$$

By the Fatou lemma, it follows

$$\int_{\mathbb{R}} |f|^2 \, d\mu_{\varphi_n,\varphi_n} \leqslant n^2 \|\varphi\|^2 < +\infty.$$

The density follows.

Let us explain why $f(\mathscr{L})\varphi_n = f_n(\mathscr{L})\varphi$. We have $f_k(\mathscr{L})\varphi_n = (f\mathbb{1}_{|f| \leqslant k}\mathbb{1}_{|f| \leqslant n})(\mathscr{L})\varphi = f_n(\mathscr{L})\varphi_k$. We can take the limit $k \to +\infty$, and we find $f(\mathscr{L})\varphi_n = f_n(\mathscr{L})\varphi$. $\qquad\square$

Proposition 8.30. — *Let us consider* $f = \mathrm{Id}_\mathbb{R}$. *Then, we have* $f(\mathscr{L}) = \mathscr{L}$.

Proof. — We must check that

$$\mathrm{Dom}\,(\mathscr{L}) = \{u \in \mathsf{H} : \int_\mathbb{R} |\lambda|^2 \, \mathrm{d}\mu_{u,u} < +\infty\}.$$

Thanks to Proposition 8.21, we have, for all $u \in \mathsf{H}$,

$$\left\|\frac{e^{i\varepsilon\mathscr{L}} - \mathrm{Id}}{\varepsilon}u\right\|^2 = \int_\mathbb{R} \left|\frac{e^{i\varepsilon\lambda} - 1}{\varepsilon}\right|^2 \, \mathrm{d}\mu_{u,u}.$$

If $u \in \mathrm{Dom}\,(\mathscr{L})$, we have $\lim_{\varepsilon\to 0} \frac{e^{i\varepsilon\mathscr{L}}-\mathrm{Id}}{\varepsilon}u = \mathscr{L}u$. Thus, by the Fatou lemma, it follows that

$$\|\mathscr{L}u\|^2 \geqslant \int_\mathbb{R} |\lambda|^2 \, \mathrm{d}\mu_{u,u}.$$

Conversely, if $\int_\mathbb{R} |\lambda|^2 \, \mathrm{d}\mu_{u,u} < +\infty$, and noticing that

$$\left|\frac{e^{i\varepsilon\lambda} - 1}{\varepsilon}\right|^2 \leqslant |\lambda|^2,$$

we get that $\left\|\frac{e^{i\varepsilon\mathscr{L}}-\mathrm{Id}}{\varepsilon}u\right\|^2$ is bounded for $\varepsilon \in (0, 1]$. Thus, $u \in \mathrm{Dom}\,(\mathscr{L})$. Note that this implies that

$$\|\mathscr{L}u\|^2 = \int_\mathbb{R} |\lambda|^2 \, \mathrm{d}\mu_{u,u}.$$

Then, we consider $f_n(\lambda) = \lambda \mathbb{1}_{|\lambda|\leqslant n}(\lambda)$ and we write, for all $u \in \mathsf{H}$,

$$\langle f_n(\mathscr{L})u, u \rangle = \int_\mathbb{R} \lambda \mathbb{1}_{|\lambda|\leqslant n}(\lambda) \, \mathrm{d}\mu_{u,u}.$$

By the Cauchy–Schwarz inequality, we have

$$\int_\mathbb{R} |\lambda| \, \mathrm{d}\mu_{u,u} \leqslant \left(\int_\mathbb{R} |\lambda|^2 \, \mathrm{d}\mu_{u,u}\right)^{\frac{1}{2}} \|u\|,$$

and thus, we can use the dominated convergence theorem to get, for all $u \in \mathrm{Dom}\,(\mathscr{L})$,

$$\langle f(\mathscr{L})u, u \rangle = \int_\mathbb{R} \lambda \, \mathrm{d}\mu_{u,u} = \langle \mathscr{L}u, u \rangle,$$

where we used the derivative of (8.8.2.2) for the last equality. The conclusion follows. $\square$

Proposition 8.31. — *If Ω is a bounded Borelian, we have, for all $u \in \mathrm{Dom}\,(\mathscr{L})$,*

$$\|\mathbb{1}_\Omega(\mathscr{L})\mathscr{L}u\| \leqslant \sup_{\lambda \in \Omega} |\lambda| \|u\|\,.$$

In particular, $\mathbb{1}_\Omega(\mathscr{L})\mathscr{L}$ can be extended as a bounded operator on H.

Proof. — For all $n \in \mathbb{N}^*$, we let $f_n(\lambda) = \mathbb{1}_{|\lambda| \leqslant n}(\lambda)$. We have that

$$\forall u \in \mathrm{Dom}\,(\mathscr{L})\,, \quad \lim_{n \to +\infty} f_n(\mathscr{L})u = \mathscr{L}u\,.$$

Now, for all $n \in \mathbb{N}^*$ and $u \in$ H,

$$\|\mathbb{1}_\Omega(\mathscr{L})f_n(\mathscr{L})u\| \leqslant \sup_{\lambda \in \Omega} |\lambda| \|u\|\,.$$

Taking the limit for $u \in \mathrm{Dom}\,(\mathscr{L})$, we get the result. $\square$

Proposition 8.32. — *In the class of Borelian functions, we have (iii)–(v). Moreover, the operator $f(\mathscr{L})$ is closed with dense domain.*

Proof. — The density comes from Lemma 8.29. For all $u, v \in \mathrm{Dom}\,(f(\mathscr{L})) = \mathrm{Dom}\,(\overline{f}(\mathscr{L}))$, we have

$$\langle f(\mathscr{L})u, v \rangle = \lim_{n \to +\infty} \langle f_n(\mathscr{L})u, v \rangle = \lim_{n \to +\infty} \langle u, \overline{f_n}(\mathscr{L})v \rangle$$

$$= \langle u, \overline{f}(\mathscr{L})v \rangle\,.$$

This shows that $\overline{f}(\mathscr{L}) \subset f(\mathscr{L})^*$. Take $v \in \mathrm{Dom}\,(f(\mathscr{L})^*)$. We have, for all $u \in \mathrm{Dom}\,(f(\mathscr{L}))$,

$$\langle f(\mathscr{L})u, v \rangle = \langle u, f(\mathscr{L})^*v \rangle\,,$$

so that

$$|\langle f(\mathscr{L})u, v \rangle| \leqslant \|f(\mathscr{L})^*v\| \|u\|\,.$$

For all $n \in \mathbb{N}$, we take $u = u_n = \mathbb{1}_{|f| \leqslant n}\varphi$ with $\varphi \in$ H (see the proof of Lemma 8.29). We get, for all $n \in \mathbb{N}$ and $\varphi \in$ H,

$$|\langle f_n(\mathscr{L})\varphi, v \rangle| \leqslant \|f(\mathscr{L})^*v\| \|\varphi\|\,,$$

and thus

$$|\langle \varphi, \overline{f_n}(\mathscr{L})v \rangle| \leqslant \|f(\mathscr{L})^*v\| \|\varphi\|\,.$$

We deduce that, for all $n \in \mathbb{N}$,

$$\int_{\mathbb{R}} |f_n|^2 \, \mathrm{d}\mu_{v,v} = \|\overline{f_n}(\mathscr{L})v\|^2 \leqslant \|f(\mathscr{L})^*v\|^2\,.$$

By the Fatou lemma, we get that $v \in \mathrm{Dom}\,(\overline{f}(\mathscr{L}))$. This proves that $f(\mathscr{L})^* = \overline{f}(\mathscr{L})$. In particular, this establishes that $f(\mathscr{L})$ is closed as the adjoint of $\overline{f}(\mathscr{L})$.

It remains to prove (iv). We have, for all $u \in \mathsf{H}$,

$$f_m(\mathscr{L})g_n(\mathscr{L})u = (f_m g_n)(\mathscr{L})u\,.$$

Then,

$$\|f_m(\mathscr{L})g_n(\mathscr{L})u\|^2 = \int_{\mathbb{R}} |f_m|^2 |g_n|^2 \,\mathrm{d}\mu_{u,u}\,,$$

so that, for all $u \in \{v \in \mathrm{Dom}\,(g(\mathscr{L})) : g(\mathscr{L})v \in \mathrm{Dom}\,(f(\mathscr{L}))\}$,

$$\liminf_{m\to+\infty} \liminf_{n\to+\infty} \int_{\mathbb{R}} |f_m|^2 |g_n|^2 \,\mathrm{d}\mu_{u,u} \leqslant \|f(\mathscr{L})g(\mathscr{L})u\|^2\,.$$

By the Fatou lemma, it follows that $u \in \mathrm{Dom}\,((fg)(\mathscr{L}))$. We have

$$f_m(\mathscr{L})g_n(\mathscr{L})u = (f_m g_n)(\mathscr{L})u\,,$$

and it remains to take the limits. $\qquad\qquad\qquad\qquad\qquad\square$

8.3.3. Characterization of the spectra. —

Proposition 8.33. *— $\lambda \in \mathrm{sp}(\mathscr{L})$ if and only if, for all $\varepsilon > 0$, $\mathbb{1}_{(\lambda-\varepsilon,\lambda+\varepsilon)}(\mathscr{L}) \neq 0$. In particular, for all $u \in \mathsf{H}$, the support of $\mu_{u,u}$ is contained in $\mathrm{sp}(\mathscr{L})$.*

Proof. — Assume that, for all $\varepsilon > 0$, we have $\mathbb{1}_{(\lambda-\varepsilon,\lambda+\varepsilon)}(\mathscr{L}) \neq 0$. Since $\mathbb{1}_{(\lambda-\varepsilon,\lambda+\varepsilon)}(\mathscr{L})$ is a non-zero projector, we can consider $u_\varepsilon \in \mathsf{H}$ such that $\|u_\varepsilon\| = 1$ and

$$\mathbb{1}_{(\lambda-\varepsilon,\lambda+\varepsilon)}(\mathscr{L})u_\varepsilon = u_\varepsilon \in \mathrm{Dom}\,(\mathscr{L})\,.$$

We write

$$\|(\mathscr{L} - \lambda)u_\varepsilon\|^2$$
$$= \|\mathbb{1}_{(\lambda-\varepsilon,\lambda+\varepsilon)}(\mathscr{L})(\mathscr{L} - \lambda)u_\varepsilon\|^2$$
$$= \int_{\lambda-\varepsilon}^{\lambda+\varepsilon} (t - \lambda)^2 \,\mathrm{d}\mu_{u_\varepsilon,u_\varepsilon}(t) \leqslant \varepsilon^2 \mu_{u_\varepsilon,u_\varepsilon}(\mathbb{R}) \leqslant \varepsilon^2\,.$$

Thus, $\lambda \in \mathrm{sp}(\mathscr{L})$. Conversely, assume that there exists $\varepsilon_0 > 0$ such that $\mathbb{1}_{(\lambda-\varepsilon_0,\lambda+\varepsilon_0)}(\mathscr{L}) = 0$. Let us consider the bounded operator R_λ defined via

$$\forall u \in \mathsf{H}\,, \quad \langle R_\lambda u, u\rangle = \int_{|\mu-\lambda|\geqslant\varepsilon_0} (\mu - \lambda)^{-1} \,\mathrm{d}\mu_{u,u}\,.$$

Remark that, for all $t \in (0, 1]$ and all $u \in \mathsf{H}$, we have

$$\left\| \frac{e^{it\mathscr{L}} - \mathrm{Id}}{t} R_\lambda u \right\|^2 = \int_{|\mu - \lambda| \geqslant \varepsilon_0} (\mu - \lambda)^{-2} \left| \frac{e^{it\lambda} - 1}{t} \right|^2 \, \mathrm{d}\mu_{u,u}$$
$$\leqslant \int_{|\mu - \lambda| \geqslant \varepsilon_0} \lambda^2 (\mu - \lambda)^{-2} \, \mathrm{d}\mu_{u,u} < +\infty.$$

Applying the criterion (7.7.3.4), we get that $R_\lambda u \in \mathrm{Dom}\,(\mathscr{L})$. With Lemma 8.14 and Proposition 8.32, we write, for all $u \in \mathsf{H}$,

$$\langle (\mathscr{L} - \lambda) R_\lambda u, u \rangle = \int_{|\mu - \lambda| \geqslant \varepsilon_0} \mathrm{d}\mu_{u,u} = \mu_{u,u}(\mathbb{R}) = \|u\|^2.$$

This shows that $(\mathscr{L} - \lambda) R_\lambda = \mathrm{Id}$. In the same way, we can get that $R_\lambda (\mathscr{L} - \lambda) = \mathrm{Id}_{\mathrm{Dom}\,(\mathscr{L})}$. Thus, we have $\lambda \in \rho(\mathscr{L})$. $\qquad \square$

Exercise 8.34. — For $z \notin \mathrm{sp}(\mathscr{L})$, consider the Borelian function $f_z(x) = (x - z)^{-1} \mathbb{1}_{|x-z| > \varepsilon}(x)$. Show that $f_z(\mathscr{L}) = (\mathscr{L} - z)^{-1}$ by using the same ideas as in the proof of Proposition 8.33.

Lemma 8.35. — *Let f be a Borelian function. If $u \in \mathrm{Dom}\,(\mathscr{L})$ satisfies $\mathscr{L}u = \lambda u$, then $f(\mathscr{L})u = f(\lambda)u$.*

Proof. — We have, for all $t \in \mathbb{R}$, $e^{it\mathscr{L}}u = e^{it\lambda}u$. Thus, for all $f \in \mathscr{S}(\mathbb{R})$, by the inverse Fourier transform, we have $f(\mathscr{L})u = f(\lambda)u$. This can be extended to $f \in \mathscr{C}^0_{\to 0}(\mathbb{R})$ by density and then to all Borelian function. Note that the formula holds for all functions f coinciding outside sets which are of zero measure for the spectral measure. $\qquad \square$

Proposition 8.36. — *An element λ belongs to the point spectrum if and only if $\mathbb{1}_{\{\lambda\}}(\mathscr{L}) \neq 0$. Moreover, $\mathbb{1}_{\{\lambda\}}(\mathscr{L})$ is the orthogonal projection on $\ker(\mathscr{L} - \lambda)$.*

Proof. — Assume that there exists $u \in \mathrm{Dom}\,(\mathscr{L})$ with $u \neq 0$ such that $\mathscr{L}u = \lambda u$. By Lemma 8.35, we have

$$\mathbb{1}_{\{\lambda\}}(\mathscr{L})u = \mathbb{1}_{\{\lambda\}}(\lambda)u = u \neq 0.$$

Conversely, assume that $\mathbb{1}_{\{\lambda\}}(\mathscr{L}) \neq 0$. Then, take $u \neq 0$ such that $\mathbb{1}_{\{\lambda\}}(\mathscr{L})u = u$. We get

$$\mathscr{L}\mathbb{1}_{\{\lambda\}}(\mathscr{L})u = \mathscr{L}u = g(\lambda)u, \qquad g(t) := t\mathbb{1}_{\{\lambda\}}(t),$$

and thus $\mathscr{L}u = \lambda u$. $\qquad \square$

Proposition 8.37. — *We have $\lambda \in \mathsf{sp}_{\mathsf{ess}}(\mathscr{L})$ if and only if, for all $\varepsilon > 0$, we have*

$$\dim \operatorname{ran} \mathbb{1}_{(\lambda-\varepsilon,\lambda+\varepsilon)}(\mathscr{L}) = +\infty.$$

Proof. — If $\lambda \notin \mathsf{sp}_{\mathsf{ess}}(\mathscr{L})$, it is an isolated eigenvalue with finite multiplicity. Then, for some $\varepsilon > 0$, we have $\mathbb{1}_{(\lambda-\varepsilon,\lambda+\varepsilon)}(\mathscr{L}) = \mathbb{1}_{\{\lambda\}}(\mathscr{L})$. By Proposition 8.36, we have

$$\operatorname{ran} \mathbb{1}_{(\lambda-\varepsilon,\lambda+\varepsilon)}(\mathscr{L}) = \operatorname{ran} \mathbb{1}_{\{\lambda\}}(\mathscr{L}) = \ker(\mathscr{L} - \lambda),$$

which has a finite dimension.

Conversely, assume that λ is not isolated with finite multiplicity (see Remark 6.18). By replacing $\mathscr{L}$ by $\mathscr{L} - \lambda$, we can always assume that $\lambda = 0$. If λ is isolated, it is an eigenvalue of infinite multiplicity (see Lemma 6.16). It is sufficient to consider an infinite orthonormal family in $\ker(\mathscr{L})$ to get that

$$\dim \operatorname{ran} \mathbb{1}_{\{0\}}(\mathscr{L}) = \dim \ker(\mathscr{L}) = +\infty.$$

Thus, we can assume that 0 is not isolated. Then, we have

$$\forall n \in \mathbb{N}^*, \qquad \exists \lambda_n \in]-1/n, 0[\cup]0, 1/n[, \qquad \lambda_n \in \mathsf{sp}(\mathscr{L}).$$

By Proposition 8.33, we have that

$$\forall n \in \mathbb{N}^*, \qquad \mathbb{1}_{(\lambda_n-|\lambda_n|/2,\lambda_n+|\lambda_n|/2)}(\mathscr{L}) \neq 0.$$

Since we have a projection, we can find $u_n \in \mathsf{H}$ such that

$$\forall n \in \mathbb{N}^*, \qquad \mathbb{1}_{(\lambda_n-|\lambda_n|/2,\lambda_n+|\lambda_n|/2)}(\mathscr{L})u_n = u_n, \qquad \|u_n\| = 1.$$

Up to extracting a subsequence (*i.e.*, an increasing function $\varphi : \mathbb{N} \to \mathbb{N}$), we can assume that the intervals

$$I_n = (\lambda_{\varphi(n)} - |\lambda_{\varphi(n)}|/2, \lambda_{\varphi(n)} + |\lambda_{\varphi(n)}|/2)$$

are disjoint. Fix any $\varepsilon > 0$. For $m \neq n$, we find

$$\begin{aligned}
\mathbb{1}_{I_n}(\mathscr{L})u_{\varphi(m)} &= \mathbb{1}_{I_n}(\mathscr{L})\mathbb{1}_{I_m}(\mathscr{L})u_{\varphi(m)} \\
&= \mathbb{1}_{\emptyset}(\mathscr{L})u_{\varphi(m)} = 0.
\end{aligned}$$

This shows that

$$u_{\varphi(m)} \in \ker \mathbb{1}_{I_n}(\mathscr{L}) \perp \operatorname{ran} \mathbb{1}_{I_n}(\mathscr{L}) \ni u_{\varphi(n)}.$$

The family $(u_{\varphi(n)})_n$ is therefore an infinite orthonormal family. For $n \geqslant n_\varepsilon$ with n_ε large enough, it is in the range of the projector $\mathbb{1}_{(-\varepsilon,\varepsilon)}(\mathscr{L})$. $\qquad\square$

8.3.4. Positive and negative parts of a self-adjoint operator. —
In this section, we consider a self-adjoint operator $\mathscr{L}$. We assume that $\mathscr{L}$ is bounded from below, *i.e.*, $\mathscr{L} \geqslant -C$. From the min-max theorem, this implies in particular that $\mathrm{sp}(\mathscr{L}) \subset [-C, +\infty)$.

Let us consider the following operators defined through the functional calculus:

$$\mathscr{L}_+ = f_+(\mathscr{L}), \quad \mathscr{L}_- = f_-(\mathscr{L}),$$

(where $f_+(\lambda) = \lambda \mathbb{1}_{[0,+\infty)}(\lambda)$ and $f_-(\lambda) = -\lambda \mathbb{1}_{(-\infty,0)}(\lambda)$) acting on their respective domains:

$$\mathrm{Dom}(\mathscr{L}_+) = \left\{ u \in \mathsf{H} : \int_{[0,+\infty)} |\lambda|^2 \, \mathrm{d}\mu_{u,u} \right\},$$

$$\mathrm{Dom}(\mathscr{L}_-) = \left\{ u \in \mathsf{H} : \int_{(-\infty,0)} |\lambda|^2 \, \mathrm{d}\mu_{u,u} \right\}.$$

Lemma 8.38. — *We have* $\mathrm{Dom}(\mathscr{L}_-) = \mathsf{H}$, *and* $\mathscr{L}_-$ *is a bounded operator on* H.

Proof. — From Proposition 8.33, we know that the support of $\mu_{u,u}$ is contained in $[-C, +\infty)$ for all $u \in \mathsf{H}$. Thus, for all $u \in \mathsf{H}$,

$$\int_{(-\infty,0)} |\lambda|^2 \, \mathrm{d}\mu_{u,u} = \int_{(-C,0)} |\lambda|^2 \, \mathrm{d}\mu_{u,u} \leqslant C^2 < +\infty.$$

Thus,

$$\mathrm{Dom}(\mathscr{L}_-) = \mathsf{H}.$$

Moreover,

$$f_-(\mathscr{L})u = \lim_{n \to +\infty} f_n(\mathscr{L})u, \quad f_n(\lambda) = f_-(\lambda) \mathbb{1}_{|f_-| \leqslant n}.$$

Therefore,

$$\langle \mathscr{L}_- u, u \rangle = \lim_{n \to +\infty} \langle f_n(\mathscr{L})u, u \rangle$$

$$= - \lim_{n \to +\infty} \int_{|f| \leqslant n} \mathbb{1}_{(-\infty,0)}(\lambda) \lambda \, \mathrm{d}\mu_{u,u}$$

$$= - \lim_{n \to +\infty} \int_{|f| \leqslant n, \, -C \leqslant \lambda < 0} \lambda \, \mathrm{d}\mu_{u,u} \leqslant C \|u\|^2.$$

This shows that $\mathscr{L}_-$ is bounded by C.   □

Lemma 8.39. — *We have* $\mathrm{Dom}(\mathscr{L}_+) = \mathrm{Dom}(\mathscr{L})$. *Moreover,*
$$\mathscr{L} = \mathscr{L}_+ - \mathscr{L}_- \,.$$

Proof. — We recall that
$$\mathrm{Dom}(\mathscr{L}) = \int_{\mathbb{R}} |\lambda|^2 \, \mathrm{d}\mu_{u,u} < +\infty \,.$$
This shows that $\mathrm{Dom}(\mathscr{L}) \subset \mathrm{Dom}(\mathscr{L}_+)$. Then, for all $u \in \mathrm{Dom}(\mathscr{L}_+)$,
$$\int_{\mathbb{R}} |\lambda|^2 \, \mathrm{d}\mu_{u,u} = \int_{[-C,+\infty)} |\lambda|^2 \, \mathrm{d}\mu_{u,u}$$
$$= \int_{(0,+\infty)} |\lambda|^2 \, \mathrm{d}\mu_{u,u} + \int_{[-C,0]} |\lambda|^2 \, \mathrm{d}\mu_{u,u} < +\infty \,.$$
Then, it remains to notice that $f_+ - f_- = \mathrm{Id}$. $\qquad\qquad \square$

Exercise 8.40. — Prove that, for all $u \in \mathrm{Dom}(\mathscr{L}_+)$, we have $u \in \mathrm{Dom}(\mathscr{L}_+^{\frac{1}{2}})$ and
$$\|\mathscr{L}_+^{\frac{1}{2}} u\|^2 = \int \lambda_+ \, \mathrm{d}\mu_{u,u} = \langle \mathscr{L}_+ u, u \rangle \,.$$

8.3.5. Decomposition of the spectral measure. —

8.3.5.1. *Lebesgue decomposition theorem.* —

Definition 8.41. — Let μ be a Borel measure on $\mathbb{R}$. We say that
   i. μ is a pure point measure when, for all Borelian set X,
$$\mu(X) = \sum_{x \in X} \mu(\{x\}) \,.$$
   ii. μ is continuous when, for all $x \in \mathbb{R}$, $\mu(\{x\}) = 0$.
   iii. μ is absolutely continuous with respect to the Lebesgue measure when all Borelian set X with Lebesgue measure zero satisfies $\mu(X) = 0$.
   iv. μ is singular with respect to the Lebesgue measure when there exists a Borelian set S_0 such that $\mu(S_0) = 0$ and $\lambda(\mathbb{R} \setminus S_0) = 0$.

Lemma 8.42. — *Consider two Borelian measures μ and ν on a topological space X. Then, μ and ν are singular if and only if* $\inf(\mu, \nu) = 0$.

Theorem 8.43 (**Lebesgue decomposition**). — *All finite Borelian measure μ can be written in a unique way as*

$$\mu = \mu_{ac} + \mu_{sing}\,,$$

where μ_{ac} is absolutely continuous with respect to λ and μ_{sing} is singular with respect to λ.

Proof. — Let us consider $\mathcal{N}$ the vector space spanned by the characteristic functions of the Borelian sets of Lebesgue measure 0. If A is a Borelian set, we let

$$\mu_{ac}(A) = \inf_{\psi \in \mathcal{N}} \int_{\mathbb{R}} |\mathbb{1}_A - \psi|^2 \, d\mu\,.$$

Notice that

$$\mu_{ac}(A) \leqslant \mu(A)\,,$$

and that, if $\lambda(A) = 0$, then $\mu_{ac}(A) = 0$ since $\mathbb{1}_A \in \mathcal{N}$. It remains to show that μ_{ac} is a measure.

Let us first prove that

$$\mu_{ac}(A) = \inf_{\psi \in \mathcal{N}} \int_{\mathbb{R}} |\mathbb{1}_A - \mathbb{1}_A \psi|^2 \, d\mu\,.$$

Since, for all $\psi \in \mathcal{N}$, we have $\mathbb{1}_A \psi \in \mathcal{N}$, we get

$$\mu_{ac}(A) \leqslant \inf_{\psi \in \mathcal{N}} \int_{\mathbb{R}} |\mathbb{1}_A - \mathbb{1}_A \psi|^2 \, d\mu\,.$$

Moreover, for all $\psi \in \mathcal{N}$, we have

$$\int_{\mathbb{R}} |\mathbb{1}_A - \psi|^2 \, d\mu = \int_{\mathbb{R}} |\mathbb{1}_A - \mathbb{1}_A \psi - \mathbb{1}_{\complement A} \psi|^2 \, d\mu$$

$$= \int_{\mathbb{R}} |\mathbb{1}_A - \mathbb{1}_A \psi|^2 \, d\mu + \int_{\mathbb{R}} |\mathbb{1}_{\complement A} \psi|^2 \, d\mu$$

$$\geqslant \int_{\mathbb{R}} |\mathbb{1}_A - \mathbb{1}_A \psi|^2 \, d\mu\,.$$

Now, consider two disjoint Borelian sets A and B. We have, for all $\psi \in \mathcal{N}$,

$$\int_{\mathbb{R}} |\mathbb{1}_{A \cup B} - \mathbb{1}_{A \cup B} \psi|^2 \, d\mu$$

$$= \int_{\mathbb{R}} |\mathbb{1}_A - \mathbb{1}_A \psi|^2 \, d\mu + \int_{\mathbb{R}} |\mathbb{1}_B - \mathbb{1}_B \psi|^2 \, d\mu \geqslant \mu_{ac}(A) + \mu_{ac}(B)\,.$$

Thus,

$$\mu_{ac}(A \cup B) \geqslant \mu_{ac}(A) + \mu_{ac}(B) \,.$$

Then, consider $\psi_1, \psi_2 \in \mathcal{N}$ and let $\psi = \mathbb{1}_A \psi_1 + \mathbb{1}_B \psi_2 \in \mathcal{N}$. We have

$$\int_{\mathbb{R}} |\mathbb{1}_{A \cup B} - \mathbb{1}_{A \cup B} \psi|^2 \, d\mu$$

$$= \int_{\mathbb{R}} |\mathbb{1}_A - \mathbb{1}_A \psi|^2 \, d\mu + \int_{\mathbb{R}} |\mathbb{1}_B - \mathbb{1}_B \psi|^2 \, d\mu$$

$$= \int_{\mathbb{R}} |\mathbb{1}_A - \mathbb{1}_A \psi_1|^2 \, d\mu + \int_{\mathbb{R}} |\mathbb{1}_B - \mathbb{1}_B \psi_2|^2 \, d\mu \,.$$

Taking the infimum in ψ_1 and ψ_2 gives

$$\mu_{ac}(A \cup B) \leqslant \mu_{ac}(A) + \mu_{ac}(B) \,.$$

The extension of this argument to a countable disjoint union is easy. Now, let us show that $\mu - \mu_{ac}$ is singular with respect to λ.

Let us now notice that, if θ is another measure such that $\theta \leqslant \mu$ and θ is absolutely continuous with respect to λ, then $\theta \leqslant \mu_{ac}$. Indeed, for all Borelian set A, we have, for all $\psi \in \mathcal{N}$,

$$\theta(A) = \int_{\mathbb{R}} |\mathbb{1}_A|^2 \, d\theta = \int_{\mathbb{R}} |\mathbb{1}_A - \psi|^2 \, d\theta \leqslant \int_{\mathbb{R}} |\mathbb{1}_A - \psi|^2 \, d\mu \,.$$

Consider another Borelian measure ν such that

$$\nu \leqslant \mu - \mu_{ac} \,, \quad \nu \leqslant \lambda \,.$$

Then, $\mu_{ac} + \nu$ is absolutely continuous with respect to λ and smaller than μ. Therefore, $\nu = 0$ and we apply Lemma 8.42.

For the uniqueness, let us write $\mu = \mu_1 + \mu_2$ with μ_1 absolutely continuous with respect to λ and μ_2 singular. Then, $\mu_1 \leqslant \mu_{ac}$ so that $\mu_{ac} - \mu_1$ is still a (finite) measure and is absolutely continuous. Since $\mu_{ac} - \mu_1 = \mu_{ac} - \mu + \mu_2$, we see that this measure is singular. Thus, $\mu_1 = \mu_{ac}$.

$\square$

Theorem 8.44. — *Any Borelian measure μ can be written in a unique way as*

$$\mu = \mu_{pp} + \mu_c \,,$$

where μ_{pp} is a pure point measure and μ_c is continuous.

This allows to write all measure μ, in a unique way,

$$\mu = \mu_{pp} + \mu_{ac} + \mu_{sc},$$

where μ_{sc} is singular and continuous.

8.3.5.2. *Remarkable subspaces.* — For all $\psi \in \mathsf{H}$, we can therefore apply the Lebesgue decomposition theorem to $\mu_{\psi,\psi}$. This suggests the following definitions.

Definition 8.45. —

$$\mathsf{H}_{ac} = \{\psi \in \mathsf{H} : \mu_{\psi,\psi} \text{ is absolutely continuous}\},$$
$$\mathsf{H}_{pp} = \{\psi \in \mathsf{H} : \mu_{\psi,\psi} \text{ is pure point}\},$$
$$\mathsf{H}_c = \{\psi \in \mathsf{H} : \mu_{\psi,\psi} \text{ is continuous}\},$$
$$\mathsf{H}_s = \{\psi \in \mathsf{H} : \mu_{\psi,\psi} \text{ is singular}\},$$
$$\mathsf{H}_{sc} = \{\psi \in \mathsf{H} : \mu_{\psi,\psi} \text{ is singular continuous}\}.$$

Proposition 8.46. — *The subsets H_{ac}, H_{pp}, H_c, H_s, and H_{sc} are closed vector spaces invariant under $\mathscr{L}$.*

Proof. — Let us consider H_{pp}. Consider $u, v \in \mathsf{H}_{pp}$ and $\lambda \in \mathbb{C}$. Let Ω be a Borelian set avoiding the (countable) atoms of $\mu_{u,u}$ and $\mu_{v,v}$. Then, $\mathbb{1}_\Omega(\mathscr{L})u = \mathbb{1}_\Omega(\mathscr{L})v = 0$. Then,

$$\mu_{u+\lambda v, u+\lambda v}(\Omega) = \langle \mathbb{1}_\Omega(\mathscr{L})(u + \lambda v), u + \lambda v \rangle = 0.$$

Thus, $u + \lambda v \in \mathsf{H}_{pp}$. Let us now consider a sequence (u_n) such that μ_{u_n,u_n} is pure point and $\lim_{n \to +\infty} u_n = u$. Let S be the (countable) union of the atoms of the μ_{u_n,u_n}. If Ω is a Borelian set avoiding S, we have $\mathbb{1}_\Omega(\mathscr{L})u_n = 0$ and then $\mathbb{1}_\Omega(\mathscr{L})u = 0$.

Let us consider H_{ac}. Consider $u, v \in \mathsf{H}_{ac}$ and $\lambda \in \mathbb{C}$. Let Ω be a Borelian set with Lebesgue measure 0. We have

$$\mu_{u+\lambda v, u+\lambda v}(\Omega) = 2\mathrm{Re}\,\langle \mathbb{1}_\Omega(\mathscr{L})u, \lambda v \rangle.$$

By the Cauchy–Schwarz inequality,

$$|\mu_{u+\lambda v, u+\lambda v}(\Omega)| \leqslant 2|\lambda|\mu_{u,u}(\Omega)^{\frac{1}{2}}\mu_{v,v}(\Omega)^{\frac{1}{2}} = 0.$$

Thus, $u + \lambda v \in \mathsf{H}_{ac}$. Let (u_n) be a sequence in H_{ac} and such that $\lim_{n \to +\infty} u_n = u$. Let Ω be a Borelian set of Lebesgue measure 0. We have

$$0 = \mu_{u_n,u_n}(\Omega) = \|\mathbb{1}_\Omega(\mathscr{L})u_n\|^2 \underset{n \to +\infty}{\longrightarrow} \mu_{u,u}(\Omega).$$

Finally, let us consider H_{sc}. Consider $u, v \in H_{sc}$ and a complex number $\lambda \in \mathbb{C}$. There exist Borelian sets S_u and S_v such that $\mu_{u,u}(S_u) = \mu_{v,v}(S_v) = 0$ and $\lambda(\mathbb{R} \setminus S_u) = \lambda(\mathbb{R} \setminus S_v) = 0$. We let $S = S_u \cap S_v$. We have $\lambda(\mathbb{R} \setminus S) = 0$. Moreover,

$$\mu_{u+\lambda v, u+\lambda v}(S) = 2\mathrm{Re}\, \langle \mathbb{1}_S(\mathscr{L})u, \lambda v \rangle$$

$$\leqslant 2|\lambda|\mu_{u,u}(S)^{\frac{1}{2}}\mu_{v,v}(S)^{\frac{1}{2}} = 0 \,.$$

We also see that $\mu_{u+\lambda v, u+\lambda v}$ is continuous. Let (u_n) be a sequence in H_{sc} and such that $\lim_{n \to +\infty} u_n = u$. We may consider (S_n) a countable family of Borelian sets such that $\mu_{u_n,u_n}(S_n) = 0$ and $\lambda(\mathbb{R} \setminus S_n) = 0$. We let $S = \bigcap_{n=0}^{+\infty} S_n$. We have $\lambda(\mathbb{R} \setminus S) = 0$. Then,

$$0 = \mu_{u_n,u_n}(S) = \|\mathbb{1}_S(\mathscr{L})u_n\|^2 \xrightarrow[n \to +\infty]{} \mu_{u,u}(S) \,.$$

We also see that $\mu_{u,u}$ is continuous.

The same kind of arguments also show that H_s and H_c are closed vector spaces.

Then, for all $u \in H$, all $t \in \mathbb{R}$, and all Borelian set Ω, we have (by using for instance the bounded functional calculus, see Propositions 8.21 and 8.26)

$$\mu_{e^{it\mathscr{L}}u, e^{it\mathscr{L}}u}(\Omega) = \|\mathbb{1}_\Omega(\mathscr{L})e^{it\mathscr{L}}u\|^2 = \|e^{it\mathscr{L}}\mathbb{1}_\Omega(\mathscr{L})u\|^2$$

$$= \|\mathbb{1}_\Omega(\mathscr{L})u\|^2 = \mu_{u,u}(\Omega) \,.$$

This shows that the spaces under consideration are invariant under $e^{it\mathscr{L}}$. If $u \in \mathrm{Dom}(\mathscr{L})$, we recall that

$$(8.8.3.3) \qquad \lim_{t \to 0} \frac{e^{it\mathscr{L}}u - u}{t} = i\mathscr{L}u \,.$$

Take $u \in \mathrm{Dom}(\mathscr{L}) \cap H_{xx}$. Since the H_{xx} are closed vector spaces, the left-hand side of (8.8.3.3) belongs to H_{xx}, and so does $i\mathscr{L}u$. $\qquad \square$

Proposition 8.47. — *We have the decomposition*

$$H = H_{ac} \overset{\perp}{\oplus} H_s \,.$$

Proof. — Since these spaces are closed, it is enough to prove that $H_s^\perp = H_{ac}$. Let $u \in H_s^\perp$. Notice that, for all $v \in H$, and all Borelian set S with Lebesgue measure 0, we have $w = \mathbb{1}_S(\mathscr{L})v \in \mathsf{H}_s$. Indeed,

$$\mu_{w,w}(\mathbb{R} \setminus S) = \|\mathbb{1}_{\mathbb{R} \setminus S}(\mathscr{L})w\|^2 = 0.$$

Thus, we have

$$\mu_{u,u}(S) = \langle u, \mathbb{1}_S(\mathscr{L})u \rangle = 0.$$

This shows that $u \in \mathsf{H}_{ac}$. Thus, $\mathsf{H}_s^{\perp} \subset \mathsf{H}_{ac}$.

Then, consider $u \in \mathsf{H}_s$. There exists a Borelian set S_0 with Lebesgue measure 0 such that $\mu_{u,u}(\mathbb{R} \setminus S_0) = 0$. This implies that $u = \mathbb{1}_{S_0}(\mathscr{L})u$ since

$$\|\mathbb{1}_{\mathbb{R} \setminus S_0}u\|^2 = \mu_{u,u}(\mathbb{R} \setminus S_0) = 0.$$

For all $v \in \mathsf{H}_{ac}$, we have

$$\|\mathbb{1}_{S_0}(\mathscr{L})v\|^2 = \mu_{v,v}(S_0) = 0.$$

Thus, $\langle u, v \rangle = 0$ and $u \in \mathsf{H}_{ac}^{\perp}$. $\qquad\square$

Proposition 8.48. *— We have the decompositions*

$$\mathsf{H} = \mathsf{H}_{pp} \overset{\perp}{\oplus} \mathsf{H}_c,$$

$$\mathsf{H}_s = \mathsf{H}_{pp} \overset{\perp}{\oplus} \mathsf{H}_{sc}.$$

Proof. — It is enough to prove that $\mathsf{H}_{pp}^{\perp} = \mathsf{H}_c$.

Let $u \in \mathsf{H}_{pp}^{\perp}$. For all $v \in \mathsf{H}_{pp}$, we have $\langle u, v \rangle = 0$. We have $v = \mathbb{1}_{\{x\}}(\mathscr{L})u \in \mathsf{H}_{pp}$. This shows that $\mu_{u,u}(\{x\}) = 0$ for all $x \in \mathsf{H}$. Thus, $u \in \mathsf{H}_c$.

Then, take $u \in \mathsf{H}_c$ and $v \in \mathsf{H}_{pp}$. Let P be the (countable) support of $\mu_{v,v}$. We have $v = \mathbb{1}_P(\mathscr{L})v$ so that

$$\langle u, v \rangle = \langle u, \mathbb{1}_P(\mathscr{L})v \rangle = \langle \mathbb{1}_P(\mathscr{L})u, v \rangle.$$

Since $u \in \mathsf{H}_c$, we have $\mathbb{1}_P(\mathscr{L})u = 0$. Thus, $\mathsf{H}_{pp} \subset \mathsf{H}_c^{\perp}$.

The second decomposition follows from the same kind of arguments. $\qquad\square$

We deduce the following general decomposition.

Theorem 8.49. *— We have*

$$\mathsf{H} = \mathsf{H}_{ac} \overset{\perp}{\oplus} \mathsf{H}_{pp} \overset{\perp}{\oplus} \mathsf{H}_{sc}.$$

Definition 8.50. *— The xx-spectrum of $\mathscr{L}$ is the spectrum of* $\mathscr{L}_{|\mathsf{H}_{xx}}$.

8.3.5.3. *Absolutely continuous spectrum.* — Let us provide the reader with some criteria to ensure that a part of the spectrum of $\mathscr{L}$ is absolutely continuous.

Proposition 8.51. — *Let $a < b$. Assume that* $\mathrm{ran}\mathbb{1}_{(a,b)}(\mathscr{L}) \subset \mathsf{H}_{ac}$. *Then,*

$$(a,b) \cap \mathsf{sp}(\mathscr{L}) \subset \mathsf{sp}_{ac}(\mathscr{L})\,.$$

Proof. — Note that $\mathscr{L}$ is isomorphic to the direct sum

$$\mathscr{L}_{|\mathrm{ran}\mathbb{1}_{(a,b)}(\mathscr{L})} \oplus \mathscr{L}_{|\mathrm{ran}\mathbb{1}_{\mathbb{R}\setminus(a,b)}(\mathscr{L})}\,.$$

If $z \in (a,b) \cap \mathsf{sp}(\mathscr{L})$, then $z \notin \mathsf{sp}(\mathscr{L}_{|\mathrm{ran}\mathbb{1}_{\mathbb{R}\setminus(a,b)}(\mathscr{L})})$ and thus $z \in \mathsf{sp}(\mathscr{L}_{|\mathrm{ran}\mathbb{1}_{(a,b)}(\mathscr{L})})$. Due to our assumption, this implies that $z \in \mathsf{sp}_{ac}(\mathscr{L})$. $\qquad\square$

Proposition 8.52. — *Let $a < b$. Assume that, for all ψ is a dense set of H, there exists $C(\psi) > 0$ such that, for all Borelian set $\Omega \subset (a,b)$, we have*

$$\langle \mathbb{1}_{\Omega}(\mathscr{L})\psi, \psi\rangle \leqslant C(\psi)|\Omega|\,.$$

Then,

$$(a,b) \cap \mathsf{sp}(\mathscr{L}) \subset \mathsf{sp}_{ac}(\mathscr{L})\,,$$

and there is no eigenvalue in (a,b).

Proof. — Let us consider a Borelian set Ω with Lebesgue measure 0. For all $\psi \in \mathsf{H}$, we let $v = \mathbb{1}_{(a,b)}(\mathscr{L})\psi$. We have

$$\mu_{v,v}(\Omega) = \langle \mathbb{1}_{\Omega}(\mathscr{L})v, v\rangle = \langle \mathbb{1}_{\Omega\cap(a,b)}(\mathscr{L})\psi, \psi\rangle\,.$$

Let us consider a sequence (ψ_n) converging to ψ and such that

$$\langle \mathbb{1}_{\Omega\cap(a,b)}(\mathscr{L})\psi_n, \psi_n\rangle = 0\,.$$

Taking the limit, it follows that $\mu_{v,v}(\Omega) = 0$. Thus, $v \in \mathsf{H}_{ac}$. We deduce that $\mathrm{ran}\mathbb{1}_{(a,b)}(\mathscr{L}) \subset \mathsf{H}_{ac}$, and we can apply Proposition 8.51. $\qquad\square$

8.4. Notes

i. One can consult [**35**, Vol. I, Chapter VII] or [**38**, Chapter 13, p. 360] for an alternative presentation of the spectral measure or the older references [**42, 18**]. The Reader is also warmly invited to discover the excellent book [**45**] where the spectral measure is defined by means of the Nevanlinna–Herglotz functions.

ii. It would be possible to construct a functional calculus by means of the « Helffer-Sjöstrand formula ». Let us explain this. We identify $\mathbb{R}^2$ with $\mathbb{C}$ by letting $z = x_1 + ix_2$. Consider a function $f \in \mathscr{C}_0^\infty(\mathbb{C})$ such that there exists $C > 0$ such that, for all $z \in \mathbb{C}$, $|\bar{\partial}f(z)| \leqslant C|\mathrm{Im}\, z|$. Then, for all $\lambda \in \mathbb{R}$,

(8.8.4.4)
$$f(\lambda) = -\frac{1}{\pi} \int_{\mathbb{R}^2} \bar{\partial}f(z)\,(z-\lambda)^{-1} \mathrm{d}x_1 \wedge \mathrm{d}x_2\,, \quad \bar{\partial} = \frac{1}{2}(\partial_1 + i\partial_2)\,.$$

Let us briefly explain why this formula holds. From a basic integrability argument, we see that the R.H.S. of (8.8.4.4) is well-defined. Moreover, we have, for some $M > 0$ such that $\mathrm{supp}\, f \subset D(0, M)$,

$$\int_{\mathbb{R}^2} \bar{\partial}f(z)\,(z-\lambda)^{-1}\,\mathrm{d}x = \lim_{\varepsilon \to 0} \int_{D(0,M)\setminus D(0,\varepsilon)} \bar{\partial}f(z)\,(z-\lambda)^{-1}\,\mathrm{d}x\,.$$

The Green–Riemann formula yields

$$\int_{D(0,M)\setminus D(0,\varepsilon)} \bar{\partial}f(z)\,(z-\lambda)^{-1}\,\mathrm{d}x$$

$$= -\int_{D(0,M)\setminus D(0,\varepsilon)} f(z)\,\bar{\partial}[(z-\lambda)^{-1}]\,\mathrm{d}x$$

$$+ \int_{\partial D(0,\varepsilon)} \frac{n}{2} f(z)(z-\lambda)^{-1}\,\mathrm{d}\sigma\,,$$

where $n = n_1 + in_2$ is the outward pointing normal (identified with a complex number), and $\mathrm{d}\sigma$ is the surface measure of the circle $\partial D(0, \varepsilon)$, i.e., $\varepsilon\,\mathrm{d}\theta$. Noticing that $n = -e^{i\theta}$, we get

$$\int_{D(0,M)\backslash D(0,\varepsilon)} \overline{\partial} f\,(z-\lambda)^{-1}\,\mathrm{d}x$$

$$= -\int_{\theta=0}^{2\pi} \frac{e^{i\theta}}{2} f(\lambda+\varepsilon e^{i\theta})(\varepsilon e^{i\theta})^{-1}\varepsilon\,\mathrm{d}\theta$$

$$= -\frac{1}{2}\int_{\theta=0}^{2\pi} f(\lambda+\varepsilon e^{i\theta})\,\mathrm{d}\theta \xrightarrow[\varepsilon\to 0]{} -\pi f(\lambda)\,.$$

Now, let us explain why formula (8.8.4.4) can be used to construct a functional calculus. Consider a function $\mathscr{C}_0^\infty(\mathbb{R})$. We can prove that there exists a function $\tilde{f}\in\mathscr{C}_0^\infty(\mathbb{C})$ such that $\tilde{f}_{|\mathbb{R}}=f$ and $|\overline{\partial}\tilde{f}|\leqslant C|\operatorname{Im} z|$. The interested reader can consult [**9**, Chapter 8] where such a $\tilde{f}$ is exhibited. We notice that

$$(8.8.4.5)\qquad f(\lambda)=-\frac{1}{\pi}\int_{\mathbb{R}^2}\overline{\partial}\tilde{f}(z)\,(z-\lambda)^{-1}\,\mathrm{d}x\,.$$

Then, it is not difficult to guess that

$$(8.8.4.6)\qquad f(\mathscr{L})=-\frac{1}{\pi}\int_{\mathbb{R}^2}\overline{\partial}\tilde{f}(z)\,(z-\mathscr{L})^{-1}\,\mathrm{d}x\,,$$

which is the « Helffer-Sjöstrand formula » (see, for instance, [**9**, Theorem 8.1]). Let us prove this formula thanks to our functional calculus. We have, for all $u\in\mathsf{H}$,

$$\langle f(\mathscr{L})u,u\rangle=\int_{\mathbb{R}} f(\lambda)\,\mathrm{d}\mu_{u,u}\,,$$

and thus, with (8.8.4.5), we get

$$\langle f(\mathscr{L})u,u\rangle=-\frac{1}{\pi}\int_{\mathbb{R}}\mathrm{d}\mu_{u,u}\int_{\mathbb{R}^2}\overline{\partial}\tilde{f}(z)\,(z-\lambda)^{-1}\,\mathrm{d}x\,.$$

By using the Fubini theorem, we get

$$\langle f(\mathscr{L})u,u\rangle=-\frac{1}{\pi}\int_{\mathbb{R}^2}\mathrm{d}x\,\overline{\partial}\tilde{f}(z)\int_{\mathbb{R}}\mathrm{d}\mu_{u,u}(z-\lambda)^{-1}\,.$$

Using Exercise 8.34, we have, for all $z\notin\mathbb{R}$,

$$\int_{\mathbb{R}}\mathrm{d}\mu_{u,u}(z-\lambda)^{-1}=\langle(z-\mathscr{L})^{-1}u,u\rangle\,.$$

Then, by continuity of the scalar product, we get, for all $u \in$ H,

$$\langle f(\mathscr{L})u, u \rangle = -\frac{1}{\pi} \left\langle \left(\int_{\mathbb{R}^2} \mathrm{d}x \, \overline{\partial} \tilde{f}(z)(z - \mathscr{L})^{-1} \right) u, u \right\rangle,$$

and the « Helffer-Sjöstrand formula » follows. Therefore, we could use (8.8.4.6) to define $f(\mathscr{L})$ when $f \in \mathscr{C}_0^\infty(\mathbb{R})$.

iii. The statement and proof of Urysohn's lemma (used in the proof of Proposition 8.23) can be found in [**37**, Lemma 2.12].

iv. The version of the monotone class theorem that we use in the proof of Proposition 8.24 is proved in [**24**, Theorem 1.1].

v. The fundamental fact that all bounded Borelian functions are uniformly approximated by sequences of step functions is established, for instance, in [**37**, Theorem 1.17].

vi. The proof of Theorem 8.12 can be found in [**37**, Theorem 2.14].

vii. The elegant proof of Theorem 8.43 is taken from [**46**]. Some insights of our presentation are due to R. Garbit.

viii. The Reader can find an alternative proof of Proposition 8.47 in [**25**, Section X.2].

CHAPTER 9

TRACE-CLASS AND HILBERT-SCHMIDT OPERATORS

We complete here our study of unbounded, bounded, and compact operators by two new classes: *trace-class* and *Hilbert–Schmidt* (H-S.) operators. The general picture is the following:

unbounded $\supset$ bounded $\supset$ compact $\supset$ H-S $\supset$ trace-class.

The trace of an operator extends to the infinite-dimensional setting the notion of the trace of a matrix. Basically, trace-class operators are compact operators for which a trace may be defined. In Quantum Physics, the trace may represent the energy of a system. We will give explicit examples, involving the Laplace operator, for which the trace can be explicitly computed or estimated.

9.1. Trace-class operators

Definition 9.1. — Let $T \in \mathcal{L}(\mathsf{H})$. We say that T is in $\mathcal{L}_1(\mathsf{H})$, the set of *trace-class* operators, when there exists a Hilbert basis $(\psi_n)_{n \in \mathbb{N}}$ such that

$$(9.9.1.1) \qquad \sum_{n=0}^{+\infty} \langle |T| \psi_n, \psi_n \rangle < +\infty.$$

Remark 9.2. — The summability condition (9.9.1.1) does not depend on the Hilbert basis. Indeed, consider another Hilbert basis $(\varphi_n)_{n \in \mathbb{N}}$. We have, via the Bessel–Parseval formula and Fubini's

© The Author(s), under exclusive license
to Springer Nature Switzerland AG 2021
C. Cheverry and N. Raymond, *A Guide to Spectral Theory*,
Birkhäuser Advanced Texts Basler Lehrbücher,
https://doi.org/10.1007/978-3-030-67462-5_9

theorem,

$$\sum_{n=0}^{+\infty}\langle|T|\psi_n,\psi_n\rangle = \sum_{n=0}^{+\infty}\||T|^{\frac{1}{2}}\psi_n\|^2 = \sum_{n=0}^{+\infty}\sum_{k=0}^{+\infty}|\langle|T|^{\frac{1}{2}}\psi_n,\varphi_k\rangle|^2$$
$$= \sum_{k=0}^{+\infty}\||T|^{\frac{1}{2}}\varphi_k\|^2 = \sum_{k=0}^{+\infty}\langle|T|\varphi_k,\varphi_k\rangle.$$

It follows that the notion of trace-class operator does not depend on the choice of the Hilbert basis allowing to test (9.9.1.1).

Remark 9.3. — When H is of finite dimension, any $T \in \mathcal{L}(H)$ is trace-class. If moreover, T is self-adjoint, the basis $(\psi_n)_{1\leqslant n\leqslant N}$ may be adjusted in such a way that ψ_n is an eigenvector of norm 1 of $|T|$, so that

$$(9.9.1.2) \qquad \sum_{n=0}^{N}\langle|T|\psi_n,\psi_n\rangle = \sum_{n=0}^{N}|\lambda_n|,$$

where the λ_n are the eigenvalues.

Lemma 9.4. — *Let $S \geqslant 0$ and V be a partial isometry, that is an isometry on the orthogonal complement of* $\ker V$*. For all Hilbert basis, we have*

$$\sum_{n=0}^{+\infty}\langle V^*SV\psi_n,\psi_n\rangle = \sum_{n=0}^{+\infty}\langle SV\psi_n,V\psi_n\rangle \leqslant \sum_{n=0}^{+\infty}\langle S\psi_n,\psi_n\rangle.$$

Proof. — First, we notice that both sides are independent of the chosen Hilbert basis. Thus, we may choose a basis adapted to the decomposition $H = \ker V \oplus \ker V^\perp$. If (φ_n) is a Hilbert basis of $\ker V^\perp$, we get

$$\sum_{n=0}^{+\infty}\langle SV\psi_n,V\psi_n\rangle = \sum_{n=0}^{+\infty}\langle SV\varphi_n,V\varphi_n\rangle.$$

Since $(V\varphi_n)$ is an orthonormal family, we can complete it into a Hilbert basis $(\tilde{\psi}_n)_{n\in\mathbb{N}}$. Then, we have

$$\sum_{n=0}^{+\infty}\langle SV\varphi_n,V\varphi_n\rangle \leqslant \sum_{n=0}^{+\infty}\langle S\tilde{\psi}_n,\tilde{\psi}_n\rangle = \sum_{n=0}^{+\infty}\langle S\psi_n,\psi_n\rangle.$$

$\square$

Definition 9.5. — For all $T \in \mathcal{L}_1(\mathsf{H})$, we let

$$\|T\|_1 = \sum_{n=0}^{+\infty} \langle |T|\psi_n, \psi_n \rangle.$$

Proposition 9.6. — $(\mathcal{L}_1(\mathsf{H}), \|\cdot\|_1)$ *is a normed vector space.*

Proof. — The invariance by multiplication by a scalar is straight-forward. Consider $T_1, T_2 \in \mathcal{L}_1(\mathsf{H})$. We write the polar decompositions

$$T_j = U_j|T_j|, \quad T_1 + T_2 = V|T_1 + T_2|.$$

We have

$$\sum_{n=0}^{N} \langle |T_1 + T_2|\psi_n, \psi_n \rangle$$

$$= \sum_{n=0}^{N} \langle V^*(T_1 + T_2)\psi_n, \psi_n \rangle$$

$$= \sum_{n=0}^{N} \langle V^*T_1\psi_n, \psi_n \rangle + \sum_{n=0}^{N} \langle V^*T_2\psi_n, \psi_n \rangle$$

$$= \sum_{n=0}^{N} \langle V^*U_1|T_1|\psi_n, \psi_n \rangle + \sum_{n=0}^{N} \langle V^*U_2|T_2|\psi_n, \psi_n \rangle.$$

By Cauchy–Schwarz, we have

$$\sum_{n=0}^{N} |\langle V^*U_j|T_j|\psi_n, \psi_n \rangle|$$

$$= \sum_{n=0}^{N} |\langle |T_j|^{\frac{1}{2}}\psi_n, |T_j|^{\frac{1}{2}}U_j^*V\psi_n \rangle|$$

$$\leqslant \sum_{n=0}^{N} \||T_j|^{\frac{1}{2}}\psi_n\|\||T_j|^{\frac{1}{2}}U_j^*V\psi_n\|$$

$$\leqslant \left(\sum_{n=0}^{N} \||T_j|^{\frac{1}{2}}\psi_n\|^2 \right)^{\frac{1}{2}} \left(\sum_{n=0}^{N} \||T_j|^{\frac{1}{2}}U_j^*V\psi_n\|^2 \right)^{\frac{1}{2}}$$

$$= \|T_j\|_1^{\frac{1}{2}} \left(\sum_{n=0}^{N} \||T_j|^{\frac{1}{2}}U_j^*V\psi_n\|^2 \right)^{\frac{1}{2}}.$$

Using two times Lemma 9.4,

$$\sum_{n=0}^{N} \||T_j|^{\frac{1}{2}} U_j^* V \psi_n\|^2 = \sum_{n=0}^{N} \langle U_j|T_j|U_j^* V \psi_n, V \psi_n \rangle \leqslant \|T_j\|_1 \,.$$

We deduce that

$$\sum_{n=0}^{N} \langle |T_1 + T_2|\psi_n, \psi_n \rangle \leqslant \|T_1\|_1 + \|T_2\|_1 \,.$$

$\square$

Proposition 9.7. — $\mathcal{L}_1(\mathsf{H})$ *is a two-sided ideal of* $\mathcal{L}(\mathsf{H})$.

Proof. — Let $T \in \mathcal{L}_1(\mathsf{H})$ and $T' \in \mathcal{L}(\mathsf{H})$. By Propositions 9.6 and 2.83, we can assume that T' is unitary. Then, $|T'T| = |T|$ and $T'T \in \mathcal{L}_1(\mathsf{H})$. Moreover, $|TT'| = |T'^{-1}TT'| = T'^{-1}|T|T'$ and T' sends any Hilbert basis onto a Hilbert basis, thus $TT' \in \mathcal{L}_1(\mathsf{H})$. $\square$

Proposition 9.8. — T *is trace-class iff* T^* *is trace-class.*

Proof. — Let T be a trace-class operator. It follows that $|T|$ is a trace-class operator. The polar decomposition $T = U|T|$ gives rise to $T^* = |T|U^*$, which is trace-class by the ideal property. Conversely, just note that $T = (T^*)^*$. □

Proposition 9.9. — *We have* $\mathcal{L}_1(\mathsf{H}) \subset \mathcal{K}(\mathsf{H})$.

Proof. — Consider $T \in \mathcal{L}_1(\mathsf{H})$. By Proposition 9.7, we have $T^*T = |T^*T| \in \mathcal{L}_1(\mathsf{H})$. Then, if $(\psi_n)_{n \in \mathbb{N}}$ is a Hilbert basis, we have

$$\sum_{n=0}^{+\infty} \langle |T^*T|\psi_n, \psi_n \rangle = \sum_{n=0}^{+\infty} \langle T\psi_n, T\psi_n \rangle = \sum_{n=0}^{+\infty} \|T\psi_n\|^2$$
$$< +\infty.$$

Then, for $N \in \mathbb{N}$, we let

$$T_N = \sum_{n=0}^{N} \langle \cdot, \psi_n \rangle T\psi_n.$$

For $M \geqslant N$, we write, for all $\psi \in \mathsf{H}$,

$$\|(T_N - T_M)\psi\| \leqslant \sum_{n=N+1}^{M} |\langle \psi, \psi_n \rangle| \|T\psi_n\|$$

$$\leqslant \left(\sum_{n=N+1}^{M} |\langle \psi, \psi_n \rangle|^2 \right)^{\frac{1}{2}} \left(\sum_{n=N+1}^{M} \|T\psi_n\|^2 \right)^{\frac{1}{2}}$$

$$\leqslant \|\psi\| \left(\sum_{n=N+1}^{\infty} \|T\psi_n\|^2 \right)^{\frac{1}{2}}.$$

It follows that

$$\|T_N - T_M\| \leqslant \left(\sum_{n=N+1}^{\infty} \|T\psi_n\|^2 \right)^{\frac{1}{2}} \xrightarrow[N \to +\infty]{} 0.$$

By the Cauchy criterion, this indicates that the sequence of finite-rank operators (T_n) converges to some bounded operator S, which is therefore compact. This limit coincides with T on a Hilbert basis and thus $T = S$ is compact. □

Lemma 9.10. — *For all* $T \in \mathcal{L}_1(\mathsf{H})$, *we have* $\|T\| \leqslant \|T\|_1$.

Proof. — The compact and self-adjoint operator $|T|$ has a spectral decomposition like

$$|T| = \sum_{n \geqslant 1} s_n \langle \cdot, \psi_n \rangle \psi_n, \qquad 0 \leqslant s_n \leqslant \| |T| \|.$$

We can complete the orthonormal family $(\psi_n)_{n \geqslant 1}$ into a Hilbert basis (by using a Hilbert basis of $\ker |T|$). In this adapted basis, we have

$$\|T\|_1 = \sum_{n=1}^{+\infty} \langle |T| \psi_n, \psi_n \rangle = \sum_{n=1}^{+\infty} \sum_{j=1}^{+\infty} \langle s_j \langle \psi_n, \psi_j \rangle \psi_j, \psi_n \rangle = \sum_{n \geqslant 1} s_n.$$

We can see in this formula a generalization of (9.9.1.2).

Now, recall that $\|T\| = \| |T| \|$. But, since $|T|$ is self-adjoint, $\| |T| \|$ coincides with its spectral radius, which is

$$\|T\| = \| |T| \| = \sup_{n \geqslant 1} s_n \leqslant \sum_{n \geqslant 1} s_n = \|T\|_1.$$

$\square$

Proposition 9.11. — $(\mathcal{L}_1(\mathsf{H}), \| \cdot \|_1)$ *is a Banach space.*

Proof. — Consider a Cauchy sequence (T_n) for the $\| \cdot \|_1$-norm. In particular, this sequence is bounded by some finite M. In view of Lemma 9.10, this sequence is also a Cauchy sequence for the norm the $\| \cdot \|$. Therefore, (T_n) converges to T in $\mathcal{L}(\mathsf{H})$. Note that

$$\forall \varepsilon > 0, \exists N \in \mathbb{N}, \forall \ell, n \geqslant N, \|T_n - T_\ell\|_1 \leqslant \varepsilon.$$

If (ψ_k) is a Hilbert basis, and using the proof of the triangle inequality,

$$\sum_{k=0}^{m} \langle |T - T_n + T_n| \psi_k, \psi_k \rangle \leqslant \sum_{k=0}^{m} \langle (|T - T_n| + |T_n|) \psi_k, \psi_k \rangle$$

$$\leqslant m\|T - T_n\| + \|T_n\|_1$$

$$\leqslant m\|T - T_n\| + M.$$

For all m, we can find n such that $m\|T - T_n\| \leqslant 1$. It follows that

$$\forall m \in \mathbb{N}^*, \qquad \sum_{k=0}^{m} \langle |T| \psi_k, \psi_k \rangle \leqslant 1 + M.$$

This implies that $T \in \mathcal{L}_1(\mathsf{H})$. Then, we write

$$\forall \varepsilon > 0, \exists N \in \mathbb{N}, \forall \ell, n \geqslant N, \qquad \sum_{k=0}^{+\infty} \langle |T_n - T_\ell| \psi_k, \psi_k \rangle \leqslant \varepsilon.$$

With the same arguments, we get

$$\sum_{k=0}^{m} \langle ((|T - T_n|) \psi_k, \psi_k \rangle \leqslant m \|T - T_\ell\| + \|T_\ell - T_n\|_1,$$

so that, for all m, taking ℓ and then n large enough, we get

$$\sum_{k=0}^{m} \langle ((|T - T_n|) \psi_k, \psi_k \rangle \leqslant \varepsilon \quad \implies \quad \|T - T_n\|_1 \leqslant \varepsilon.$$

The Cauchy sequence (T_n) does converge to $T \in \mathcal{L}_1(\mathsf{H})$. $\square$

9.2. Hilbert–Schmidt operators

9.2.1. Definition and first properties. —

Definition 9.12. — We say that $T \in \mathcal{L}(\mathsf{H})$ is an Hilbert–Schmidt operator if $|T|^2 = T^*T$ is in $\mathcal{L}_1(\mathsf{H})$. In this case, we write $T \in \mathcal{L}_2(\mathsf{H})$.

Remark 9.13. — By Proposition 9.7, we have $\mathcal{L}_1(\mathsf{H}) \subset \mathcal{L}_2(\mathsf{H})$.

Proposition 9.14. — $\mathcal{L}_2(\mathsf{H})$ *is a vector space.*

Proof. — If $T_1, T_2 \in \mathcal{L}_2(\mathsf{H})$, then for any Hilbert basis $(\psi_n)_{n \in \mathbb{N}}$,

$$\|T_j^* T_j\|_1 = \sum_{n=0}^{+\infty} \langle T_j^* T_j \psi_n, \psi_n \rangle = \sum_{n=0}^{+\infty} \|T_j \psi_n\|^2 < +\infty,$$

and thus

$$\sum_{n=0}^{+\infty} \|(T_1 + T_2) \psi_n\|^2 \leqslant 2 \sum_{j=1}^{2} \sum_{n=0}^{+\infty} \|T_j \psi_n\|^2 < +\infty.$$

$\square$

Proposition 9.15. — $\mathcal{L}_2(\mathsf{H})$ *is a two-sided ideal of* $\mathcal{L}(\mathsf{H})$.

Proof. — Let $T \in \mathcal{L}_2(H)$ and U be a unitary operator. If (ψ_n) is a Hilbert basis, we have

$$\sum_{n \geqslant 0} \|UT\psi_n\|^2 = \sum_{n \geqslant 0} \|T\psi_n\|^2 < +\infty.$$

Thus, $UT \in \mathcal{L}_2(H)$.

Moreover, $(U\psi_n)_{n \in \mathbb{N}}$ is a Hilbert basis so that

$$\sum_{n \geqslant 0} \|T(U\psi_n)\|^2 = \sum_{n \geqslant 0} \|T\psi_n\|^2 < +\infty.$$

$\square$

Proposition 9.16. — $T \in \mathcal{L}_2(H)$ *if and only if* $T^* \in \mathcal{L}_2(H)$.

Proof. — The polar decomposition $T = U|T|$ gives rise to $T^* = |T|U^*$, as well as

$$(T^*)^*T^* = (|T|U^*)^*|T|U^* = U|T|^2 U^*,$$

which is trace-class by the ideal property. Conversely, just note that $T = (T^*)^*$.

$\square$

Proposition 9.17. — *We have* $\mathcal{L}_2(H) \subset \mathcal{K}(H)$.

Proof. — This fact has been established in the proof of Proposition 9.9, which was only based on the information $T^*T \in \mathcal{L}_1(H)$.

$\square$

Proposition 9.18. — *Let* $T_1, T_2 \in \mathcal{L}_2(H)$. *Consider a Hilbert basis* $(\psi_n)_{n \in \mathbb{N}}$. *Then,*

$$\sum_{n \geqslant 0} |\langle T_2^* T_1 \psi_n, \psi_n \rangle| < +\infty.$$

The sum of the series

$$\sum_{n \geqslant 0} \langle T_2^* T_1 \psi_n, \psi_n \rangle$$

is independent of the chosen Hilbert basis. More precisely, if $(\varphi_n)_{n \in \mathbb{N}}$ *is another Hilbert basis, we have*

$$\sum_{n \geqslant 0} \langle T_2^* T_1 \psi_n, \psi_n \rangle = \sum_{n \geqslant 0} \langle T_1 T_2^* \varphi_n, \varphi_n \rangle = \sum_{n \geqslant 0} \langle T_2^* T_1 \varphi_n, \varphi_n \rangle.$$

Proof. — Because T_1 and T_2 are in $\mathcal{L}_2(\mathsf{H})$, we have

$$\sum_{n=0}^{N} |\langle T_2^* T_1 \psi_n, \psi_n \rangle| = \sum_{n=0}^{N} |\langle T_1 \psi_n, T_2 \psi_n \rangle|$$

$$= \sum_{n=0}^{N} |\langle T_1 \psi_n, \sum_{k=0}^{+\infty} \langle \varphi_k, T_2 \psi_n \rangle \varphi_k|$$

$$= \sum_{n=0}^{N} \sum_{k=0}^{+\infty} |\langle T_1 \psi_n, \varphi_k \rangle| \, |\langle \varphi_k, T_2 \psi_n \rangle|$$

$$\leqslant \frac{1}{2} \sum_{n=0}^{N} \sum_{k=0}^{+\infty} \left(|\langle T_1 \psi_n, \varphi_k \rangle|^2 + |\langle \varphi_k, T_2 \psi_n \rangle|^2 \right)$$

$$\leqslant \frac{1}{2} \sum_{n=0}^{N} \left(\|T_1 \psi_n\|^2 + \|T_2 \psi_n\|^2 \right)$$

$$\leqslant \frac{1}{2} \left(\|T_1^* T_1\|_1 + \|T_2^* T_2\|_1 \right) < +\infty.$$

Now, consider another Hilbert basis $(\varphi_n)_{n \in \mathbb{N}}$ and write again the Bessel–Parseval formula

$$\sum_{n \geqslant 0} \langle T_2^* T_1 \psi_n, \psi_n \rangle = \sum_{n \geqslant 0} \sum_{k \geqslant 0} \langle T_1 \psi_n, \varphi_k \rangle \langle \varphi_k, T_2 \psi_n \rangle.$$

The obtained double series is absolutely convergent (due to the above summation argument). Moreover, by the Fubini theorem, we get

$$\sum_{n \geqslant 0} \langle T_2^* T_1 \psi_n, \psi_n \rangle = \sum_{k \geqslant 0} \sum_{n \geqslant 0} \langle T_1 \psi_n, \varphi_k \rangle \langle \varphi_k, T_2 \psi_n \rangle$$

$$= \sum_{k \geqslant 0} \sum_{n \geqslant 0} \langle T_2^* \varphi_k, \psi_n \rangle \langle \psi_n, T_1^* \varphi_k \rangle$$

$$= \sum_{k \geqslant 0} \langle T_2^* \varphi_k, T_1^* \varphi_k \rangle = \sum_{k \geqslant 0} \langle T_1 T_2^* \varphi_k, \varphi_k \rangle.$$

This shows the independence. $\qquad \square$

Proposition 9.19. — *Let* $T_1, T_2 \in \mathcal{L}_2(\mathsf{H})$. *Then,* $T_1 T_2 \in \mathcal{L}_1(\mathsf{H})$.

Proof. — We start with the polar decomposition $|T_1 T_2| = U^* T_1 T_2$. By Proposition 9.15, we know that $U^* T_1 \in \mathcal{L}_2(\mathsf{H})$.

Then, by Proposition 9.16, we get that $T_1^* U \in \mathcal{L}_2(\mathsf{H})$. Proposition 9.18 implies that

$$\sum_{n \geqslant 0} \langle (T_1^* U)^* T_2 \psi_n, \psi_n \rangle = \sum_{n \geqslant 0} \langle |T_1 T_2| \psi_n, \psi_n \rangle < +\infty,$$

which guarantees that $T_1 T_2 \in \mathcal{L}_1(\mathsf{H})$. $\square$

9.2.2. Trace of a trace-class operator. —

Proposition 9.20 (**Trace of a trace-class operator**)
   Let $T \in \mathcal{L}_1(\mathsf{H})$ and $(\psi_n)_{n \in \mathbb{N}}$ be a Hilbert basis. Then, the series

$$\operatorname{Tr} T := \sum_{n=0}^{+\infty} \langle T \psi_n, \psi_n \rangle$$

is absolutely convergent and independent of the chosen Hilbert basis.

Proof. — We write $T = U|T| = (U|T|^{\frac{1}{2}})|T|^{\frac{1}{2}}$ and apply Proposition 9.18. $\square$

Proposition 9.21. — *The application $\mathcal{L}_1(\mathsf{H}) \ni T \mapsto \operatorname{Tr} T \in \mathbb{C}$ is a linear form. Moreover, for all $T \in \mathcal{L}_1(\mathsf{H})$, $\operatorname{Tr} T^* = \overline{\operatorname{Tr} T}$.*

Proposition 9.22. — *The application $\mathcal{L}_2(\mathsf{H}) \times \mathcal{L}_2(\mathsf{H}) \ni (A, B) \mapsto \operatorname{Tr}(AB^*) \in \mathbb{C}$ is a scalar product on $\mathcal{L}_2(\mathsf{H})$. The associated norm, called Hilbert–Schmidt norm, is denoted by $\| \cdot \|_2$. Moreover, the application $\mathcal{L}_2(\mathsf{H}) \ni T \mapsto T^* \in \mathcal{L}_2(\mathsf{H})$ is unitary.*

Proposition 9.23. — *For all $T \in \mathcal{L}_1(\mathsf{H})$,*

$$\|T\| \leqslant \|T\|_2 \leqslant \|T\|_1.$$

Proof. — Consider first the case when $T = T^* \geqslant 0$ and write

$$T = \sum_{n \geqslant 0} s_n \langle \cdot, \psi_n \rangle \psi_n, \qquad 0 \leqslant s_n.$$

We have already seen that

$$\|T\|_1 = \sum_{n \geqslant 0} s_n.$$

In the same way, we get

$$\|T\|_2^2 = \sum_{n \geqslant 0} s_n^2 \,.$$

The inequality is then proved since

$$\max_{n \geqslant 0} s_n \leqslant \left(\sum_{n \geqslant 0} s_n^2 \right)^{\frac{1}{2}} \leqslant \sum_{n \geqslant 0} s_n \,.$$

In the general case, we have

$$\|T\| = \| \, |T| \, \| \leqslant \| |T| \|_2 = \|T\|_2 \leqslant \| \, |T| \, \|_1 = \|T\|_1 \,.$$

$\square$

Proposition 9.24. — $(\mathcal{L}_2(\mathsf{H}), \| \cdot \|_2)$ *is a Hilbert space.*

Proposition 9.25. — *For all $T_1, T_2 \in \mathcal{L}_2(\mathsf{H})$,*

$$\|T_1 T_2\|_2 \leqslant \|T_1\| \|T_2\|_2 \leqslant \|T_1\|_2 \|T_2\|_2 \,.$$

Proposition 9.26. — *Let $T_1, T_2 \in \mathcal{L}_1(\mathsf{H})$. Then, $T_2 T_1$ and $T_1 T_2$ are in $\mathcal{L}_1(\mathsf{H})$, and we have $\mathrm{Tr}\,(T_2 T_1) = \mathrm{Tr}\,(T_1 T_2)$. Moreover,*

$$\|T_1 T_2\|_1 \leqslant \|T_1\| \|T_2\|_1 \,.$$

Proof. — This is a consequence of Propositions 9.19 and 9.18. The inequality follows from the polar decomposition $T_2 = U|T_2|$ and

$$\begin{aligned}
\|T_1 T_2\|_1 &= \|T_1 U |T_2|^{\frac{1}{2}} |T_2|^{\frac{1}{2}} \|_1 \\
&\leqslant \|T_1 U |T_2|^{\frac{1}{2}} \|_2 \| |T_2|^{\frac{1}{2}} \|_2 \\
&\leqslant \|T_1 U\| \| \, |T_2|^{\frac{1}{2}} \|_2^2 \leqslant \|T_1\| \| \, |T_2|^{\frac{1}{2}} \|_2^2 \,.
\end{aligned}$$

$\square$

Proposition 9.27. — *For all $T_1 \in \mathcal{L}_1(\mathsf{H})$ and $T_2 \in \mathcal{L}(\mathsf{H})$, we have $\mathrm{Tr}\,(T_2 T_1) = \mathrm{Tr}\,(T_1 T_2)$ and $|\mathrm{Tr}\,(T_1 T_2)| \leqslant \|T_1\|_1 \|T_2\|$.*

Proof. — For the cyclicity of the trace, by Propositions 2.83 and 9.21, it is enough to establish the formula when T_2 is unitary. In this case, we have

$$\mathrm{Tr}(T_2 T_1) = \sum_{n=0}^{+\infty} \langle T_1 \psi_n, T_2^* \psi_n \rangle \,.$$

Using the new Hilbert basis $\varphi_n = T_2\psi_n$, we get

$$\mathrm{Tr}(T_2T_1) = \sum_{n=0}^{+\infty} \langle T_1T_2\psi_n, \psi_n \rangle = \mathrm{Tr}(T_1T_2).$$

For the inequality, assume first that $T_1 = T_1^* \geqslant 0$, and write its spectral decomposition

$$T_1 = \sum_{n \geqslant 0} s_n \langle \cdot, \psi_n \rangle \psi_n.$$

Recall that $\sum_{n \geqslant 1} s_n = \|T_1\|_1$. By using an adapted Hilbert basis, we have

$$\mathrm{Tr}\,(T_2T_1) = \sum_{n=0}^{+\infty} \langle T_1T_2\psi_n, \psi_n \rangle = \sum_{n=0}^{+\infty} \langle \sum_{j \geqslant 0} s_j \langle T_2\psi_n, \psi_j \rangle \psi_j, \psi_n \rangle$$

$$= \sum_{n \geqslant 0} s_n \langle T_2\psi_n, \psi_n \rangle,$$

and thus

$$|\mathrm{Tr}\,(T_2T_1)| \leqslant \|T_2\| \|T_1\|_1.$$

If T_1 is not non-negative, we write $T_1 = U|T_1|$ and get

$$|\mathrm{Tr}\,(T_2T_1)| = |\mathrm{Tr}\,((T_2U)|T_1|)| \leqslant \|T_2U\| \||T_1|\|_1 \leqslant \|T_2\| \|T_1\|_1.$$

$\square$

9.3. A fundamental example

Let us consider a non-empty open set $\Omega \subset \mathbb{R}^d$. We consider the classical separable Hilbert space $\mathsf{H} = \mathsf{L}^2(\Omega)$. For $K \in \mathsf{L}^2(\Omega \times \Omega)$, and all $\psi \in \mathsf{L}^2(\Omega)$, we let

$$T_K\psi(x) = \int_\Omega K(x,y)\psi(y)\,\mathrm{d}y.$$

The application T_K is clearly a bounded operator from H to H. By Cauchy–Schwarz inequality, we have

$$\|T_K\| := \sup_{\|\psi\|_{\mathsf{L}^2(\Omega)} \leqslant 1} \|T_K\psi\|_{\mathsf{L}^2(\Omega)} \leqslant \|K\|_{\mathsf{L}^2(\Omega \times \Omega)}.$$

Its adjoint satisfies $T_K^* = T_{\check{K}}$ with $\check{K}(x,y) = \overline{K(y,x)}$.

Proposition 9.28. — *We have* $T_K \in \mathcal{L}_2(\mathsf{H})$ *and* $\|T_K\|_2 = \|K\|_{\mathsf{L}^2(\Omega \times \Omega)}$.

Proof. — Consider $(\psi_n)_{n \geqslant 0}$ a Hilbert basis of $\mathsf{L}^2(\Omega)$. Letting

$$\varphi_{m,n}(x, y) = \psi_m(x)\psi_n(y),$$

we see that $(\varphi_{m,n})_{(m,n) \in \mathbb{N}^2}$ is a Hilbert basis of $\mathsf{L}^2(\Omega \times \Omega)$. Thus, we can write

$$K = \sum_{(m,n) \in \mathbb{N}^2} k_{m,n}\varphi_{m,n}, \qquad \sum_{(m,n) \in \mathbb{N}^2} |k_{m,n}|^2 = \|K\|_{\mathsf{L}^2(\Omega \times \Omega)}^2.$$

In fact,

$$k_{m,n} = \langle K, \varphi_{m,n}\rangle_{\mathsf{L}^2(\Omega \times \Omega)} = \langle T_K \psi_m, \psi_n\rangle_{\mathsf{L}^2(\Omega)},$$

where we used the Fubini theorem. Since

$$\|T_K \psi_m\|_{\mathsf{L}^2(\Omega)}^2 = \sum_{n \geqslant 0} |\langle T_K \psi_m, \psi_n\rangle_{\mathsf{L}^2(\Omega)}|^2,$$

we get that T_K is Hilbert–Schmidt and $\|T_K\|_2^2 = \|K\|_{\mathsf{L}^2(\Omega \times \Omega)}^2$. □

Proposition 9.29. — *The application* $\mathsf{L}^2(\Omega \times \Omega) \ni K \mapsto T_K \in \mathcal{L}_2(\mathsf{H})$ *is unitary.*

Proof. — The application $K \mapsto T_K$ is an isometry. In particular, it is injective with closed range. So is its adjoint $T_{\check{K}}$, with $\check{K}(x, y) = \overline{K(y, x)}$. Therefore, $K \mapsto T_K$ is bijective. □

Proposition 9.30. — *Consider* $T \in \mathcal{L}_1(\mathsf{H})$. *The polar decomposition of T can be written as* $T = U|T|^{\frac{1}{2}}|T|^{\frac{1}{2}}$, *and we may write* $T = AB$ *with* $A, B \in \mathcal{L}_2(\mathsf{H})$. *Writing* $A = T_a$ *and* $B = T_b$, *with* $a, b \in \mathsf{L}^2(\Omega \times \Omega)$, *we have*

$$\operatorname{Tr} T = \int_\Omega t(x, x)\, \mathrm{d}x,$$

where

$$t(x, y) = \int_\Omega a(x, z)b(z, y)\, \mathrm{d}z.$$

Proof. — We have, with Proposition 9.29,

$$\operatorname{Tr} T = \operatorname{Tr}(AB) = \langle A, B^*\rangle_2 = \langle a, \check{b}\rangle_{\mathsf{L}^2(\Omega \times \Omega)}$$

$$= \int_{\Omega \times \Omega} a(x, y)b(y, x)\, \mathrm{d}x\, \mathrm{d}y.$$

From Fubini's theorem, we deduce that

$$\operatorname{Tr} T = \int_\Omega \left(\int_\Omega a(x,y)b(y,x)\, dy \right) dx \,.$$

$\square$

Let us give a simple, but non completely trivial, example.

Proposition 9.31. — *Let $\mathscr{L}$ be the Dirichlet Laplacian on the interval $I = (0,1)$. Consider the compact self-adjoint operator $T = \mathscr{L}^{-1}$. Then, T is actually trace-class and*

$$\operatorname{Tr} T = \frac{1}{6} \,, \qquad \|T\|_2^2 = \frac{1}{90} \,.$$

In particular,

$$\sum_{n \geqslant 1} n^{-2} = \frac{\pi^2}{6} \qquad \sum_{n \geqslant 1} n^{-4} = \frac{\pi^4}{90} \,.$$

Proof. — Consider the Hilbert basis $(\varphi_n)_{n \geqslant 1}$ of eigenfunctions of T, associated with the eigenvalues $(\lambda_n)_{n \geqslant 1}$. Explicitly (see Lemma 1.8), we have

$$\varphi_n(x) = \sqrt{n} \sin(n\pi x)\,, \qquad \lambda_n = (n\pi)^{-2} \,.$$

From the explicit expression of the eigenvalues, we see that $T \in \mathcal{L}_2(\mathsf{H})$ with $\mathsf{H} = \mathsf{L}^2(I)$. Let us find the kernel K of T. Consider $f \in \mathsf{L}^2(I)$. Let us try to find $u \in \mathsf{H}_0^1(I) \cap \mathsf{H}^2(I)$ such that

$$u'' = -f \,.$$

We can write this equation in the form

$$U' = \begin{pmatrix} 0 & 1 \\ 0 & 0 \end{pmatrix} U + \begin{pmatrix} 0 \\ f \end{pmatrix} \qquad U = \begin{pmatrix} u \\ u' \end{pmatrix} .$$

Consider the following independent solutions of the homogeneous equation

$$U_1 = \begin{pmatrix} x \\ 1 \end{pmatrix}, \qquad U_2 = \begin{pmatrix} x - 1 \\ 1 \end{pmatrix} .$$

Letting $u_1(x) = x$ and $u_2(x) = x - 1$, we notice that $u_1(0) = u_2(1) = 0$. Then, we look for u is in the form

$$u(x) = \alpha(x)u_1(x) + \beta(x)u_2(x) \,,$$

with

$$[U_1, U_2] \begin{pmatrix} \alpha' \\ \beta' \end{pmatrix} = \begin{pmatrix} 0 \\ -f \end{pmatrix},$$

or, equivalently,

$$\alpha'(x) = (x-1)f(x), \qquad \beta'(x) = -xf(x).$$

Since $u(0) = u(1) = 0$, we get $\alpha(1) = \beta(0) = 0$ so that

$$\alpha(x) = \int_1^x (y-1)f(y)\,\mathrm{d}y, \qquad \beta(x) = -\int_0^x yf(y)\,\mathrm{d}y.$$

Thus,

$$u(x) = \int_0^1 \left(\mathbb{1}_{[x,1]}(y)x(1-y) + \mathbb{1}_{[0,x]}(y)y(1-x) \right) f(y)\,\mathrm{d}y.$$

The kernel is given by

$$K(x,y) = \mathbb{1}_{[x,1]}(y)x(1-y) + \mathbb{1}_{[0,x]}(y)y(1-x).$$

We can check that $K \in \mathscr{C}^0([0,1]^2, \mathbb{R})$. A computation gives

$$\|T\|_2^2 = \int_{[0,1]^2} |K(x,y)|^2\,\mathrm{d}x\,\mathrm{d}y = \frac{1}{90}.$$

Note that, in $\mathsf{L}^2(I \times I)$,

$$K(x,y) = \sum_{(m,n)\in\mathbb{N}^* \times \mathbb{N}^*} k_{mn}\varphi_m(x)\varphi_n(y),$$

with

$$k_{mn} = \int_{[0,1]^2} K(x,y)\varphi_m(x)\varphi_n(y)\,\mathrm{d}x\,\mathrm{d}y.$$

By the Fubini theorem,

$$k_{mn} = \langle T\varphi_m, \varphi_n \rangle = \lambda_m \delta_{mn},$$

so that

$$K(x,y) = \sum_{n\geqslant 1} \lambda_n \varphi_n(x)\varphi_n(y).$$

In fact, for all fixed $y \in [0,1]$, this series is convergent in $\mathsf{L}^2(I)$. Let us explain why the convergence is also true in the $\mathsf{L}^\infty(I)$ sense.

Notice that $T^{\frac{1}{2}} \in \mathcal{L}_2(\mathsf{H})$ so that it has a kernel $\tilde{K}$ in $\mathsf{L}^2(I \times I)$, and, for almost all fixed $y \in I$, in $\mathsf{L}^2(I_x)$,

$$\tilde{K}(\cdot, y) = \sum_{n\geqslant 1} \sqrt{\lambda_n}\varphi_n(y)\varphi_n(\cdot).$$

In particular,

$$\int_0^1 |\tilde{K}(x,y)|^2 \, dx = \sum_{n \geqslant 1} \lambda_n \varphi_n(y)^2 \,,$$

and also, for almost all $x \in I$,

$$\int_0^1 |\tilde{K}(x,y)|^2 \, dy = \sum_{n \geqslant 1} \lambda_n \varphi_n(x)^2 \,.$$

Now, consider

$$\left| \sum_{n=M}^N \sqrt{\lambda_n} \varphi_n(y) \sqrt{\lambda_n} \varphi_n(x) \right|$$

$$\leqslant \left(\sum_{n=M}^N \lambda_n \varphi_n(y)^2 \right)^{\frac{1}{2}} \left(\sum_{n=M}^N \lambda_n \varphi_n(x)^2 \right)^{\frac{1}{2}}$$

$$\leqslant \left(\int_0^1 |\tilde{K}(x,y)|^2 \, dy \right)^{\frac{1}{2}} \left(\sum_{n=M}^N \lambda_n \varphi_n(y)^2 \right)^{\frac{1}{2}} \xrightarrow[M,N \to +\infty]{} 0 \,.$$

Therefore, by continuity of K and of the eigenfunctions, we have, for all $(x,y) \in [0,1]^2$,

$$K(x,y) = \sum_{n \geqslant 1} \lambda_n \varphi_n(x) \varphi_n(y) \,.$$

We take $x = y$ and integrate to get

$$\int_0^1 K(x,x) \, dx = \sum_{n \geqslant 1} \lambda_n = \text{Tr} \, \lambda_n \,.$$

We have

$$\int_0^1 K(x,x) \, dx = \int_0^1 x(1-x) \, dx = \frac{1}{6} \,.$$

$\square$

9.4. Local traces of the Laplacian

The Laplace operator $-\Delta$ on $\mathbb{R}^d$, equipped with its natural domain, is not bounded, and thus its trace (or its Hilbert–Schmidt norm) is not defined. Nevertheless, we may give a « local » sense to such traces. To do so, we can restrict the action of the operator to a finite region by means of cutoff functions, *i.e.*, consider $\varphi(-\Delta)\varphi$.

9.4.1. The case of $\mathbb{R}^d$. — Let $h > 0$ (a small parameter). We consider here the semi-classical operator $\mathscr{L}_h^{\mathbb{R}^d} = -h^2\Delta - 1$ acting on $\mathsf{L}^2(\mathbb{R}^d)$.

***Definition 9.32.* —** We define the (unitary) Fourier transform by

$$\mathscr{F}\psi(\xi) = (2\pi)^{-\frac{d}{2}} \int_{\mathbb{R}^d} e^{-ix\xi}\psi(x)\,\mathrm{d}x\,.$$

Given some operator $\mathscr{L}$, the notion of positive and negative parts $\mathscr{L}_\pm$ of $\mathscr{L}$ is explained in Section 8.3.4.

***Lemma 9.33.* —** *In the sense of quadratic forms, we have*

$$(\mathscr{L}_h^{\mathbb{R}^d})_- \leqslant \gamma_h\,, \qquad \gamma_h = \mathbb{1}_{\mathbb{R}^-}(\mathscr{L}_h^{\mathbb{R}^d})\,.$$

Consider $\varphi \in \mathscr{C}_0^\infty(\mathbb{R}^d)$. Then,

$$\varphi(\mathscr{L}_h^{\mathbb{R}^d})_-\varphi \leqslant \varphi\gamma_h\varphi\,.$$

***Lemma 9.34.* —** *We have*

$$\mathscr{F}\mathscr{L}_h\mathscr{F}^{-1} = h^2\xi^2 - 1\,.$$

In particular,

$$\gamma_h = \mathscr{F}^{-1}\mathbb{1}_{\mathbb{R}^-}(h^2\xi^2 - 1)\mathscr{F}\,,$$

and

$$\gamma_h\psi(x) = (2\pi)^{-d}\int_{\mathbb{R}^{2d}} e^{i(x-y)\xi}\mathbb{1}_{\mathbb{R}^-}(h^2\xi^2 - 1)\psi(y)\,\mathrm{d}y\,\mathrm{d}\xi$$

$$= (2\pi)^{-\frac{d}{2}}\int_{\mathbb{R}^d} \mathrm{d}y\psi(y)\mathscr{F}^{-1}(\mathbb{1}_{\mathbb{R}^-}(h^2\xi^2 - 1))(x - y)\,.$$

***Proposition 9.35.* —** *Consider $\varphi \in \mathscr{C}_0^0(\mathbb{R}^d)$. Then,*

(i) *The bounded operator $\varphi\gamma_h$ is Hilbert–Schmidt and*

$$\|\varphi\gamma_h\|_2^2 = \frac{\omega_d}{(2\pi h)^d}\|\varphi\|^2 \, .$$

(ii) *The bounded self-adjoint operator $\varphi\gamma_h\varphi$ is trace-class and*

(9.9.4.3) $$\mathrm{Tr}(\varphi\gamma_h\varphi) = \frac{\omega_d}{(2\pi h)^d}\|\varphi\|^2 \, .$$

Proof. — For the first item, we notice that the kernel K of $\varphi\gamma_h$ is given by

$$K(x,y) = (2\pi)^{-\frac{d}{2}}\varphi(x)\mathscr{F}^{-1}(\mathbb{1}_{\mathbb{R}_-}(h^2\xi^2 - 1))(x - y) \, .$$

From the Parseval formula, we see that $K \in \mathsf{L}^2(\mathbb{R}^{2d})$ and

$$\|K\|_{\mathsf{L}^2(\mathbb{R}^{2d})}^2 = (2\pi)^{-d}\|\varphi\|^2 \int_{\mathbb{R}^d} \mathrm{d}\xi \mathbb{1}_{\mathbb{R}_-}(h^2\xi^2 - 1) \, .$$

For the second item, it is sufficient to notice that

$$\varphi\gamma_h\varphi = \varphi\gamma_h\gamma_h\varphi = (\varphi\gamma_h)(\varphi\gamma_h)^* \, ,$$

and

$$\mathrm{Tr}(\varphi\gamma_h\varphi) = \|\varphi\gamma_h\|_2^2 \, .$$

$\square$

Corollary 9.36. — *For all $\varphi \in \mathscr{C}_0^\infty(\mathbb{R}^d)$, the operator $\varphi(\mathscr{L}_h^{\mathbb{R}^d})_-\varphi$ is trace-class. Moreover, $(\varphi\mathscr{L}_h^{\mathbb{R}^d}\varphi)_-$ is also trace-class and*

$$\mathrm{Tr}(\varphi\mathscr{L}_h^{\mathbb{R}^d}\varphi)_- \leqslant \mathrm{Tr}(\varphi(\mathscr{L}_h^{\mathbb{R}^d})_-\varphi) \, .$$

Moreover,

$$\mathrm{Tr}(\varphi(\mathscr{L}_h^{\mathbb{R}^d})_-\varphi) = (2\pi)^{-d}h^{-d}\|\varphi\|^2 \int_{\mathbb{R}^d} (\xi^2 - 1)_- \, \mathrm{d}\xi \, .$$

Proof. — The first part of the statement follows from Lemma 9.33 and Proposition 9.35. For the second part, we consider a Hilbert basis $(\psi_j)_{j\geqslant 1}$ such that $(\psi_j)_{j\in J}$ is a Hilbert basis of the negative (Hilbert) subspace of $\varphi\mathscr{L}_h^{\mathbb{R}^d}\varphi$. Then, for all $j \in J$,

$$\begin{aligned}
\langle \psi_j, (\varphi\mathscr{L}_h^{\mathbb{R}^d}\varphi)_-\psi_j\rangle &= -\langle \psi_j, (\varphi\mathscr{L}_h^{\mathbb{R}^d}\varphi)\psi_j\rangle \\
&\leqslant \langle \psi_j, (\varphi(\mathscr{L}_h^{\mathbb{R}^d})_-\varphi)\psi_j\rangle \, ,
\end{aligned}$$

and, for all $j \in \mathbb{N}^* \setminus J$, $\langle \psi_j, (\varphi \mathscr{L}_h^{\mathbb{R}^d} \varphi)_- \psi_j \rangle = 0$. This shows that $\sum_{j \geqslant 1} \langle \psi_j, (\varphi \mathscr{L}_h^{\mathbb{R}^d} \varphi)_- \psi_j \rangle$ is convergent, that the non-negative operator $(\varphi \mathscr{L}_h^{\mathbb{R}^d} \varphi)_-$ is trace-class, and the inequality follows. Then, we write

$$\varphi(\mathscr{L}_h^{\mathbb{R}^d})_- \varphi = \varphi(\mathscr{L}_h^{\mathbb{R}^d})_-^{\frac{1}{2}} \left(\varphi(\mathscr{L}_h^{\mathbb{R}^d})_-^{\frac{1}{2}} \right)^* .$$

The kernel of $\varphi(\mathscr{L}_h^{\mathbb{R}^d})_-^{\frac{1}{2}}$ is

$$(2\pi)^{-\frac{d}{2}} \varphi(x) \mathscr{F}^{-1}((h^2\xi^2 - 1)_-^{\frac{1}{2}})(x - y) .$$

Therefore, it is Hilbert–Schmidt and

$$\|\varphi(\mathscr{L}_h^{\mathbb{R}^d})_-^{\frac{1}{2}}\|_2^2 = (2\pi)^{-d} h^{-d} \|\varphi\|^2 \int_{\mathbb{R}^d} (\xi^2 - 1)_- \, d\xi .$$

$\square$

9.4.2. The case of $\mathbb{R}_+^d$. —

Let us now consider the operator $\mathscr{L}_h^{\mathbb{R}_+^d} = -h^2 \Delta - 1$ acting on $L^2(\mathbb{R}_+^d)$ with Dirichlet boundary condition on $x_d = 0$.

9.4.2.1. *Computation of the local trace.* —

Proposition 9.37. — *For all* $\varphi \in \mathscr{C}_0^0(\mathbb{R}^d)$,

$$\mathrm{Tr}(\varphi(\mathscr{L}_h^{\mathbb{R}_+^d})_- \varphi)$$
$$= 2(2\pi)^{-d} h^{-d} \int_{\mathbb{R}_+^d} \varphi^2(x) \int_{\mathbb{R}^d} (\xi^2 - 1)_- \sin^2(h^{-1} x_d \xi_d) \, d\xi \, dx .$$

Proof. — Let us diagonalize $\mathscr{L}_h^{\mathbb{R}_+^d}$. For that purpose, let us consider the application $\mathscr{T} : L^2(\mathbb{R}_+^d) \to L^2(\mathbb{R}^d)$ defined by

$$\mathscr{T} = \frac{1}{\sqrt{2}} \mathscr{F} \circ S ,$$

where S is defined by $S\psi(x) = \psi(x)$ when $x_d > 0$ and $S\psi(x) = -\psi(x)$ when $x_d \leqslant 0$. The operator $\mathscr{T}$ is an isometry and $\mathscr{T} : L^2(\mathbb{R}_+^d) \to L_{\mathrm{odd}}^2(\mathbb{R}^d)$ is bijective and $\mathscr{T}^{-1} = \sqrt{2} \mathscr{F}^{-1}$ where we have used $\mathscr{F} : L_{\mathrm{odd}}^2(\mathbb{R}^d) \to L_{\mathrm{odd}}^2(\mathbb{R}^d)$. We have

$$\mathscr{L}_h^{\mathbb{R}_+^d} = \mathscr{T}^{-1}(h^2 |\xi|^2 - 1) \mathscr{T} .$$

In fact, $\mathscr{T}$ can be related to the « sine Fourier transform »:

$$
\begin{aligned}
\mathscr{T}\psi(x) &= \frac{1}{\sqrt{2}}(2\pi)^{-\frac{d}{2}} \int_{\mathbb{R}^d} e^{-ix\xi} S\psi(x)\,\mathrm{d}x \\
&= -i\sqrt{2}(2\pi)^{-\frac{d}{2}} \int_{\mathbb{R}^d_+} e^{-ix'\xi'} \sin(x_d\xi_d)\psi(x)\,\mathrm{d}x .
\end{aligned}
$$

Notice that

$$
(\mathscr{L}_h^{\mathbb{R}^d_+})^{\frac{1}{2}} = \mathscr{T}^{-1}(h^2|\xi|^2-1)_-^{\frac{1}{2}}\mathscr{T} .
$$

In particular,

$$
(\mathscr{L}_h^{\mathbb{R}^d_+})^{\frac{1}{2}}\psi(x)
$$

$$
= -\frac{2i}{(2\pi)^d} \int_{\mathbb{R}^d} e^{ix\xi}(h^2|\xi|^2-1)_-^{\frac{1}{2}} \int_{\mathbb{R}^d_+} e^{-iy'\xi'} \sin(y_d\xi_d)\psi(y)\,\mathrm{d}y\,\mathrm{d}\xi
$$

$$
= -\frac{2i}{(2\pi)^d} \int_{\mathbb{R}^d_+} \mathrm{d}y\,\psi(y) \int_{\mathbb{R}^d} e^{ix\xi} e^{-iy'\xi'} \sin(y_d\xi_d)(h^2|\xi|^2-1)_-^{\frac{1}{2}}\,\mathrm{d}\xi .
$$

Thus, the kernel of $(\mathscr{L}_h^{\mathbb{R}^d_+})^{\frac{1}{2}}\varphi$ is

$$
-\frac{2i}{(2\pi)^d}\varphi(y) \int_{\mathbb{R}^d} e^{ix\xi} e^{-iy'\xi'} \sin(y_d\xi_d)(h^2|\xi|^2-1)_-^{\frac{1}{2}}\,\mathrm{d}\xi .
$$

The squared L^2-norm of this kernel is

$$
\begin{aligned}
&\frac{4}{(2\pi)^{2d}} \int_{\mathbb{R}^d_+} \mathrm{d}y\,|\varphi(y)|^2 \int_{\mathbb{R}^d_+} \mathrm{d}x\,\rho(x,y) \\
&= \frac{2}{(2\pi)^d} \int_{\mathbb{R}^d_+} \mathrm{d}y\,|\varphi(y)|^2 \int_{\mathbb{R}^d} \mathrm{d}x\,\rho(x,y) \\
&= \frac{2}{(2\pi)^d} \int_{\mathbb{R}^d_+} \mathrm{d}y\,|\varphi(y)|^2 \int_{\mathbb{R}^d} \mathrm{d}\xi\,\sin^2(y_d\xi_d)(h^2|\xi|^2-1)_- ,
\end{aligned}
$$

where the Parseval formula is used and

$$
\begin{aligned}
\rho(x,y) &= \left| \int_{\mathbb{R}^d} e^{ix\xi} e^{-iy'\xi'} \sin(y_d\xi_d)(h^2|\xi|^2-1)_-^{\frac{1}{2}}\,\mathrm{d}\xi \right|^2 \\
&= \left| \int_{\mathbb{R}^d} e^{i(x-y)\xi} e^{iy_d\xi_d} \sin(y_d\xi_d)(h^2|\xi|^2-1)_-^{\frac{1}{2}}\,\mathrm{d}\xi \right|^2 .
\end{aligned}
$$

This shows that $(\mathscr{L}_h^{\mathbb{R}^d_+})^{\frac{1}{2}}\varphi$ is a Hilbert–Schmidt operator, that $\varphi(\mathscr{L}_h^{\mathbb{R}^d_+})_-\varphi$ is trace-class, and

$$\mathrm{Tr}(\varphi(\mathscr{L}_h^{\mathbb{R}^d_+})_-\varphi)$$
$$= \frac{2}{(2\pi)^d}\int_{\mathbb{R}^d_+}\mathrm{d}y|\varphi(y)|^2\int_{\mathbb{R}^d}\mathrm{d}\xi\,\sin^2(y_d\xi_d)(h^2|\xi|^2-1)_-\,.$$

$\square$

In fact, one can estimate the asymptotic behavior of this « local » trace $\mathrm{Tr}(\varphi(\mathscr{L}_h^{\mathbb{R}^d_+})_-\varphi)$.

Lemma 9.38. — *Let us consider the function defined for $t \geqslant 0$ by*

$$J(t) = \int_{\mathbb{R}^d}(\xi^2-1)_-\cos(2t\xi_d)\,\mathrm{d}\xi\,.$$

Then, for all $t \geqslant 0$,

$$J(t) = C_0\,\mathrm{Re}\,K(t)\,, \quad K(t) = \int_{-1}^1 e^{2iut}(1-u^2)^{\frac{d+1}{2}}\,\mathrm{d}u\,,$$

with

$$C_0 = \int_{\mathbb{B}^{d-1}}(1-|v|^2)\,\mathrm{d}v\,.$$

Moreover, $J(t)\underset{t\to+\infty}{=}\mathscr{O}(t^{-\frac{d+1}{2}-1}).$

Proof. — We have

$$J(t) = \int_{-1}^1\mathrm{d}\xi_d\cos(2\xi_dt)\int_{|\xi'|^2\leqslant 1-|\xi_d|^2}(1-\xi_d^2-|\xi'|^2)\,\mathrm{d}\xi'\,.$$

By using a rescaling,

$$\int_{|\xi'|^2\leqslant 1-|\xi_d|^2}(1-\xi_d^2-|\xi'|^2)\,\mathrm{d}\xi' = (1-\xi_d^2)^{\frac{d+1}{2}}\int_{\mathbb{B}^{d-1}}(1-|v|^2)\,\mathrm{d}v\,.$$

Let us now consider K. We let $\delta = \frac{d+1}{2}$. By integrating by parts $\lfloor\delta\rfloor$ times, we can write

$$K(t) = t^{-\lfloor\delta\rfloor}\int_{-1}^1 e^{2iut}k(u)(1-u^2)^{\delta-\lfloor\delta\rfloor}\,\mathrm{d}u\,,$$

where k is a polynomial.

If $\delta \in \mathbb{N}$, another integration by parts yields

$$K(t) = \mathcal{O}(t^{-\lfloor \delta \rfloor - 1}),$$

which is the desired estimate.

If not, we have $\delta - \lfloor \delta \rfloor > 0$ and we can integrate by parts:

$$K(t) = -it^{-\lfloor \delta \rfloor - 1} \int_{-1}^{1} e^{2iut} u k(u)(1 - u^2)^{\delta - \lfloor \delta \rfloor - 1} \, du,$$

where we used that $(1 - u^2)^{\delta - \lfloor \delta \rfloor - 1} \in L^1((-1, 1))$ since $-1 < \delta - \lfloor \delta \rfloor - 1 < 0$. Now, we can write, for some smooth function $\tilde{k}$,

$$\int_0^1 e^{2iut} u k(u)(1 - u^2)^{\delta - \lfloor \delta \rfloor - 1} \, du$$

$$= \int_0^1 e^{2iut} \tilde{k}(u)(1 - u)^{\delta - \lfloor \delta \rfloor - 1} \, du,$$

and also, for some smooth function $\check{k}$,

$$\int_0^1 e^{2iut} u k(u)(1 - u^2)^{\delta - \lfloor \delta \rfloor - 1} \, du$$

$$= e^{2it} \int_0^1 e^{-2ivt} \check{k}(v) v^{\delta - \lfloor \delta \rfloor - 1} \, dv.$$

Note that

$$\int_0^1 e^{-2ivt} \check{k}(v) v^{\delta - \lfloor \delta \rfloor - 1} \, dv = \check{k}(0) \int_0^1 e^{-2ivt} v^{\delta - \lfloor \delta \rfloor - 1} \, dv$$

$$+ \int_0^1 e^{-2ivt}(\check{k}(v) - \check{k}(0)) v^{\delta - \lfloor \delta \rfloor - 1} \, dv.$$

We have

$$\int_0^1 e^{-2ivt} v^{\delta - \lfloor \delta \rfloor - 1} \, dv = t^{\lfloor \delta \rfloor - \delta} \int_0^t e^{-2iv} v^{\delta - \lfloor \delta \rfloor - 1} \, dv$$

$$= \mathcal{O}(t^{\lfloor \delta \rfloor - \delta}),$$

where we used that the last integral is convergent (by using integration by parts). We can write, for some smooth function r,

$$\check{k}(v) - \check{k}(0) = vr(v),$$

so that

$$\int_0^1 e^{-2ivt}(\check{k}(v) - \check{k}(0))v^{\delta - \lfloor\delta\rfloor - 1} \, dv = \int_0^1 e^{-2ivt} r(v) v^{\delta - \lfloor\delta\rfloor} \, dv$$
$$= \mathscr{O}(t^{-1}) = \mathscr{O}(t^{\lfloor\delta\rfloor - \delta}).$$

We deduce that

$$\int_0^1 e^{2iut} u k(u)(1 - u^2)^{\delta - \lfloor\delta\rfloor - 1} \, du = \mathscr{O}(t^{\lfloor\delta\rfloor - \delta}).$$

In the same way,

$$\int_{-1}^0 e^{2iut} u k(u)(1 - u^2)^{\delta - \lfloor\delta\rfloor - 1} \, du = \mathscr{O}(t^{\lfloor\delta\rfloor - \delta}).$$

Thus,

$$K(t) = \mathscr{O}(t^{-\lfloor\delta\rfloor - 1 + \lfloor\delta\rfloor - \delta}),$$

and the conclusion follows. □

Proposition 9.39. — *Consider $\varphi \in \mathscr{C}_0^1(\mathbb{R}^d)$. Then, with L_d as in (9.9.5.7), we have*

$$h^d \mathrm{Tr}(\varphi(\mathscr{L}_h^{\mathbb{R}_+^d})_- \varphi)$$
$$= L_d \|\varphi\|_{L^2(\mathbb{R}_+^d)}^2 - \frac{L_{d-1}}{4} h \int_{\mathbb{R}^{d-1}} |\varphi(x', 0)|^2 \, dx' + \mathscr{O}(h^2 \|\nabla\varphi^2\|_\infty).$$

Proof. — Let us write

$$(2\pi h)^d \mathrm{Tr}(\varphi(\mathscr{L}_h^{\mathbb{R}_+^d})_- \varphi)$$
$$= \int_{\mathbb{R}_+^d} \varphi^2(x) \int_{\mathbb{R}^d} (\xi^2 - 1)_- (1 - \cos(2h^{-1} x_d \xi_d)) \, d\xi \, dx.$$

Let us now consider the (absolutely convergent) integral

$$I(h) = \int_{\mathbb{R}_+^d} \varphi^2(x) \int_{\mathbb{R}^d} (\xi^2 - 1)_- \cos(2h^{-1} x_d \xi_d) \, d\xi \, dx.$$

We have
(9.9.4.4)

$$I(h) = h \int_{\mathbb{R}_+^d} \varphi^2(x', ht) \int_{\mathbb{R}^d} (\xi^2 - 1)_- \cos(2t\xi_d) \, d\xi \, dx' \, dt,$$

and thus

$$(9.9.4.5) \qquad I(h) = h \int_Q \left(\int_0^{+\infty} \varphi^2(x', ht) J(t) \, dt \right) dx',$$

where Q is a compact subset of $\mathbb{R}^{d-1}$. We write, uniformly with respect to $x' \in Q$,

$$|\varphi^2(x', ht) - \varphi^2(x', 0)| \leqslant \|\nabla \varphi^2\|_\infty ht.$$

Therefore,

$$\left| \int_0^{+\infty} \varphi^2(x', ht) J(t) \, dt - \int_0^{+\infty} \varphi^2(x', 0) J(t) \, dt \right|$$
$$\leqslant \|\nabla \varphi^2\|_\infty h \int_0^{+\infty} t J(t) \, dt.$$

This shows that

$$\left| I(h) - h \int_Q dx' \int_0^{+\infty} \varphi^2(x', 0) J(t) \, dt \right|$$
$$\leqslant |Q| h^2 \|\nabla \varphi^2\|_\infty \int_0^{+\infty} t J(t) \, dt.$$

We notice that

$$\int_{\mathbb{R}} (\xi^2 - 1)_- \cos(2t\xi_d) \, d\xi_d = \operatorname{Re} \int_{\mathbb{R}} (\xi^2 - 1)_- e^{2it\xi_d} \, d\xi_d.$$

Then, by using the inverse Fourier transform,

$$\int_{\mathbb{R}} \int_{\mathbb{R}} (\xi^2 - 1)_- e^{2it\xi_d} \, d\xi_d \, dt = \frac{1}{2}(2\pi)(|\xi'|^2 - 1)_-.$$

Thus,

$$(2\pi)^{-d} \int_0^{+\infty} J(t) \, dt = \frac{L_{d-1}}{4}.$$

$\square$

9.5. Notes

Proposition 9.39 is part of a rather long story. It started with Weyl in [**47**] and the asymptotic expansion of the counting function

$$N_\Omega(h) = |\Omega| \frac{C_d}{h^d} + o(h^{-d}), \qquad C_d = \frac{\omega_d}{(2\pi)^d}.$$

Here, $N_\Omega(h) = |\{k \geqslant 1 : h^2\lambda_k < 1\}|$. Under the geometric assumption that Ω has no « periodic point » and when Ω is smooth, V. Ivrii exhibited the second term of the asymptotic expansion in [**23**] (see also its translation):

$$(9.9.5.6) \quad N_\Omega(h) = |\Omega|C_d h^{-d} - \frac{1}{4}|\partial\Omega|C_{d-1}h^{-d+1} + o(h^{-d+1}).$$

In general, Weyl's asymptotic expansions can be obtained by means of microlocal technics. The reader can consult [**9**] where it is proved, for instance, that

$$|\{\lambda_k(h) < 1\}| = (2\pi h)^{-d}\int_{a(x,\xi)<1} dx\ d\xi + o(h^{-d+1}),$$

where the $(\lambda_k(h))_{k\geqslant 1}$ are the eigenvalues of an elliptic pseudo-differential operator defined by

$$\mathrm{Op}_h^{\mathrm{W}}(a)\psi(x)$$
$$= (2\pi h)^{-d}\int_{\mathbb{R}^{2d}} e^{i\langle x-y,\eta\rangle/h} a\left(\frac{x+y}{2},\eta\right)\psi(y)\,dy\,d\eta.$$

A very good introduction to semi-classical/microlocal analysis is the book by Zworski [**49**].

Sometimes (especially in old references), the Weyl asymptotics is written in terms of a large parameter $\lambda = h^{-\frac{1}{2}}$. The expansion (9.9.5.6) can be rewritten as

$$|\{k \geqslant 1 : \lambda_k < \lambda\}| =: N(\lambda)$$
$$= |\Omega|C_d\lambda^{\frac{d}{2}} - \frac{1}{4}C_{d-1}|\partial\Omega|\lambda^{\frac{d-1}{2}} + o(\lambda^{\frac{d-1}{2}}).$$

Note that

$$\int_0^\lambda N(u)\,du = |\Omega|\frac{2C_d}{d+2}\lambda^{\frac{d}{2}+1} - \frac{1}{4}\frac{2C_{d-1}}{d+1}|\partial\Omega|\lambda^{\frac{d+1}{2}} + o(\lambda^{\frac{d+1}{2}}),$$

and also, by definition of the counting function,

$$\int_0^\lambda N(u)\,du = -T(\lambda) + \lambda N(\lambda) = \sum_{k=1}^{N(\lambda)}(\lambda_k - \lambda)_-,$$

with

$$T(\lambda) = \sum_{k=1}^{N(\lambda)}\lambda_k.$$

Coming back to h, we deduce the following theorem, under Ivrii's assumptions.

Theorem 9.40. —

$$\mathrm{Tr}(-h^2\Delta - 1)_- = L_d|\Omega|h^{-d} - \frac{1}{4}L_{d-1}|\partial\Omega|h^{-d+1} + o(h^{-d+1}),$$

where

(9.9.5.7) $$L_d = (2\pi)^{-d}\int_{\mathbb{R}^d}(\xi^2 - 1)_-\,\mathrm{d}\xi = (2\pi)^{-d}\frac{2\omega_d}{d+2}.$$

Proposition 9.39 can be used as a step in a (direct) proof of Theorem 9.40 (see [**13**]).

CHAPTER 10

SELECTED APPLICATIONS OF THE FUNCTIONAL CALCULUS

The aim of this chapter is to illustrate how useful the functional calculus can be. In particular, we prove a version of Lieb's Variational Principle (to estimate traces of operators by means of density matrices). Then, we prove Stone's formula that relates the spectral projections to the resolvent. We use it to provide a sufficient condition for the spectrum to be absolutely continuous in a convenient spectral interval. Finally, we give a concise presentation of the celebrated Mourre estimates: how a positive commutator may be used to prove absolute continuity? During the analysis, we establish a version of the Limiting Absorption Principle. The core of the investigation will rely on elementary coercivity estimates for non-self-adjoint operators. It is somehow in the spirit of the Lax–Milgram theorem.

10.1. Lieb's Variational Principle

10.1.1. Statement. —

Definition 10.1 (**Density matrix**). — A density matrix on a Hilbert space H is a trace-class self-adjoint operator γ such that

$$0 \leqslant \gamma \leqslant 1 .$$

Let us consider a self-adjoint operator $\mathscr{L}$ bounded from below. We can write (see Section 8.3.4)

$$\mathscr{L} = \mathscr{L}_+ - \mathscr{L}_- ,$$

where $\mathscr{L}_\pm = \mathbb{1}_{\mathbb{R}_\pm}(\mathscr{L})\mathscr{L}$. Since $\mathscr{L}$ is bounded from below, the operator $\mathscr{L}_-$ is bounded. In particular, if γ is trace-class, $\gamma\mathscr{L}_-$ is trace-class.

Lemma 10.2. — *Consider a trace-class operator γ in the form domain $\mathrm{Dom}\,(\mathscr{Q})$ associated with the operator $\mathscr{L}$. We may consider a Hilbert basis (ψ_j) such that $(\psi_j) \subset \mathrm{Dom}\,\mathscr{Q}$. The quantity*

$$\sum_{j\geqslant 0}\langle\gamma\mathscr{L}_+^{\frac{1}{2}}\psi_j, \mathscr{L}_+^{\frac{1}{2}}\psi_j\rangle - \sum_{j\geqslant 0}\langle\gamma\mathscr{L}_-^{\frac{1}{2}}\psi_j, \mathscr{L}_-^{\frac{1}{2}}\psi_j\rangle$$

is well-defined (possibly $+\infty$) and independent of the choice of (ψ_j). We denote it by $\mathrm{Tr}(\gamma\mathscr{L})$.

Proof. — Let us write[1]

$$\gamma = \sum_{k\geqslant 0}\gamma_k|\varphi_k\rangle\langle\varphi_k|\,,$$

and notice that, by the properties of trace-class operators,

$$\sum_{j\geqslant 0}\langle\gamma\mathscr{L}_-^{\frac{1}{2}}\psi_j, \mathscr{L}_-^{\frac{1}{2}}\psi_j\rangle = \sum_{j\geqslant 0}\langle\gamma\mathscr{L}_-^{\frac{1}{2}}\varphi_j, \mathscr{L}_-^{\frac{1}{2}}\varphi_j\rangle\,.$$

Then, by the Bessel–Parseval formula and Fubini,

$$\sum_{j\geqslant 0}\langle\gamma\mathscr{L}_+^{\frac{1}{2}}\psi_j, \mathscr{L}_+^{\frac{1}{2}}\psi_j\rangle = \sum_{j\geqslant 0}\sum_{k\geqslant 0}\langle\gamma\mathscr{L}_+^{\frac{1}{2}}\psi_j, \varphi_k\rangle\langle\varphi_k, \mathscr{L}_+^{\frac{1}{2}}\psi_j\rangle$$

$$= \sum_{j\geqslant 0}\sum_{k\geqslant 0}\gamma_k|\langle\mathscr{L}_+^{\frac{1}{2}}\psi_j, \varphi_k\rangle|^2$$

$$= \sum_{k\geqslant 0}\gamma_k\|\mathscr{L}_+^{\frac{1}{2}}\varphi_k\|^2\,.$$

$\square$

Proposition 10.3 (Variational Principle). — *Assume that the Hilbert space $\mathrm{ran}\mathbb{1}_{\mathbb{R}_-}(\mathscr{L})$ has a Hilbert basis made of eigenfunctions of $\mathscr{L}$. We have*

$$\inf_{0\leqslant\gamma\leqslant 1}\mathrm{Tr}(\gamma\mathscr{L}) = -\mathrm{Tr}(\mathscr{L}_-)\,,$$

---

[1]The notation $|\varphi_k\rangle\langle\varphi_k|$ is the physical notation for the projection on φ_k.

where the infimum is taken over the density matrices valued in Dom $(\mathcal{Q})$. *Here,* $\mathrm{Tr}(\mathcal{L}_-)$ *can be infinite.*

Proof. — Let us consider $(\varphi_j)_{j \geqslant 1}$ a Hilbert basis of $\mathrm{ran}\mathbb{1}_{\mathbb{R}_-}(\mathcal{L})$ associated with (negative) eigenvalues $(E_j)_{j \geqslant 1}$ of $\mathcal{L}$. For all $N \geqslant 1$, consider

$$\gamma = \sum_{j=1}^{N} |\varphi_j\rangle\langle\varphi_j| \, .$$

The projection γ is trace-class and valued in Dom $(\mathcal{L})$. Moreover,

$$\mathrm{Tr}(\gamma\mathcal{L}) = \sum_{j=1}^{N} E_j \, .$$

Thus,

$$\inf_{0 \leqslant \gamma \leqslant 1} \mathrm{Tr}(\gamma\mathcal{L}) \leqslant \sum_{j=1}^{N} E_j \, ,$$

so that

$$\inf_{0 \leqslant \gamma \leqslant 1} \mathrm{Tr}(\gamma\mathcal{L}) \leqslant \sum_{j=1}^{+\infty} E_j = -\mathrm{Tr}(\mathcal{L}_-) \, .$$

Conversely, consider a trace class operator γ valued in the form domain. We may consider a Hilbert basis (ψ_j) valued in the form domain as in the proof of Lemma 10.2. We have

$$\mathrm{Tr}(\gamma\mathcal{L}) = \sum_{j=1}^{+\infty} \gamma_j \|\mathcal{L}_+^{\frac{1}{2}}\psi_j\|^2 - \sum_{j=1}^{+\infty} \gamma_j \|\mathcal{L}_-^{\frac{1}{2}}\psi_j\|^2 = \sum_{j=1}^{+\infty} \gamma_j \mathcal{Q}(\psi_j) \, ,$$

where the series converges in $\mathbb{R} \cup \{+\infty\}$. We have

$$\mathrm{Tr}(\gamma\mathcal{L}) \geqslant - \sum_{j=1}^{+\infty} \gamma_j \|\mathcal{L}_-^{\frac{1}{2}}\psi_j\|^2 \geqslant - \sum_{j=1}^{+\infty} \|\mathcal{L}_-^{\frac{1}{2}}\psi_j\|^2$$

$$= - \sum_{j=1}^{+\infty} \mathcal{Q}_-(\psi_j) = -\mathrm{Tr}(\mathcal{L}_-) \, .$$

$\square$

10.1.2. Illustration. — Let us recall Corollary 9.36. We would like to compare $\mathrm{Tr}(\varphi \mathscr{L}_h^{\mathbb{R}^d} \varphi)_-$ to $\mathrm{Tr}(\varphi(\mathscr{L}_h^{\mathbb{R}^d})_-\varphi)$. The Lieb Variational Principle can help us to do so (at least in the limit $h \to 0$).

Proposition 10.4. — *There exist $C, h_0 > 0$ such that, for all $h \in (0, h_0)$, and for all function $\varphi \in \mathscr{C}_0^\infty(\mathbb{R}^d)$, we have*

$$|\mathrm{Tr}(\varphi \mathscr{L}_h^{\mathbb{R}^d} \varphi)_- - \mathrm{Tr}(\varphi(\mathscr{L}_h^{\mathbb{R}^d})_-\varphi)| \leqslant Ch^{2-d}\|\nabla\varphi\|^2 .$$

Proof. — Consider $\chi \in \mathscr{C}_0^\infty(\mathbb{R}^d)$ such that $\varphi\chi = \varphi$. Since $\chi\gamma_h\chi$ is trace-class and $0 \leqslant \chi\gamma_h\chi \leqslant 1$ (see Section 9.4.1), the Variational Principle provides us with

$$(10.10.1.1) \qquad -\mathrm{Tr}(\varphi \mathscr{L}_h^{\mathbb{R}^d} \varphi)_- \leqslant \mathrm{Tr}((\varphi \mathscr{L}_h^{\mathbb{R}^d} \varphi)(\chi\gamma_h\chi)) .$$

We write

$$\chi\gamma_h\chi = \sum_{j \geqslant 1} \mu_j |\psi_j\rangle\langle\psi_j| ,$$

so that

$$\chi\gamma_h\chi\psi_j = \mu_j\psi_j .$$

Note that, if $\varphi \in \mathscr{C}_0^\infty(\mathbb{R}^d)$,

$$\langle\psi_j, (\varphi \mathscr{L}_h^{\mathbb{R}^d} \varphi)\psi_j\rangle = \langle\psi_j, (\mathscr{L}_h^{\mathbb{R}^d}\varphi^2 + [\varphi, \mathscr{L}_h^{\mathbb{R}^d}]\varphi)\psi_j\rangle$$
$$= \langle\psi_j, (\varphi^2\mathscr{L}_h^{\mathbb{R}^d} + \varphi[\mathscr{L}_h^{\mathbb{R}^d}, \varphi])\psi_j\rangle ,$$

so that

$$2\langle\psi_j, (\varphi \mathscr{L}_h^{\mathbb{R}^d} \varphi)\psi_j\rangle$$
$$= \langle\psi_j, \left(\varphi^2\mathscr{L}_h^{\mathbb{R}^d} + \mathscr{L}_h\varphi^2 - [\varphi, [\varphi, \mathscr{L}_h^{\mathbb{R}^d}]]\right)\psi_j\rangle ,$$

and thus

$$\langle\psi_j, (\varphi\mathscr{L}_h^{\mathbb{R}^d} \varphi)\psi_j\rangle = \mathrm{Re}\,\langle\psi_j, \left(\varphi^2\mathscr{L}_h^{\mathbb{R}^d} - \frac{1}{2}[\varphi, [\varphi, \mathscr{L}_h^{\mathbb{R}^d}]]\right)\psi_j\rangle ,$$

since

$$[\varphi, [\varphi, \mathscr{L}_h^{\mathbb{R}^d}]] = -2h^2|\nabla\varphi|^2 .$$

Thus, if $\varphi \in \mathscr{C}_0^\infty(\mathbb{R}^d)$,

$$\mathscr{Q}_h(\varphi\psi_j) = \langle\psi_j, (\varphi\mathscr{L}_h^{\mathbb{R}^d} \varphi)\psi_j\rangle$$
$$= \mathrm{Re}\,\langle\psi_j, \varphi^2\mathscr{L}_h^{\mathbb{R}^d}\psi_j\rangle + h^2\langle\psi_j, |\nabla\varphi|^2\psi_j\rangle .$$

This formula can be extended to $\varphi \in \mathscr{C}_0^1(\mathbb{R}^d)$. We have

$$\mu_j\langle\psi_j, (\varphi\mathscr{L}_h^{\mathbb{R}^d}\varphi)\psi_j\rangle$$
$$= \mathrm{Re}\,\langle\psi_j, \varphi^2\mathscr{L}_h^{\mathbb{R}^d}(\chi\gamma_h\chi)\psi_j\rangle + h^2\langle\psi_j, |\nabla\varphi|^2(\chi\gamma_h\chi)\psi_j\rangle\,.$$

Since $\mathscr{L}_h^{\mathbb{R}^d}$ is a local operator, we have

$$\mu_j\langle\psi_j, (\varphi\mathscr{L}_h^{\mathbb{R}^d}\varphi)\psi_j\rangle$$
$$= -\mathrm{Re}\,\langle\psi_j, \varphi^2\chi(\mathscr{L}_h^{\mathbb{R}^d})_-\chi)\psi_j\rangle + h^2\langle\psi_j, |\nabla\varphi|^2\chi\gamma_h\chi\psi_j\rangle\,.$$

Now, we observe that $\varphi^2\chi(\mathscr{L}_h^{\mathbb{R}^d})_-\chi)$ and $|\nabla\varphi|^2\chi\gamma_h\chi$ are trace-class. In particular, the series $\sum_{j\geqslant 1}\mu_j\langle\psi_j, (\varphi\mathscr{L}_h^{\mathbb{R}^d}\varphi)\psi_j\rangle$ is convergent and

$$\sum_{j\geqslant 1}\mu_j\langle\psi_j, (\varphi\mathscr{L}_h^{\mathbb{R}^d}\varphi)\psi_j\rangle$$
$$= -\mathrm{Tr}(\varphi^2\chi(\mathscr{L}_h^{\mathbb{R}^d})_-\chi) + h^2\mathrm{Tr}(|\nabla\varphi|^2\chi\gamma_h\chi)\,.$$

By using the cyclicity of the trace, we also get

$$\sum_{j\geqslant 1}\mu_j\langle\psi_j, (\varphi\mathscr{L}_h^{\mathbb{R}^d}\varphi)\psi_j\rangle$$
$$= -\mathrm{Tr}(\varphi(\mathscr{L}_h^{\mathbb{R}^d})_-\varphi) + h^2\mathrm{Tr}(|\nabla\varphi|\gamma_h|\nabla\varphi|)\,.$$

This shows that the left-hand side does not depend on the choice of the diagonalizing Hilbert basis $(\psi_j)_{j\geqslant 1}$. By definition, the left-hand side is $\mathrm{Tr}(\varphi\mathscr{L}_h^{\mathbb{R}^d}\varphi(\chi\gamma_h\chi))$, and we have proved that

$$\mathrm{Tr}(\varphi\mathscr{L}_h^{\mathbb{R}^d}\varphi(\chi\gamma_h\chi)) = -\mathrm{Tr}(\varphi(\mathscr{L}_h^{\mathbb{R}^d})_-\varphi) + h^2\mathrm{Tr}(|\nabla\varphi|\gamma_h|\nabla\varphi|)\,.$$

By (10.10.1.1), this implies that

$$\mathrm{Tr}(\varphi\mathscr{L}_h^{\mathbb{R}^d}\varphi)_- \geqslant \mathrm{Tr}(\varphi(\mathscr{L}_h^{\mathbb{R}^d})_-\varphi) - h^2\mathrm{Tr}(|\nabla\varphi|\gamma_h|\nabla\varphi|)\,.$$

With Proposition 9.35, we get

$$\mathrm{Tr}(\varphi\mathscr{L}_h^{\mathbb{R}^d}\varphi)_- \geqslant \mathrm{Tr}(\varphi(\mathscr{L}_h^{\mathbb{R}^d})_-\varphi) - Ch^{2-d}\|\nabla\varphi\|^2\,.$$

$\square$

10.2. Stone's formula

10.2.1. Statement. — Let $\mathscr{L}$ be a self-adjoint operator.

***Proposition 10.5* (Stone's formula).** — *Consider $a, b \in \mathbb{R}$ such that $a < b$. We have, for all $u \in \mathsf{H}$,*

$$\lim_{\varepsilon \to 0^+} \frac{1}{2i\pi} \int_{[a,b]} \left((\mathscr{L} - (\lambda + i\varepsilon))^{-1} - ((\mathscr{L} - (\lambda - i\varepsilon))^{-1} \right) u \, \mathrm{d}\lambda$$

$$= \frac{1}{2} \left(\mathbb{1}_{[a,b]}(\mathscr{L}) + \mathbb{1}_{(a,b)}(\mathscr{L}) \right) u .$$

Proof. — For $\varepsilon > 0$, we introduce, for all $x \in [a, b]$,

$$f_\varepsilon(x) = \frac{1}{2i\pi} \int_{[a,b]} \left((x - (\lambda + i\varepsilon))^{-1} - ((x - (\lambda - i\varepsilon))^{-1} \right) \mathrm{d}\lambda ,$$

and we notice that, for all $x \in [a, b]$,

$$f_\varepsilon(x) = \frac{1}{\pi} \left(\arctan \left(\frac{b - x}{\varepsilon} \right) - \arctan \left(\frac{a - x}{\varepsilon} \right) \right) ,$$

so that

$$\lim_{\varepsilon \to 0^+} f_\varepsilon(x) = g(x) := \frac{1}{2} \left(\mathbb{1}_{[a,b]}(x) + \mathbb{1}_{(a,b)}(x) \right) ,$$

and $|f_\varepsilon(x)| \leqslant 1$. Since, for all $u \in \mathsf{H}$,

$$\|(f_\varepsilon(\mathscr{L}) - g(\mathscr{L}))u\|^2 = \int_{\mathbb{R}} |f_\varepsilon(\lambda) - g(\lambda)|^2 \, \mathrm{d}\mu_{u,u} ,$$

we get, by dominated convergence,

$$\lim_{\varepsilon \to 0^+} f_\varepsilon(\mathscr{L})u = \frac{1}{2} \left(\mathbb{1}_{[a,b]}(\mathscr{L}) + \mathbb{1}_{(a,b)}(\mathscr{L}) \right) u .$$

By using Riemannian sums and Exercise 8.34, we get, for $\varepsilon > 0$,

$$f_\varepsilon(\mathscr{L})$$
$$= \frac{1}{2i\pi} \int_{[a,b]} \left((\mathscr{L} - (\lambda + i\varepsilon))^{-1} - ((\mathscr{L} - (\lambda - i\varepsilon))^{-1} \right) \mathrm{d}\lambda ,$$

and the conclusion follows. $\square$

10.2.2. A criterion for absolute continuity. — The Stone formula may be used as follows.

Proposition 10.6. — *For all $z \in \mathbb{C} \setminus \mathrm{sp}(\mathscr{L})$, we consider the resolvent $R(z) = (\mathscr{L} - z)^{-1}$. Assume that, for all ψ in a dense set D of H, there exists $C(\psi) > 0$ such that*

$$\sup_{\varepsilon \in (0,1)} \sup_{\mu \in (a,b)} \langle \mathrm{Im}\, R(\mu + i\varepsilon)\psi, \psi \rangle \leqslant C(\psi).$$

Then, the spectrum of $\mathscr{L}$ in (a, b) is absolutely continuous. In particular, there is no eigenvalue in (a, b).

Proof. — Consider a Borelian set Ω such that $\Omega \subset (a, b)$. We have, by construction of the Lebesgue measure,

$$|\Omega| = \inf \Big\{ \sum_{j \in \mathbb{N}} |I_j|\,,\ \text{with, for all } j \in \mathbb{N},$$

$$I_j \subset (a, b) \text{ open bounded interval}\,, \Omega \subset \bigcup_{j \in \mathbb{N}} I_j \Big\}.$$

Let us consider such a family (I_j). We write $I_j = (c_j, d_j)$. The Stone formula gives

$$\frac{1}{2}\Big(\mathbb{1}_{[c_j, d_j]}(\mathscr{L}) + \mathbb{1}_{(c_j, d_j)}(\mathscr{L}) \Big)$$

$$= \lim_{\varepsilon \to 0} \frac{1}{2i\pi} \int_{c_j}^{d_j} 2i\mathrm{Im}\,(R(\mu + i\varepsilon))\, \mathrm{d}\mu.$$

Then, for all $\psi \in \mathsf{D}$,

$$\langle \mathbb{1}_{(c_j, d_j)}(\mathscr{L})\psi, \psi \rangle \leqslant \frac{|I_j|}{\pi} \sup_{\varepsilon \in (0,1)} \sup_{\mu \in (a,b)} \langle \mathrm{Im}\, R(\mu + i\varepsilon)\psi, \psi \rangle$$

$$\leqslant C(\psi)|I_j|.$$

Since $\mathbb{1}_\Omega \leqslant \sum_{j \in \mathbb{N}} \mathbb{1}_{I_j}$, we get

$$\langle \mathbb{1}_\Omega(\mathscr{L})\psi, \psi \rangle \leqslant C(\psi) \sum_{j \in \mathbb{N}} |I_j|.$$

Taking the infimum, we get

$$\langle \mathbb{1}_\Omega(\mathscr{L})\psi, \psi \rangle \leqslant C(\psi)|\Omega|.$$

The conclusion follows by using Proposition 8.52.   □

10.3. Elementary Mourre's theory and Limiting Absorption Principle

10.3.1. Mourre estimates. —

10.3.1.1. *Assumptions.* — We select two intervals I and J with $I \subset\subset J$ and J bounded. We consider two self-adjoint operators $\mathscr{L}$ and $\mathscr{A}$. We assume that

$$\mathbb{1}_J(\mathscr{L})\mathscr{B}\mathbb{1}_J(\mathscr{L}) \geqslant c_0\mathbb{1}_J(\mathscr{L}), \quad \mathscr{B} := [\mathscr{L}, i\mathscr{A}], \quad c_0 > 0,$$

as well as

  (i) $[\mathscr{L}, \mathscr{A}](\mathscr{L} + i)^{-1}$ is bounded.
  (ii) $(\mathscr{L} + i)^{-1}[[\mathscr{L}, \mathscr{A}], \mathscr{A}]$ is bounded.

10.3.1.2. *Coercivity estimates.* — Consider $\varepsilon \geqslant 0$, $\operatorname{Re} z \in I$, $\operatorname{Im} z \geqslant 0$. We define, on $\operatorname{Dom}\mathscr{L}$,

$$\mathscr{L}_{z,\varepsilon} := \mathscr{L} - z - i\varepsilon\mathscr{B}.$$

Proposition 10.7 **(Mourre estimates).** — *There exist positive constants* $\varepsilon_0, C_1, C_2, c_0', c_0'', c, c'$ *such that, for all* $\varepsilon \in (0, \varepsilon_0)$ *and* $z \in I \times [0, +\infty)$*, we have the following coercivity estimates.*

  (a) *For all* $u \in \operatorname{Dom}(\mathscr{L})$,

$$\|\mathscr{L}_{z,\varepsilon}u\| \geqslant c\|(\mathscr{L} + i)\mathbb{1}_{J^c}(\mathscr{L})u\| - C_1\varepsilon\|\mathbb{1}_J(\mathscr{L})u\|,$$

  where $J^c = \complement J = \mathbb{R} \setminus J$.
  (b) *For all* $u \in \operatorname{Dom}(\mathscr{L})$,

$$\|\mathscr{L}_{z,\varepsilon}u\| \geqslant c_0'\varepsilon\|(\mathscr{L} + i)\mathbb{1}_J(\mathscr{L})u\|.$$

  (c) *For all* $u \in \operatorname{Dom}(\mathscr{L})$,

$$\|\mathscr{L}_{z,\varepsilon}u\| \geqslant c'\|(\mathscr{L} + i)\mathbb{1}_{J^c}(\mathscr{L})u\|.$$

  (d) *In particular* $\mathscr{L}_{z,\varepsilon}$ *is bijective and*

$$\|(\mathscr{L} + i)\mathscr{L}_{z,\varepsilon}^{-1}\| \leqslant C_2\varepsilon^{-1}.$$

  Moreover, for $\varepsilon = 0$*,* $\mathscr{L}_{z,\varepsilon}$ *is also bijective when* $\operatorname{Im} z > 0$*.*
  (e)

$$\forall u \in \operatorname{Dom}(\mathscr{L}), \quad |\langle \mathscr{L}_{z,\varepsilon}u, u\rangle| + \|\mathscr{L}_{z,\varepsilon}u\|^2 \geqslant c_0''\varepsilon\|u\|^2.$$

Proof. —   (a) The first step is to reduce the discussion to the case $|\text{Im } z| \leqslant C$ for some constant C depending only on the bounds available on $[\mathscr{L}, \mathscr{A}](\mathscr{L} + i)^{-1}$ and J. To this end, observe that (for ε small enough and $|\text{Im } z| \geqslant C$):

$$
\begin{aligned}
&\|\mathscr{L}_{z,\varepsilon} u\| \\
&= \|(\mathscr{L} - z)u + \varepsilon[\mathscr{L}, \mathscr{A}](\mathscr{L} + i)^{-1}(\mathscr{L} + i)u\| \\
&\geqslant \|(\mathscr{L} - z)u\| - C\varepsilon\|(\mathscr{L} + i)u\| \\
&\geqslant \tfrac{1}{\sqrt{2}}\left(\|(\mathscr{L} - \text{Re } z)u\| + |\text{Im } z|\|u\| - \sqrt{2}C\varepsilon\|(\mathscr{L} + i)u\|\right) \\
&\geqslant (\tfrac{1}{\sqrt{2}} - C\varepsilon)\|(\mathscr{L} + i)u\| + \tfrac{1}{\sqrt{2}}(|\text{Im } z| - |(-i - \text{Re } z)|)\|u\| \\
&\geqslant (\tfrac{1}{\sqrt{2}} - C\varepsilon)\|(\mathscr{L} + i)u\| + \tfrac{1}{\sqrt{2}}(|\text{Im } z| - C)\|u\| \\
&\geqslant c\|(\mathscr{L} + i)\mathbb{1}_{J^c}(\mathscr{L})u\|.
\end{aligned}
$$

The second step deals with the case when z is bounded. By (i),

$$
\begin{aligned}
\|\mathbb{1}_{J^c}(\mathscr{L})\mathscr{L}_{z,\varepsilon} u\| &\geqslant \|\mathbb{1}_{J^c}(\mathscr{L})(\mathscr{L} - z)u\| - \varepsilon\|\mathscr{B}u\| \\
&\geqslant \|\mathbb{1}_{J^c}(\mathscr{L})(\mathscr{L} - z)u\| - C\varepsilon\|(\mathscr{L} + i)u\|.
\end{aligned}
$$

By using the orthogonal decomposition of the last term, and since J is bounded, we have

$$
\begin{aligned}
\|(\mathscr{L} + i)u\| &\leqslant C\|(\mathscr{L} + i)\mathbb{1}_{J^c}(\mathscr{L})u\| + C\|(\mathscr{L} + i)\mathbb{1}_J(\mathscr{L})u\| \\
&\leqslant C\|(\mathscr{L} + i)\mathbb{1}_{J^c}(\mathscr{L})u\| + \tilde{C}\|\mathbb{1}_J(\mathscr{L})u\|.
\end{aligned}
$$

Then, since z lies in a bounded set, we have

$$
\begin{aligned}
&\|\mathbb{1}_{J^c}(\mathscr{L})\mathscr{L}_{z,\varepsilon} u\| \\
&\geqslant \|\mathbb{1}_{J^c}(\mathscr{L})(\mathscr{L} - z)u\| - C\varepsilon\|\mathbb{1}_J(\mathscr{L})u\| - C\varepsilon\|(\mathscr{L} + i)\mathbb{1}_{J^c}(\mathscr{L})u\| \\
&\geqslant (1 - C\varepsilon)\|\mathbb{1}_{J^c}(\mathscr{L})(\mathscr{L} - z)u\| - C\varepsilon\|\mathbb{1}_J(\mathscr{L})u\| - C\varepsilon\|\mathbb{1}_{J^c}(\mathscr{L})u\|.
\end{aligned}
$$

Since $|\lambda - z|$ is bounded from below when $\lambda \in J^c$ and $z \in I \times [0, +\infty[$, we have

$$
\|\mathbb{1}_{J^c}(\mathscr{L})u\| \leqslant C\|\mathbb{1}_{J^c}(\mathscr{L})(\mathscr{L} - z)u\|,
$$

and thus, we can deduce (a).

(b) By (i) and (a),

$$
\begin{aligned}
&- \text{Im} \langle \mathscr{L}_{z,\varepsilon} u, \mathbb{1}_J(\mathscr{L})u \rangle \\
&\quad = \text{Im } z\|\mathbb{1}_J(\mathscr{L})u\|^2 + \varepsilon\langle \mathbb{1}_J(\mathscr{L})\mathscr{B}\mathbb{1}_J(\mathscr{L})u, u \rangle \\
&\qquad\qquad + \varepsilon\text{Re} \langle \mathscr{B}\mathbb{1}_{J^c}(\mathscr{L})u, \mathbb{1}_J(\mathscr{L})u \rangle,
\end{aligned}
$$

and

$$-\operatorname{Im}\langle \mathscr{L}_{z,\varepsilon}u, \mathbb{1}_J(\mathscr{L})u\rangle$$
$$\geqslant \operatorname{Im} z\|\mathbb{1}_J(\mathscr{L})u\|^2 + c_0\varepsilon\|\mathbb{1}_J(\mathscr{L})u\|^2$$
$$- C\varepsilon\|\mathbb{1}_J(\mathscr{L})u\|\|(\mathscr{L}+i)\mathbb{1}_{J^c}(\mathscr{L})u\|,$$

so that

$$(10.10.3.2) \quad -\operatorname{Im}\langle \mathscr{L}_{z,\varepsilon}u, \mathbb{1}_J(\mathscr{L})u\rangle$$
$$\geqslant \operatorname{Im} z\|\mathbb{1}_J(\mathscr{L})u\|^2 + (c_0\varepsilon - C\varepsilon^2)\|\mathbb{1}_J(\mathscr{L})u\|^2$$
$$- C\varepsilon\|\mathbb{1}_J(\mathscr{L})u\|\|\mathscr{L}_{z,\varepsilon}u\|.$$

By Cauchy–Schwarz, this gives (for ε small enough)

$$(1+C\varepsilon)\|\mathscr{L}_{z,\varepsilon}u\| \geqslant (\operatorname{Im} z + \tilde{c}_0\varepsilon)\|\mathbb{1}_J(\mathscr{L})u\| \geqslant \tilde{c}_0\varepsilon\|\mathbb{1}_J(\mathscr{L})u\|.$$

Since J is bounded, we deduce (b).

(c) It is sufficient to combine (a) and (b).

(d) From (b) and (c), since the operator $\mathscr{L}+i$ is injective, we see that $\|\mathscr{L}_{z,\varepsilon}u\| = 0$ only if $\|\mathbb{1}_J(\mathscr{L})u\| = 0$ and $\|\mathbb{1}_{J^c}(\mathscr{L})u\| = 0$ so that $u = 0$. More precisely, we infer that $\mathscr{L}_{z,\varepsilon}$ is injective with closed range (and so is the adjoint since the above estimates (a), (b), (c) are also true when $\mathscr{L}_{z,\varepsilon}$ is replaced by $\mathscr{L}_{z,\varepsilon}^*$). Therefore, $\mathscr{L}_{z,\varepsilon}$ is bijective and $\|(\mathscr{L}+i)\mathscr{L}_{z,\varepsilon}^{-1}\| \leqslant C\varepsilon^{-1}$. The case $\varepsilon = 0$ can be obtained by improving the estimates (as indicated above) when $\operatorname{Im} z > 0$.

(e) Notice that, with (10.10.3.2),

$$-\operatorname{Im}\langle \mathscr{L}_{z,c}u, u\rangle + \operatorname{Im}\langle \mathscr{L}_{z,\varepsilon}u, \mathbb{1}_{J^c}(\mathscr{L})u\rangle$$
$$= -\operatorname{Im}\langle \mathscr{L}_{z,\varepsilon}u, \mathbb{1}_J(\mathscr{L})u\rangle$$
$$\geqslant c_0'\varepsilon\|\mathbb{1}_J(\mathscr{L})u\|^2 - C\varepsilon\|\mathbb{1}_J(\mathscr{L})u\|\|\mathscr{L}_{z,\varepsilon}u\|.$$

By Cauchy–Schwarz,

$$|\langle \mathscr{L}_{z,\varepsilon}u, u\rangle| + \|\mathscr{L}_{z,\varepsilon}u\|\|\mathbb{1}_{J^c}(\mathscr{L})u\|$$
$$\geqslant c_0'\varepsilon\|\mathbb{1}_J(\mathscr{L})u\|^2 - C\varepsilon\|\mathbb{1}_J(\mathscr{L})u\|\|\mathscr{L}_{z,\varepsilon}u\|,$$

so that, by (b) and (c), the conclusion follows.

$$\square$$

10.3.2. Limiting Absorption Principle and consequence. —

Lemma 10.8. — *For all bounded self-adjoint operator* $\mathscr{C}$,

$$\|\mathscr{L}_{z,\varepsilon}^{-1}\mathscr{C}\| \leqslant C\varepsilon^{-\frac{1}{2}}(1 + \|\mathscr{C}\mathscr{L}_{z,\varepsilon}^{-1}\mathscr{C}\|^{\frac{1}{2}}),$$

and

$$\|\mathscr{C}\mathscr{L}_{z,\varepsilon}^{-1}\| \leqslant C\varepsilon^{-\frac{1}{2}}(1 + \|\mathscr{C}\mathscr{L}_{z,\varepsilon}^{-1}\mathscr{C}\|^{\frac{1}{2}}).$$

Proof. — Insert $u = \mathscr{L}_{z,\varepsilon}^{-1}\mathscr{C}\varphi$ in (e), and the first estimate follows. The second one follows by noticing that (e) is also true when $\mathscr{L}_{z,\varepsilon}$ is replaced by $\mathscr{L}_{z,\varepsilon}^*$ and letting $u = \left(\mathscr{L}_{z,\varepsilon}^{-1}\right)^*\mathscr{C}\varphi$. $\qquad\square$

Proposition 10.9 (Limiting Absorption Principle)

For all bounded self-adjoint operator $\mathscr{C}$ such that $\mathscr{C}\mathscr{A}$ and $\mathscr{A}\mathscr{C}$ are bounded, we have

$$\exists C, \varepsilon_0 > 0, \forall \varepsilon \in (0, \varepsilon_0),$$

$$\sup_{\operatorname{Im} z > 0, \operatorname{Re} z \in J} \|\mathscr{C}(\mathscr{L} - z - i\varepsilon\mathscr{B})^{-1}\mathscr{C}\| \leqslant C,$$

and

$$\sup_{\operatorname{Im} z > 0, \operatorname{Re} z \in J} \|\mathscr{C}(\mathscr{L} - z)^{-1}\mathscr{C}\| \leqslant C.$$

Proof. — We set $F = \mathscr{C}\mathscr{L}_{z,\varepsilon}^{-1}\mathscr{C}$ and we have, taking the derivative w.r.t. ε,

$$iF' = i\mathscr{C}\mathscr{L}_{z,\varepsilon}^{-1}\left(\frac{d}{d\varepsilon}\mathscr{L}_{z,\varepsilon}\right)\mathscr{L}_{z,\varepsilon}^{-1}\mathscr{C} = \mathscr{C}\mathscr{L}_{z,\varepsilon}^{-1}\mathscr{B}\mathscr{L}_{z,\varepsilon}^{-1}\mathscr{C}.$$

On the other hand, by construction, we have

$$\mathscr{B} = [\mathscr{L}, i\mathscr{A}] = [\mathscr{L}_{z,\varepsilon} + z + i\varepsilon\mathscr{B}, i\mathscr{A}] = [\mathscr{L}_{z,\varepsilon}, i\mathscr{A}] - \varepsilon[\mathscr{B}, \mathscr{A}].$$

There remains

$$iF' = \mathscr{C}\mathscr{L}_{z,\varepsilon}^{-1}[\mathscr{L}_{z,\varepsilon}, i\mathscr{A}]\mathscr{L}_{z,\varepsilon}^{-1}\mathscr{C}$$
$$- i\varepsilon\mathscr{C}(\mathscr{L} + i)\mathscr{L}_{z,\varepsilon}^{-1}(\mathscr{L} + i)^{-1}[[\mathscr{L}, \mathscr{A}], \mathscr{A}]\mathscr{L}_{z,\varepsilon}^{-1}\mathscr{C}.$$

Since $\mathscr{A}\mathscr{C}$ and $\mathscr{C}\mathscr{A}$ are bounded, we have

$$\|\mathscr{C}\mathscr{L}_{z,\varepsilon}^{-1}[\mathscr{L}_{z,\varepsilon}, \mathscr{A}]\mathscr{L}_{z,\varepsilon}^{-1}\mathscr{C}\| \leqslant \|\mathscr{C}\mathscr{A}\mathscr{L}_{z,\varepsilon}^{-1}\mathscr{C}\| + \|\mathscr{C}\mathscr{L}_{z,\varepsilon}^{-1}\mathscr{A}\mathscr{C}\|$$
$$\leqslant C\|\mathscr{L}_{z,\varepsilon}^{-1}\mathscr{C}\| + C\|\mathscr{C}\mathscr{L}_{z,\varepsilon}^{-1}\|.$$

Thanks to (ii) and (d), we have

$$\|F'\| \leqslant C\|\mathscr{L}_{z,\varepsilon}^{-1}\mathscr{C}\| + C\|\mathscr{C}\mathscr{L}_{z,\varepsilon}^{-1}\| \leqslant \tilde{C}\varepsilon^{-\frac{1}{2}}(1 + \|F\|^{\frac{1}{2}}).$$

We have $\|F\| \leqslant C\varepsilon^{-1}$ and thus, by integrating, $\|F\| \leqslant C|\ln \varepsilon|$. Therefore, F is bounded by using again the differential inequality.

$\square$

Choosing $\mathscr{C} = \mathscr{A}^{-1}$, using the density of $\mathrm{Dom}\,(\mathscr{A})$, and applying Proposition 8.51, we deduce that the spectrum of $\mathscr{L}$ in J is a.c. and that there is no eigenvalue in J.

10.3.3. Example of Mourre estimates. — We want to provide here the Reader with a paradigmatic example of the Mourre method. In the literature, this example is sometimes called the Virial Theorem. Consider

$$\mathscr{L} = -\partial_x^2 + V(x)\,,$$

where V is (real) non-negative, and smooth. We assume that $V(x)$, $xV'(x)$, and $x^2V''(x)$ are bounded. The natural domain of $\mathscr{L}$ is $\mathrm{H}^2(\mathbb{R}^d)$. We let

$$\mathscr{A} = -\frac{i}{2}(x\partial_x + \partial_x x) = -ix\partial_x - \frac{i}{2}\,.$$

Let us now inspect the Mourre assumptions. A computation gives

$$[\mathscr{L}, \mathscr{A}] = -i(-2\partial_x^2 - xV'(x)) = -i(2\mathscr{L} - 2V - xV'(x))\,.$$

In particular,

$$\mathscr{B} = [\mathscr{L}, i\mathscr{A}] = 2(\mathscr{L} - V) - xV'(x) = 2\mathscr{L} + W(x)\,,$$

with

$$W(x) = -2V(x) - xV'(x)\,.$$

Then

$$
\begin{aligned}
[[\mathscr{L}, \mathscr{A}], \mathscr{A}] \\
&= -2i[\mathscr{L}, \mathscr{A}] + 2[V, x\partial_x] + [xV'(x), x\partial_x] \\
&= 2(2\mathscr{L} - 2V - xV'(x)) - 2xV'(x) - x(\cdot V'(\cdot))'(x) \\
&= 4\mathscr{L} - 4V - 5xV'(x) - x^2V''(x)\,.
\end{aligned}
$$

From the assumptions on V, we deduce that the assumptions on the commutators are satisfied. Let us now turn to the « positive commutator assumption ». Consider $E_0 > 0$ and $\eta > 0$. We let $J = [E_0 - \eta, E_0 + \eta]$. Let us notice that

$$\langle \mathscr{B}\mathbb{1}_J(\mathscr{L})\psi, \mathbb{1}_J(\mathscr{L})\psi \rangle \geqslant \langle (2(E_0 - \eta) + W)\mathbb{1}_J(\mathscr{L})\psi, \mathbb{1}_J(\mathscr{L})\psi \rangle\,.$$

Therefore, we see that the positivity of

$$2(E_0 - V(x)) - xV'(x)$$

is a key of the positive commutator assumption. Thus, we assume that V satisfies

(10.10.3.3) $2(E_0 - V(x)) - xV'(x) \geqslant c_0 > 0 \,.$

Up to shrinking η, we get the positive commutator.

This shows that the spectrum of $\mathscr{L}$ lying in $(E_0 - \eta, E_0 + \eta)$ is absolutely continuous and that there is no eigenvalue in this window. The inequality (10.10.3.3) is satisfied for all $E_0 \in (\|V\|_\infty, +\infty)$ as soon as $xV'(x) \leqslant 0$ (which is satisfied for all even V having a maximum at 0 and being non-increasing on $(0, +\infty)$). When $V = 0$, we recover that the (positive) spectrum of $-\partial_x^2$ is absolutely continuous. This fact can be directly proved by noticing that, for all $f \in \mathscr{C}_0^0(\mathbb{R})$,

$$\langle f(\mathscr{L})u, u \rangle = \langle f(\xi^2)\hat{u}, \hat{u} \rangle = \int_\mathbb{R} f(\xi^2)|\hat{u}(\xi)|^2 \, d\xi \,,$$

where $\hat{u}$ denotes the unitary Fourier transform of u. By using a change of variable, we get

$$\int_\mathbb{R} f(\xi^2)|\hat{u}(\xi)|^2 \, d\xi = \int_0^{+\infty} f(\lambda)\frac{|\hat{u}(\sqrt{\lambda})|^2 + |\hat{u}(-\sqrt{\lambda})|^2}{2\sqrt{\lambda}} \, d\lambda \,.$$

This shows that, for all $u \in \mathsf{L}^2(\mathbb{R})$,

$$d\mu_{u,u} = \mathbb{1}_{[0,+\infty)}(\lambda)\frac{|\hat{u}(\sqrt{\lambda})|^2 + |\hat{u}(-\sqrt{\lambda})|^2}{2\sqrt{\lambda}} \, d\lambda \,,$$

which implies that $d\mu_{u,u}$ is absolutely continuous with respect to $d\lambda$.

10.4. Notes

i. Section 10.1 has been inspired by the Ph. D. dissertation of S. Gottwald [15, Appendix C] (see also [16]), and many discussions with S. Fournais and T. Østergaard-Sørensen.

ii. Section 10.3 is inspired by the original paper [32] and also [5, Section 4.3], but provides a different presentation centered around coercivity estimates. This section has benefited of discussions with É. Soccorsi.

CHAPTER A

REMINDERS OF FUNCTIONAL ANALYSIS

This appendix contains various prerequisites of functional analysis that are needed in this book.

A.1. Hahn–Banach theorem

Let E and F be two normed vector spaces on $\mathbb{K} = \mathbb{R}$ or $\mathbb{K} = \mathbb{C}$, equipped, respectively, with the norms $\|\cdot\|_E$ and $\|\cdot\|_F$. We recall that the space of bounded operators $T : E \longrightarrow F$ is denoted by $\mathcal{L}(E, F)$. Recall that the topological dual E' of E is $\mathcal{L}(E, \mathbb{K})$.

Theorem A.1 (**Analytic Hahn Banach Theorem**)
  Let $G \subset E$ be a subspace of E, and $S : G \longrightarrow \mathbb{K}$ be a bounded operator. Then, there exists a bounded operator $T : E \longrightarrow \mathbb{K}$ such that:

$$\forall x \in G, \qquad T(x) = S(x),$$

and

$$\|T\| = \sup_{\|x\|_E \leqslant 1,\, x \in E} \|T(x)\| = \|S\| = \sup_{\|x\|_E \leqslant 1,\, x \in G} \|S(x)\|.$$

The following corollary is used several times in this book.

Corollary A.2. — *Let E be a normed vector space. Then,*

$$\forall x \in E, \qquad \|x\|_E = \max_{\|T\| \leqslant 1,\, T \in E'} \|T(x)\|.$$

© The Author(s), under exclusive license
to Springer Nature Switzerland AG 2021
C. Cheverry and N. Raymond, *A Guide to Spectral Theory*,
Birkhäuser Advanced Texts Basler Lehrbücher,
https://doi.org/10.1007/978-3-030-67462-5

Proof. — For $x = 0$, this is obvious. Now, fix any $x \in E \setminus \{0\}$, and consider the subspace $G_x = \{tx; t \in \mathbb{K}\} \subset E$, as well as the application $S_x : G_x \longrightarrow \mathbb{K}$ given by

$$S_x(tx) = t\|x\|_E.$$

We have $S_x \in G'$ and $\|S_x\| = 1$. Since

$$\|T(x)\| \leqslant \|x\|_E \|T\|,$$

it is clear that

$$\sup_{\|T\| \leqslant 1, T \in E'} \|T(x)\| \leqslant \|x\|_E.$$

By the Hahn–Banach theorem, we can find $T_x \in E'$ such that $T_{x|G_x} \equiv S_x$ and $\|T_x\| = 1$. It follows that

$$\|T_x(x)\| = \|S_x(x)\| = \|x\|_E \leqslant \sup_{\|T\| \leqslant 1, T \in E'} \|T(x)\|.$$

Therefore, the supremum equals $\|x\|_E$, and it is a maximum achieved for $T = T_x$. □

A.2.   Baire theorem and its consequences

In this section, we recall the various important consequences of the Baire theorem.

Theorem A.3 (Baire Theorem). — *Let (X, d) be a complete metric space. Consider a sequence $(U_n)_{n \in \mathbb{N}}$ of dense open sets in X. Then, the intersection $\bigcap_{n \in \mathbb{N}} U_n$ is dense in X.*

Proof. — Fix any $a = x_0 \in X$ and any $\varepsilon = \varepsilon_0 > 0$. Since the set U_1 is dense, we know that $U_1 \cap B(x_0, \varepsilon_0/2[\neq \emptyset$. In particular, we can find x_1 and $0 < \varepsilon_1 < \varepsilon_0/2$ such that

$$x_1 \in U_1, \qquad d(x_0, x_1) < \varepsilon_0/2, \qquad B(x_1, \varepsilon_1] \subset U_1.$$

We can repeat this operation with the couple $(x_1, \varepsilon_1) \in X \times \mathbb{R}_+^*$, and so on. This yields a sequence $((x_n, \varepsilon_n))_n$ with $(x_n, \varepsilon_n) \in X \times \mathbb{R}_+^*$ satisfying

$$x_n \in U_n, \qquad d(x_{n-1}, x_n) < 2^{-n}\varepsilon_{n-1}, \qquad B(x_n, \varepsilon_n] \subset U_n,$$

with $0 < \varepsilon_n < \varepsilon_{n-1}/2$. Notice that, for all $(p, q) \in \mathbb{N}^2$ with $p \leqslant q$,

$$d(x_p, x_q) \leqslant \sum_{n=p+1}^{q} d(x_{n-1}, x_n) \leqslant 2^{1-p} \varepsilon_p \,,$$

which implies that $(x_n)_n$ is a Cauchy sequence. Since X is complete, it converges to some $x \in X$. Passing to the limit $(q \to +\infty)$ in the above inequality, we get

$$\forall p \in \mathbb{N}^*, \qquad d(x_p, x) \leqslant \sum_{n=p}^{+\infty} d(x_n, x_{n+1}) \leqslant 2^{1-p} \varepsilon_p \leqslant \varepsilon_p \,.$$

Thus, we have $x \in B(x_p, \varepsilon_p] \subset U_p$ and :

$$d(x, a) \leqslant d(x, x_1) + d(x_1, x_0) < \varepsilon_1 + (\varepsilon_0/2) < \varepsilon_0 = \varepsilon \,.$$

As stated, some $x \in \bigcap_{n \in \mathbb{N}} U_n$ can be selected at any distance $\varepsilon > 0$ from $a \in X$. $\qquad\square$

A rather straightforward consequence of the Baire theorem is the following.

Theorem A.4 (Uniform Boundedness Principle)
   Let E and F be vector spaces. Consider a family $(T_j)_{j \in J}$ of bounded operators $T_j : E \longrightarrow F$ such that

$$\forall x \in E, \qquad \sup_{j \in J} \|T_j(x)\| < +\infty \,.$$

Then,

$$\sup_{j \in J} \|T_j\| < +\infty \,.$$

With a little work, we can show that the Baire theorem implies the Open Mapping Theorem.

Theorem A.5 (Open Mapping Theorem). — *Let E and F be Banach spaces. Consider a surjective bounded operator $T : E \to F$. Then, T transforms open sets into open sets.*

A straightforward consequence is the following.

Theorem A.6 (Banach Isomorphism Theorem)
   Let E and F be Banach spaces. Consider a bijective bounded operator $T : E \to F$. Then, T^{-1} is bounded.

The previous theorem implies the Closed Graph Theorem.

Theorem A.7 **(Closed Graph Theorem).** — *Let E and F be Banach spaces. Consider a bounded operator $T : E \to F$. Then, the operator T is bounded if and only if its graph*

$$\Gamma(T) := \{(x, Tx), x \in E\}$$

is closed for the product topology on $E \times F$.

A.3. Ascoli Theorem

According to the Bolzano–Weierstrass theorem, any bounded sequence of real numbers has a convergent subsequence. This theorem can easily be extended to finite-dimensional $\mathbb{K}$-vector spaces. In infinite dimension, especially in functional spaces, we have first to introduce a reasonable notion of *boundedness*.

Consider a compact topological space $(X, \mathscr{T})$ and a complete metric space (E, d). Let $\mathscr{F} \subset \mathscr{C}^0(X, E)$. Typically, the Reader can imagine that $X = [0, 1]$, that $\mathscr{T}$ is the topology induced by the absolute value, and that $(E, d) = (\mathbb{C}, |\cdot|)$.

Remark A.8. — Remember the following definitions and facts:

   (a) A part of a metric space is precompact when, for all $\varepsilon > 0$, it may be covered by a finite number of balls of radius $\varepsilon > 0$.

   (b) In a complete metric space, being *precompact* is equivalent to having a *compact closure* (sometimes called *relative compactness*).

   (c) In finite dimension, *precompact* is equivalent to *bounded*.

Definition A.9 **(Pointwise precompactness)**
     The set $\mathscr{F}$ is pointwise precompact when, for all $x \in X$, the set $\mathscr{F}(x)$ is precompact in (E, d).

When dealing with sequences of functions and uniform convergence, there is no direct extension of the Bolzano–Weierstrass theorem. Just consider $f_n(x) = \sin(nx)$ on $[0, 2\pi]$ with $n \in \mathbb{N}$ to obtain a counter-example. Some additional conditions are imposed.

Definition A.10 (Equicontinuity). — The set $\mathscr{F}$ is equicontinuous when for all $x \in X$ and all $\varepsilon > 0$, there exists $O_x \in \mathscr{T}$ such that

$$y \in O_x \Longrightarrow \quad \forall f \in \mathscr{F}, \quad d(f(y), f(x)) \leqslant \varepsilon.$$

Theorem A.11 (Ascoli Theorem). — *The set $\mathscr{F}$ is equicontinuous and pointwise precompact if and only if $\mathscr{F}$ has a compact closure in $\mathscr{C}^0(X, E)$ (this means that, from any sequence in $\mathscr{F}$, we can extract a uniformly convergent sequence).*

Proof. — (i) *Necessary condition.* Let $\mathscr{F}$ be a precompact subset of $\mathscr{C}^0(X, E)$. Let $\varepsilon > 0$. There exist $f_1, \cdots, f_N$ such that

$$\mathscr{F} \subset \bigcup_{i=1}^{N} B_{d_\infty}(f_i, \varepsilon).$$

The finite part $\{f_1, \cdots, f_N\}$ is equicontinuous. Let $x \in X$. Consider O_x being the open set given by the continuity of the f_i. If $f \in \mathscr{F}$, there exists $i \in \{1, \cdots, N\}$ such that $d_\infty(f, f_i) < \varepsilon$. For all $y \in O_x$,

$$d(f(x), f(y))$$
$$\leqslant d(f(x), f_i(x)) + d(f_i(x), f_i(y)) + d(f_i(y), f(y)) \leqslant 3\varepsilon.$$

The application $\mathscr{F} \ni f \mapsto f(x) \in E$ is continuous. It sends $\mathscr{F}$ onto a compact set of E. Thus, $\mathscr{F}(x)$ is included in a compact and is precompact.

(ii) *Sufficient condition.* We assume that, for all $\varepsilon > 0$, and all $x \in X$,

$$\exists O_x \in \mathcal{T}, \forall f \in A, \forall y \in O_x \Longrightarrow d(f(x), f(y)) \leqslant \varepsilon.$$

We have $X = \bigcup_{x \in X} O_x$. From the compactness of X, we can find finitely many x_i such that

$$X = \bigcup_{i=1}^{n} O_{x_i}.$$

Consider

$$C = \bigcup_{i=1}^{n} \mathscr{F}(x_i).$$

The set $C \subset E$ is precompact. Let $\varepsilon > 0$. We can find finitely many $c_j \in E$ such that

$$C \subset \bigcup_{j=1}^{m} B_d(c_j, \varepsilon).$$

If $\phi : \{1, \cdots n\} \to \{1, \cdots, m\}$, we set

$$L_\phi = \{f \in \mathscr{C}^0(X, E) : \forall i \in \{1, \cdots, n\}, \forall y \in O_{x_i} :$$
$$d(f(y), c_{\phi(i)}) \leqslant 2\varepsilon\}.$$

Let $f \in \mathscr{F}$. For all $i \in \{1, \cdots, n\}$, there exists O_{x_i} such that

$$\forall y \in O_{x_i}, \quad d(f(y), f(x_i)) \leqslant \varepsilon.$$

Since $f(x_i) \in C$, there exists $j_i \in \{1, \cdots m\}$ such that $d(f(x_i), c_{j_i}) < \varepsilon$. Therefore $\mathscr{F}$ is covered by the finite union of the L_ϕ. The diameter of each L_ϕ is less than 4ε.

$\square$

A.4. Sobolev spaces

In this book, we often use, in the examples, a rough notion of distribution. The aim of this section is just to define weak derivatives in $\mathsf{L}^p(\Omega)$, without entering into the general theory of distributions.

Definition A.12. — Let Ω be an open set of $\mathbb{R}^d$ and $p \in [1, +\infty]$. We denote

$$\mathsf{W}^{1,p}(\Omega) = \Big\{u \in \mathsf{L}^p(\Omega) : \forall j \in \{1, \ldots, d\}, \exists f_j \in \mathsf{L}^p(\Omega),$$

$$\forall v \in \mathscr{C}_0^\infty(\Omega), \int_\Omega u \partial_j v \, \mathrm{d}x = - \int_\Omega f_j v \, \mathrm{d}x\Big\}.$$

Remark A.13. — By using standard density arguments, we can show that the f_j are unique and, when $u \in \mathsf{W}^{1,p}(\Omega)$, we let $\partial_j u := f_j$, for all $j \in \{1, \ldots, d\}$.

When $p = 2$, we use the classical notation $\mathsf{H}^1(\Omega) = W^{1,2}(\Omega)$.

Definition A.14. — Let Ω be an open set of $\mathbb{R}^d$, and $(m, p) \in \mathbb{N} \times [1, +\infty]$. We denote

$$W^{m,p}(\Omega) = \Big\{ u \in L^p(\Omega) : \forall \alpha \in \mathbb{N}^d \text{ with } |\alpha| \leqslant m, \exists f_\alpha \in L^p(\Omega),$$

$$\forall v \in \mathscr{C}_0^\infty(\Omega), \int_\Omega u \partial^\alpha v \, \mathrm{d}x = (-1)^{|\alpha|} \int_\Omega f_\alpha v \, \mathrm{d}x \Big\}.$$

Remark A.15. — The f_α are unique, and we let $f_\alpha = \partial^\alpha u$. For $u \in W^{m,p}(\Omega)$, we let

$$\|u\|_{W^{m,p}(\Omega)} = \left(\sum_{|\alpha| \leqslant m} \|\partial^\alpha u\|^p_{L^p(\Omega)} \right)^{\frac{1}{p}}.$$

When $p = 2$, we use the classical notation $H^m(\Omega) = W^{m,2}(\Omega)$, and we recall the characterization, using the Fourier transform $\mathscr{F}$,

$$u \in H^m(\mathbb{R}^d) \Leftrightarrow \mathscr{F}u \in \{v \in L^2(\mathbb{R}^d) : \langle \xi \rangle^m v \in L^2(\mathbb{R}^d)\}.$$

We can check that $W^{m,p}(\Omega) \subset \mathscr{D}'(\Omega)$. In this book, we only meet functions in Sobolev spaces, but we can use the convenient language of the distributions.

Definition A.16. — We say that a sequence $(T_n) \subset W^{m,p}(\Omega)$ converges to $T \in W^{m,p}(\Omega)$ in the sense of distributions when

$$\forall \varphi \in \mathscr{D}(\Omega) := \mathscr{C}_0^\infty(\Omega), \quad \langle T_n, \varphi \rangle_{\mathscr{D}'(\Omega), \mathscr{D}(\Omega)}$$

$$:= \int_\Omega T_n \varphi \, \mathrm{d}x \xrightarrow[n \to +\infty]{} \langle T, \varphi \rangle_{\mathscr{D}'(\Omega), \mathscr{D}(\Omega)}.$$

A.5. Notes

i. A proof of the Hahn–Banach theorem can be found in [**38**, Chapter 3].

ii. Various consequences of the Baire Theorem are proved in [**38**, Chapter 2].

iii. Our version and proof of the Ascoli Theorem are adaptations of [**8**, Chapter VII, Section 5].

BIBLIOGRAPHY

[1] F.V. Atkinson, The normal solubility of linear equations in normed spaces. Mat. Sbornik N.S. **28**(70), 3–14 (1951)

[2] S. Bochner, Integration von Funktionen, deren Werte die Elemente eines Vektorraumes sind. Fundamenta Mathematicae **20**(1), 262–176 (1933)

[3] H. Brezis, *Analyse fonctionnelle*. Collection Mathématiques Appliquées pour la Maîtrise. [Collection of Applied Mathematics for the Master's Degree]. Masson, Paris, 1983. Théorie et applications. [Theory and applications]

[4] R. Courant, D. Hilbert, *Methods of Mathematical Physics*, vol. I (Interscience Publishers Inc, New York, N.Y., 1953)

[5] H.L. Cycon, R.G. Froese, W. Kirsch, B. Simon, *Schrödinger Operators with Application to Quantum Mechanics and Global Geometry*, study edn., Texts and Monographs in Physics (Springer, Berlin, 1987)

[6] E.B. Davies, *Spectral Theory and Differential Operators*, 42nd edn., Cambridge Studies in Advanced Mathematics (Cambridge University Press, Cambridge, 1995)

[7] E.B. Davies, *Linear Operators and Their Spectra*, 106th edn., Cambridge Studies in Advanced Mathematics (Cambridge University Press, Cambridge, 2007)

© The Author(s), under exclusive license
to Springer Nature Switzerland AG 2021
C. Cheverry and N. Raymond, *A Guide to Spectral Theory*,
Birkhäuser Advanced Texts Basler Lehrbücher,
https://doi.org/10.1007/978-3-030-67462-5

[8] J. Dieudonné, *Foundations of Modern Analysis*, vol. X, Pure and Applied Mathematics (Academic Press, New York-London, 1960)

[9] M. Dimassi, J. Sjöstrand, *Spectral Asymptotics in the Semi-Classical Limit*, vol. 268, London Mathematical Society Lecture Note Series (Cambridge University Press, Cambridge, 1999)

[10] D.E. Edmunds, W.D. Evans, *Spectral Theory and Differential Operators*, Oxford Mathematical Monographs (Oxford University Press, Oxford, 2018). Second edition of [MR0929030]

[11] M. Einsiedler, T. Ward, *Functional Analysis, Spectral Theory, and Applications*, Graduate Texts in Mathematics (Springer, Cham, 2017)

[12] P. Enflo, A counterexample to the approximation problem in Banach spaces. Acta Math. **130**, 309–317 (1973)

[13] R.L. Frank, L. Geisinger, Two-term spectral asymptotics for the Dirichlet Laplacian on a bounded domain, in *Mathematical Results in Quantum Physics* (World Sci. Publ., Hackensack, NJ, 2011), pp. 138–147

[14] I. Fredholm, Sur une classe d'équations fonctionnelles. Acta Math. **27**(1), 365–390 (1903)

[15] S. Gottwald, *Two-term spectral asymptotics for the Dirichlet pseudo-relativistic kinetic energy operator on a bounded domain*. PhD thesis, Ludwig Maximilian Universität (2017)

[16] S. Gottwald, Two-term spectral asymptotics for the Dirichlet pseudo-relativistic kinetic energy operator on a bounded domain. Ann. Henri Poincaré **19**(12), 3743–3781 (2018)

[17] V.V. Grushin, Les problèmes aux limites dégénérés et les opérateurs pseudo-différentiels, in *Actes du Congrès International des Mathématiciens (Nice, 1970), Tome 2*, pp. 737–743 (1971)

[18] P.R. Halmos, *Introduction to Hilbert Space and the Theory of Spectral Multiplicity* (AMS Chelsea Publishing, Providence, RI, 1998). Reprint of the second (1957) edition

[19] H. Hanche-Olsen, H. Holden, The Kolmogorov-Riesz compactness theorem. Expo. Math. **28**(4), 385–394 (2010)

[20] B. Helffer, *Spectral Theory and Its Applications*, vol. 139, Cambridge Studies in Advanced Mathematics (Cambridge University Press, Cambridge, 2013)

[21] D. Hilbert, *Grundzüge einer allgemeinen Theorie der linearen Integralgleichungen* (Vieweg+Teubner Verlag, Wiesbaden, 1989), pp. 8–171

[22] P.D. Hislop, I.M. Sigal, *Introduction to Spectral Theory*, vol. 113, Applied Mathematical Sciences (Springer, New York, 1996). With applications to Schrödinger operators

[23] V.J. Ivriĭ, The second term of the spectral asymptotics for a Laplace-Beltrami operator on manifolds with boundary. Funktsional. Anal. i Prilozhen. **14**(2), 25–34 (1980)

[24] O. Kallenberg, *Foundations of Modern Probability*, Probability and its Applications (New York) (Springer, New York, 1997)

[25] T. Kato, *Perturbation Theory for Linear Operators*, Classics in Mathematics (Springer, Berlin, 1995). Reprint of the 1980 edition

[26] A. Kolmogoroff, Über Kompaktheit der Funktionenmengen bei der Konvergenz im Mittel. Nachrichten von der Gesellschaft der Wissenschaften zu Göttingen, Mathematisch-Physikalische Klasse **60–63**, 1931 (1931)

[27] D. Krejčiřík, P. Siegl, Elements of spectral theory without the spectral theorem, in *Non-Selfadjoint Operators in Quantum Physics* (Wiley, Hoboken, NJ, 2015), pp. 241–291

[28] C.S. Kubrusly, *Spectral Theory of Operators on Hilbert Spaces* (Birkhäuser/Springer, New York, 2012)

[29] P.D. Lax, A.N. Milgram, Parabolic equations, in *Contributions to the theory of partial differential equations*, Annals of Mathematics Studies, no 33 (Princeton University Press, Princeton, N. J., 1954), pp. 167–190

[30] J. Leray, *Étude de diverses équations intégrales non linéaires et de quelques problèmes que pose l'hydrodynamique*. NUMDAM, [place of publication not identified] (1933)

[31] P. Lévy-Bruhl, *Introduction à la théorie spectrale* (Sciences Sup, Dunod, 2003)

[32] E. Mourre, Absence of singular continuous spectrum for certain selfadjoint operators. Comm. Math. Phys. **78**(3), 391–408 (1980/81)

[33] L. Nirenberg, On elliptic partial differential equations. Ann. Scuola Norm. Sup. Pisa Cl. Sci. (3) **13**, 115–162 (1959)

[34] N. Raymond, Bound states of the magnetic Schrödinger operator, in *EMS Tracts in Mathematics*, vol. 27 (European Mathematical Society (EMS), Zürich, 2017)

[35] M. Reed, B. Simon, *Methods of Modern Mathematical Physics. I-IV* (Academic Press, New York-London)

[36] M. Riesz, Sur les ensembles compacts de fonctions sommables. Acta Szeged Sect. Math. **6**, 136–142 (1933)

[37] W. Rudin, *Real and Complex Analysis*, 3rd edn. (McGraw-Hill Book Co., New York, 1987)

[38] W. Rudin, *Functional Analysis*, 2nd edn., International Series in Pure and Applied Mathematics (McGraw-Hill Inc, New York, 1991)

[39] E. Schmidt, Über die Auflösung linearer Gleichungen mit Unendlich vielen unbekannten. Rendiconti del Circolo Matematico di Palermo (1884–1940) **25**(1), 53–77 (1908)

[40] J. Sjöstrand, Operators of principal type with interior boundary conditions. Acta Math. **130**, 1–51 (1973)

[41] J. Sjöstrand, M. Zworski, Elementary linear algebra for advanced spectral problems, vol. 57, pp. 2095–2141 (2007). Festival Yves Colin de Verdière

[42] M.H. Stone, *Linear Transformations in Hilbert Space*, vol. 15, American Mathematical Society Colloquium Publications (American Mathematical Society, Providence, RI, 1990). Reprint of the 1932 original

[43] J.D. Tamarkin, On the compactness of the space L_p. Bull. Amer. Math. Soc. **38**(2), 79–84 (1932)

[44] M.E. Taylor, *Partial Differential Equations I. Basic Theory*, vol. 115, 2nd edn., Applied Mathematical Sciences (Springer, New York, 2011)

[45] G. Teschl, *Mathematical Methods in Quantum Mechanics*, vol. 157, 2nd edn., Graduate Studies in Mathematics (American Mathematical Society, Providence, RI, 2014). With applications to Schrödinger operators

[46] T. Titkos, A simple proof of the Lebesgue decomposition theorem. Amer. Math. Monthly **122**(8), 793–794 (2015)

[47] H. Weyl, Das asymptotische Verteilungsgesetz der Eigenwerte linearer partieller Differentialgleichungen (mit einer Anwendung auf die Theorie der Hohlraumstrahlung). Math. Ann. **71**(4), 441–479 (1912)

[48] K. Yosida, *Functional Analysis*, Classics in Mathematics (Springer, Berlin, 1995). Reprint of the sixth (1980) edition

[49] M. Zworski, *Semiclassical Analysis*, vol. 138, Graduate Studies in Mathematics (American Mathematical Society, Providence, RI, 2012)

INDEX

© The Author(s), under exclusive license to Springer Nature Switzerland AG 2021
C. Cheverry and N. Raymond, *A Guide to Spectral Theory*, Birkhäuser Advanced Texts Basler Lehrbücher, https://doi.org/10.1007/978-3-030-67462-5